as integrated perception is developed you can anticipate
↑ certainty + specificity
+ consistency of results

75 - Bauer 1984
prosopagnosia - ↑ GSR w/out recognition
= must be: if I do look at flowers

76 - SC involved in defensive behavior
+ PAG, very close / proximate automatic

77 - 'looming' may result in basic defensive behavior
→ tailgater

86 - Privileged Perception
Priority

87 ERP studies offer millisec ...

88 - PAG is 'automatic' - ie it responds regardless of other task loads
- the amyg is only 'somewhat' automatic

93 - author's bias left-out the fact, until now, that fast processing in amyg (30-60) is automatic; it is slower processes (80-120) that are affected by other factors (bias!)
'intercranial' (pre surgical direct contact) = also showed amyg early = auto

95 - emo stim + potent than neutral + are more unaided

135 - very nice "motivation" definition

199 - Menon - salience network

205 - > 600 connections of amyg to other areas known in 2001

240 - higher functions of brain may depend on signals/input from subcortical to cortical structures rather than the other way around

impoverished perception

The Cognitive-Emotional Brain

The Cognitive-Emotional Brain

From Interactions to Integration

Luiz Pessoa

The MIT Press
Cambridge, Massachusetts
London, England

MIT Press books may be purchased at special quantity discounts for business or sales promotional use. For information, please email special_sales@mitpress.mit.edu.

This book was set in Syntax and Times New Roman by Toppan Best-set Premedia Limited, Hong Kong. Printed and bound in the United States of America.

Library of Congress Cataloging-in-Publication Data

Pessoa, Luiz.
The cognitive-emotional brain : from interactions to integration / Luiz Pessoa.
pages cm
Includes bibliographical references and index.
ISBN 978-0-262-01956-9 (hardcover : alk. paper)
1. Brain. 2. Emotions and cognition. 3. Neuropsychology. I. Title.
QP376.P443 2013
612.8'2—dc23
2013001551

10 9 8 7 6 5 4 3 2

For Amélie and Mrim

Contents

Preface

To me, the idea that a given brain region "computes" something has always sounded odd. I've always thought the brain's mechanisms and processes operate in fairly large circuits, or as Ralph Adolphs put it in our 2010 *Nature Reviews Neuroscience* article, "the substrate of brain function is not so much to be found within neurons as within networks." So the notion that the "emotional" and "cognitive" brains are separate seemed even stranger to me. Indeed, to attempt to describe interactions and integration processes that intertwine the two was the seed idea for this book.

In a very real sense, the debate of localized versus distributed accounts of brain function is central to the history of neuroscience. The disputes of Santiago Ramón y Cajal and Camillo Golgi, which are foundational to neuroscience as a discipline, are a great example. Golgi was a staunch holist with regard to brain function, and believed that his studies showed that axons were densely intertwined, forming a "reticulum"—a continuous network of nerve cells, as proposed by Joseph von Gerlach and others. Cajal traced the paths of axons and mapped the structure of neuronal cell bodies in astonishing detail. Objecting to the idea that axons or dendrites formed a physically linked network, he believed in individual nervous "elements," each of which was "an absolutely autonomous canton" (Cajal, as quoted in Finger 1994, 47). Many other fascinating disputes in neuroscience revolve around the question of localization and the search for the "right scale" or "right units." In the end, even though certain ideas can be discarded as incorrect (such as the idea of a reticulum espoused by Golgi or the type of holism espoused by Karl Lashley), localized and distributed ideas are like yin and yang, two interdependent components. That is not to say that both views are equally valid, or equally productive. But there are at least some grounds for adopting either one.

We all have our biases, of course—in this case, toward either localized or distributed. Mine are probably easier to trace than most. When I was finishing my undergraduate degree in computer science in Brazil, I chose to do my "undergraduate thesis" (somewhat like an honors project in the United States) on artificial neural networks. Feeling less and less enamored of the "artificial," I ended up getting a Ph.D. at Boston University's recently created Department of Cognitive and Neural Systems, named by Steve Grossberg and others in the department for a field that was growing rapidly in the 1980s and 1990s—essentially, computational neuroscience. So it's perhaps not surprising that, after my studies there, I viewed most brain mechanisms through the prism of "neural *network* computations."

In 2008, I published "On the Relationship between Emotion and Cognition." Sometime after that, on a blog, I found one person loving my paper and another hating it and saying that it was trivial, that I was breaking through an open door (or something to that effect) by asserting that emotion and cognition can't be separated in the brain. I think that helped me decide that I didn't want to simply make a book out of the arguments in my paper. I thought it would be more interesting to discuss ways in which emotion and cognition interact and are integrated. To have "the question" fade into the background, if you will. And that's what I've tried to do here.

I also felt it would be valuable to have at least one chapter that discussed the problem of structure-function mappings. Even though that's the hardest problem in brain research, because neuroscience is so heavily skewed toward empirical work, many of the best people in the field don't "waste" their time writing about it but focus instead on individual experiments (or series of experiments) that are carefully designed and executed. Of course, many others have made important contributions to this question; in fact, too many to cite here. A beautiful treatment of structure-function mappings was given by Tim Shallice, not only in his classic book *From Neuropsychology to Mental Structure* (1998), but also in his more recent book with Richard Cooper, *The Organization of Mind* (2011), an even bolder title. Another of my favorite treatments was developed by M.-Marsel Mesulam (2000). Closer to the question of emotion and cognition interactions and integration, my favorite example is *The Feeling of What Happens* by Antonio Damasio (1999). And Helen Barbas's 1995 article in *Neuroscience and Biobehavioral Reviews*, emphasizing the great potential for interactions afforded by anatomy, has also been very inspirational. If *The Cognitive-Emotional Brain* is seen to continue in the tradition of these works, I will consider it a success.

Acknowledgments

The hardest part of writing a book might well be writing the acknowledgments. To be sure, you can't list everyone starting from your eighth grade physics teacher (fifth grade math teacher?), but you want to do justice to your intellectual debt. So I thank Ricardo Gattass for being so welcoming and letting me audit graduate neurobiology courses at the Federal University's Biophysics Institute in Rio. Ricardo also played no small role in my going to the National Institute of Mental Health (NIMH) in Bethesda. He'd worked with Leslie Ungerleider for many years and somehow convinced her that it would be a good idea to make me part of her lab. I thank Steve Grossberg, Ennio Mingolla, and all the others in my graduate department at Boston University for providing one of the most extraordinary intellectual environments I've ever witnessed. I thank Evan Thompson for sharing his knowledge with a graduate student who had absolutely no philosophical training. I thank Jorge Moll for being open to collaborating and sharing his ideas about the "moral brain." Jorge almost single-handedly started doing functional magnetic resonance imaging in Brazil at a time when MRI machines were a lot less user friendly. Words can't do justice to how grateful I am to Leslie Ungerleider, who provided immense support and opened all sorts of doors for me (and so many others). Having "Pessoa and Ungerleider" papers among my publications gives me a great sense of pride and joy. Leslie was also extremely open to investigating "affective vision" and its interaction with attention, even though that wasn't something she had investigated before. My fondest memories of NIMH were meetings with Leslie when we would pore over data and dissect them every possible way to understand their "deep meaning" (and late-hour emails with sudden ideas of how to think of the data in another way).

I've been fortunate, too, for all that I've learned from graduate students, postdocs, and other researchers in my lab. Srikanth Padmala has been absolutely instrumental in developing several of the ideas described in the book;

he played a crucial role in weeding out many an idea that made little sense, though I didn't see it at the time.

Several people gave me extremely valuable feedback on chapters of the book, including Alan Anticevic, Sandra Langeslag, Agnes Moors, India Morrison, Srikanth Padmala, Phil Spechler, and Rebecca Todd (I'm sure I'm forgetting someone, sorry). I am very grateful to Luis Guardado and to my postdoc Brenton McMenamin for reading and commenting on the entire book. It is my good fortune to have worked with Ralph Adolphs on the ideas described in chapter 2 and with Michael Anderson (and John Kinnison from my lab) on meta-analysis of neuroimaging data, whose results, briefly described in chapter 8, illustrate the kinds of findings this approach can give. I'm very thankful to Phil Spechler for all his patience and help with figures and references and to Susan Buckley and Bob Prior from MIT Press for their patient support, especially as I was bringing *The Cognitive-Emotional Brain* to completion.

Finally, I thank my daughter, Amélie, and my wife, Mrim, for always being there for me. You mean the world to me.

1 Introduction

To explicate the uses of the Brain seems as difficult a task as to paint the Soul, of which it is commonly said, that it understands all things but itself.
—Thomas Willis, *The Anatomy of the Brain and Nerves* (1664)

The nineteenth century witnessed the first "modern" attempts at tracing the neural correlates of mental functions. The contributions of physiology giants such as Eduard Hitzig, Gustav Fritsch, David Ferrier, and Hermann Munk and of neuropsychology luminaries such as Paul Broca and Heinrich Lissauer now form the foundation of systems neuroscience. Not long after those first attempts, advances were made in understanding the brain basis of emotion. For example, Philip Bard related sham rage to the posterior hypothalamus, strengthening the suggestion that this structure contributed to autonomic reactions accompanying the expression of emotions—ideas discussed by Harvey Cushing, who called the hypothalamus "the mainspring of primitive existence: vegetative, emotional and reproductive" (as quoted in Card, Swanson, and Moore 2008).

James Papez (1937) was the first to propose a circuit theory of emotion with wider impact. Like Walter Cannon and Philip Bard before him, Papez considered the hypothalamus to be a central component of the circuit responsible for emotional expression—as in a cat's biting, clawing, and hissing behavior—with several other regions also playing important roles, such as the cingulate cortex, a region that served a pivotal function in emotional experience. Indeed, according to Papez (1937, 737), the cingulate gyrus allowed "environmental experiences [to be] endowed with an emotional consciousness." Other structures of the circuit included the mammillary body (situated just below the thalamus) and the hippocampus.

More important than the specific components of Papez's proposal was the idea that a specialized circuit was responsible for emotion in the brain. Such "mechanism of emotion," as Papez called it, was modified and extended by

Paul MacLean (1949, 1952), who posited the existence of a "limbic system." Since MacLean, the limbic system, which comprises regions around the midline of the brain (hence "limbic," pertaining to edge or border), has become synonymous with "emotional brain." Although the limbic system concept has been criticized, sometimes quite vigorously (Brodal 1981; LeDoux 1996), its basic tenet—that a specific brain circuit constitutes the emotional brain—has shaped thinking about emotion and the brain. Notably, a corollary of this idea is that cognition resides elsewhere in the brain.

Such treatment of emotion as distinct from cognition is not surprising, of course. In the West, the separation between the two goes back at least to ancient Greece. For example, in several chariot allegories, Plato's *Phaedrus* juxtaposes the charioteer, representing intellect (and reason), with the horse, representing the soul's irrational passions. Similar allegories are also found in older Indian Upanishad texts:

> Know the self as a rider in a chariot,
> and the body, as simply the chariot.
> Know the intellect as the charioteer,
> and the mind, as simply the reins.
> The senses, they say, are the horses,
> and sense objects are the paths around them. . . .
> When a man lacks understanding,
> and his mind is never controlled;
> His senses do not obey him,
> as bad horses, a charioteer.
>
> (Olivelle 1996, 238–239)

In the 1990s, the status of emotion research was elevated to "mainstream" (e.g., Panksepp 1998). The expansion of this research area accelerated further in the 2000s with the growth of cognitive and affective neuroscience more broadly. In these past two decades, we have considerably expanded our knowledge of emotion and its relationship to other mental domains, including perception and cognition. Behaviorally, interactions of emotion with perception, attention, and memory were described in considerable detail. Neuropsychological work revealed that lesions of the brain that compromised emotion also have important effects on decision making, a traditionally cognitive function. Separately, brain anatomy started to stress "parallel, distributed processing" and, more recently, both anatomical and functional studies have emphasized "network" organization. Neuroimaging, perhaps rightly criticized at times for being a neo-phrenological approach, has contributed by showing that brain areas are engaged during many task conditions and contexts. Taken together, that emotion interacts with cognition has become a fairly well accepted notion.

On the Relationship between Emotion and Cognition

I have previously defended the idea that emotion and cognition cannot be dissociated in the brain because "affective" brain regions participate in cognition, on the one hand, and "cognitive" brain regions participate in emotion, on the other (Pessoa 2008). Which is to say that emotion and cognition systems are not modular—indeed, they are not even weakly modular, a theme echoed throughout *The Cognitive-Emotional Brain*, albeit, for the most part, implicitly. For example, in discussing the amygdala—the quintessential emotional region—and its functions, chapter 2 describes how the contributions of this region are reflected during attention and decision making, two typical cognitive functions. And in making the case against a popular notion in affective neuroscience, namely, that a subcortical pathway (or "low road") conveys emotional information to the amygdala in a way that does not require attention and is independent of awareness, chapter 3 highlights how emotion is *not* separate from cognition in the brain. These are but two of the instances in the book that inform the relationship between emotion and cognition. More than making the case *against* dissociating the two, one of my goals in this book is to move beyond the debate and to illustrate the many ways in which emotion and cognition interact and in fact are integrated. In this way, the question, in a sense, recedes into the background.

Does One Need to Define Emotion and Cognition to Study Them?

A few points should be made regarding the use of the terms "emotion" and "cognition." They will be used *descriptively* to refer to paradigms, task conditions, or "processes" that are closer to the traditional, intended meanings of emotion and cognition. Because I do not view emotion and cognition as separate domains, typically, I will not use the terms in quotation marks.

Given that *The Cognitive-Emotional Brain* is about interactions between emotion and cognition, readers may feel that the terms need to be carefully defined.[1] How else can interactions be meaningfully studied and, more critically, understood? Merriam-Webster (online) defines "cognitive" as "of, relating to, being, or involving conscious intellectual activity (as thinking, reasoning, or remembering)." It defines "emotion" as "a conscious mental reaction (as anger or fear) subjectively experienced as strong feeling usually directed toward a specific object and typically accompanied by physiological and behavioral changes in the body."

When cognition is studied, it seems that researchers are not strongly bothered by the absence of a formal definition—cognition is what we do when we think, reason, remember, and the like. The same cannot be said about emotion. In this case, it appears essential to distinguish multiple meanings, and researchers are compelled to advocate their own, specific framework. Some investigators use definitions that incorporate the concepts of drive and motivation: "Emotions are states elicited by rewards and punishers" (Rolls 2005, 11). Others favor the view that emotions are involved in the conscious (or unconscious) evaluation of events—that is, appraisals (Arnold 1960). Some approaches focus on basic emotions (e.g., fear and anger; Ekman 1992), others on an extended set of emotions, including moral ones (e.g., pride and envy; Haidt 2003; Moll et al. 2005). Furthermore, several researchers emphasize the link between emotions and the body (see Damasio 1994, 1999).

To precisely define emotion and cognition here would be to attempt to draw what I believe to be an artificial distinction between them. I do not think that we need to define them in order to study them. In this context, a quote from V. S. Ramachandran (2012) is relevant:

> I remember a seminar [Francis] Crick was giving on consciousness at the Salk Institute here in La Jolla. He'd barely started when a gentleman in attendance raised a hand and said, "But Doctor Crick, you haven't even bothered to *define* the word consciousness before embarking on this." Crick's response was memorable: "I'd remind you that there was never a time in the history of biology when a bunch of us sat around the table and said, 'Let's first *define* what we mean by life.' We just went out there and discovered what it was—a double helix. We leave matters of semantic hygiene to you philosophers."

Even though I do not think that "life" can be equated with "double helix," there is something to be said about not insisting on defining terms that are the object of study.

In a textbook on emotion, Keith Oatley, Dacher Keltner, and Jennifer Jenkins (2006) summarize eight proposed definitions by leading theorists. For example, Karen Barrett and Joseph Campos (1987) conceive of emotions as "bidirectional processes of establishing, maintaining, and/or disrupting significant relationships between an organism and the (external or internal) environment." But, recognizing the difficulty of adopting a *single* definition, Oatley, Keltner, and Jenkins (2006, 29; italics added) state:

> Before the age of science, definitions used to be all-important for understanding. Now we know that it is more important to understand deeply how a particular process works. Think of it this way. Several hundred years ago the motions of the planets were defined as circular. It was not until the understandings offered by Newton, that planetary motions depended on the forces of gravity and inertia acting at right angles to each other, that it was understood that planetary motions were better understood as ellipses. *Understanding had to precede proper definition.*[2]

In the spirit of the discussion above, I will not define emotion and cognition further here. I believe that it is more fruitful to describe mental phenomena in terms of cognitive-emotional interactions, as in, for example, "What are the cognitive-emotional interactions that are important in understanding a certain complex behavior X?"; "What are the cognitive-emotional interactions needed to advance our understanding of mental health problem Y?"; and so on. As discussed by B. Scott Kelso and David Engstrøm (2006), it may be time to stop describing concepts in terms of dichotomies (see also Newell 1973) and to adopt a vocabulary that views concepts as complementary pairs that mutually define each other and, critically, do *not* exclude each other. Thus Kelso and Engstrøm propose that light should be understood in terms of complementary pairs of "wave~particles" (quanta). In this vein, behavior should be understood in terms of "emotion~cognition."

Organization of the Book

The amygdala is center stage to a significant portion of studies of the brain basis of emotion. Chapter 1 reviews findings that reveal how many of its functions contribute to attention and decision making, critical components of cognitive functions. The amygdala is also believed to signal the presence of affective visual stimuli in a manner that is "automatic," that is, independent of attention and awareness. This property is purported to rely on a subcortical pathway that conveys visual signals to the amygdala. Chapter 3 provides an in-depth evaluation of this proposition and describes an alternative framework to understand the visual processing of emotion-laden information—the "multiple waves" model. Chapter 4 explicitly addresses the issue of automaticity of emotional visual processing. After a decade of much back-and-forth on this question, I suggest that the debate will not be resolved based solely on empirical data—conceptual issues are needed to advance research.

Chapters 5–7 form a unit. Chapter 5 describes interactions between emotion and cognition. In the past decade, much research has addressed how reward is processed in the brain. This literature has also spurred a body of studies investigating how *motivation* influences perception and cognitive function, covered in chapter 6. Chapter 7 describes the "dual competition" model, which proposes that the impact of emotional *and* motivational processing on behavior can be viewed as influencing competition processes at the levels of *both* perception and executive function.

Chapter 8 broadens the discussion to describe the mapping between structure and function in the brain, challenging the notion that we can understand function in terms of brain regions. Although I advocate a *network* perspective,

I also discuss how many of the challenges to understanding how function is implemented in terms of brain *regions* are not dissolved when functions are described in terms of networks of these regions. Chapter 9 briefly illustrates anatomical features of several regions traditionally considered to be crucial to emotional processing and highlights their connectivity properties, which support distributed, "network" computation. Finally, the conclusion both draws conclusions and discusses questions that will be important to advance our understanding of the cognitive-emotional brain in the coming years. To help readers less familiar with anatomical or other cognitive science terms, I have included a glossary, together with figures showing the approximate locations of many of the brain regions discussed in the book.

Notes

1. More than one of the reviewers of "On the Relationship between Emotion and Cognition" (Pessoa 2008) were perplexed by my not precisely defining the terms.

2. In fairness, I should mention that later Oatley, Keltner, and Jenkins (2006, 29; italics added) state: "But let us also distinguish emotions from several other related processes, in what might be called the *affective realm*."

2 Amygdala: From Determining "What Is It?" to Determining "What's to Be Done?"

As our knowledge of the connections of the amygdala has expanded, it has become apparent that the earlier view that it is primarily involved in the control of visceral and autonomic function is incomplete. . . . These widespread interconnections with diverse parts of the brain simply do not fit with a narrow functional role for the amygdaloid complex. They support, rather, the behavioral and clinical observations which suggest that the amygdaloid complex should be included among the structures which are responsible for the elaboration of higher cognitive function.
—David Amaral and Joseph Price (1984)

The amygdala is a fascinating, complex structure that lies at the center of much of our current thinking about emotion. It is well known for its importance in fear conditioning but is also involved in a surprisingly broad array of functions, spanning both negative and positive dimensions.

As often the case when considering a brain area, a natural inclination is to attempt to unravel *the* function that it implements. In the case of the amygdala, historically, this function has been linked to fear; in fact, Arne Öhman and Susan Mineka (2001) suggested that the amygdala be considered a "fear module" and posited the existence of a subcortical pathway that expediently conveys information to it (a notion evaluated at length in chapter 2). More generally, the limitations of the fear module idea have been addressed by others, such as David Sander, Jordan Grafman, and Tiziana Zalla (2003), who rejected the idea that amygdala function is centered on "negative arousing emotions."

This chapter will review data that indicate that the amygdala participates in several processes that determine what a stimulus is and what the organism should therefore do—or, more simply, "What is it?" and "What's to be done?" It will highlight important ways in which the amygdala's functions go beyond emotion as traditionally conceived, reflecting a discernible trend toward viewing this structure not simply in terms of its fear-related mechanisms, but also in terms of a wider array of processes (for a comprehensive review of thinking on the amygdala, see Whalen and Phelps 2009; Aggleton 1992, 2000; for recent reviews, see also Morrison and Salzman 2010; Salzman and Fuzi 2010). As the initial quotation from Amaral and Price 1984 makes clear, however, these ideas are not new.

Amygdala Definition and Anatomy

Anatomically, the amygdala is a complex structure containing more than a dozen richly interconnected nuclei. Based on a range of information, including connectivity and distribution of neurotransmitters, some have questioned whether this collection of nuclei should be considered a functional-anatomical unit. Indeed, Larry Swanson and Gorica Petrovich (1998) entitled their provocative article "What Is the Amygdala?" (see also Heimer et al. 2008; Holland and Gallagher 1999). Even though this debate is far from settled (Aggleton and Saunders 2000), it is useful to consider at least two subdivisions of this brain area: the "basolateral amygdala" (involving the lateral, basal, and accessory basal nuclei) and the "central amygdala" (involving the central nucleus; Cardinal et al. 2002; Davis and Whalen 2001; Heimer et al. 2008; figure 2.1). The basolateral amygdala appears to be responsible for Pavlovian learning and the representation of value, whereas the central amygdala contributes to several attentional functions. The two subdivisions also exhibit connectivity patterns that are quite distinct from each other. Thus the basolateral amygdala receives substantial sensory information from cortex and is richly connected with parietal, cingulate, insular, and prefrontal cortex. In contrast, the central amygdala, at times viewed as a "controller of the brainstem" (Cardinal et al. 2002), uses its widespread projections to the hypothalamus and brainstem nuclei to coordinate behavioral, autonomic, and neuroendocrine responses. The basolateral amygdala structurally resembles the layered cytoarchitecture of cortex, whereas the central amygdala exhibits a more simplified cytoarchitecture with incipient lamination—indeed, in terms of embryological origin, the amygdala appears to consist of two distinct groups of nuclei (Swanson 2000).

The value of functional-anatomical subdivisions notwithstanding (Heimer et al. 2008; Swanson 2003; Swanson and Petrovich 1998), the bulk of neuroimaging research on the human amygdala employs methods that cannot resolve anatomical nuclei.[1] Furthermore, many lesion studies are simply too coarse; this is the case in many nonhuman animal studies as well. Most importantly, the division of labor between the basolateral and central amygdala is not clear cut: both of these subdivisions are involved in attentional processes, for example.

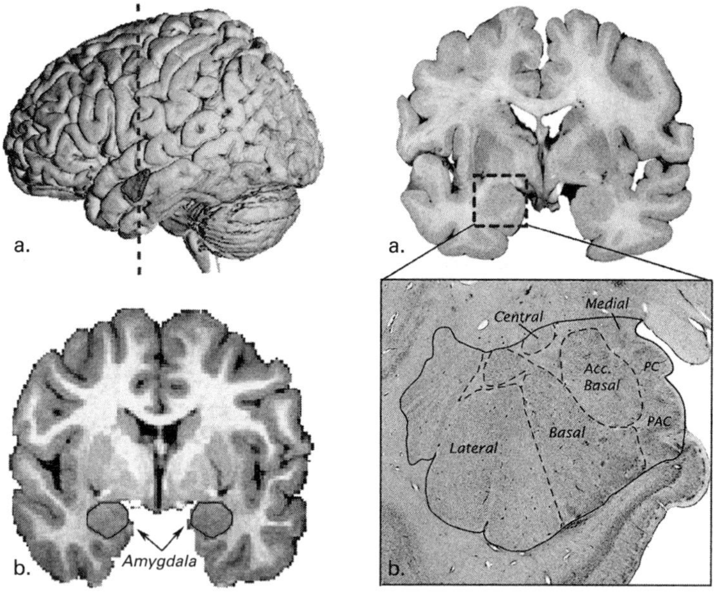

Figure 2.1
Human amygdala. (a, left) Lateral view of three-dimensional MRI scan of human brain with amygdala outlined; line represents location of coronal section in panel b (left). (b, left) Coronal section showing position of left and right amygdala. (a, right) Coronal section with box around left amygdala. (b, right) Enlarged, Nissl-stained section of left amygdala nuclei from panel a (right). Acc, accessory; PC, posterior cortical; PAC, periamygdaloid cortex. Reproduced with permission from Schumann, Bauman, and Amaral 2011.

Arousal, Vigilance, Salience, and Novelty

Early stimulation studies of the amygdala in nonhuman animals revealed a consistent "attention response," consisting of a rapid arrest of all activities in progress (e.g., licking and walking in the cat) followed by orienting movements (this behavior was also termed the "arrest response" in Kaada 1951). For instance, Holger Ursin and Birger Kaada (1960) noted that the stimulated animal usually raised its head and looked in an inquisitive manner (figure 2.2). The same amygdala stimulation produced EEG "desynchronization," as determined by both scalp electrodes and electrodes implanted into cortex of the frontal, temporal, parietal, and occipital brain (Ursin and Kaada 1960). Evidence of cortical desynchronization is particularly interesting since it has long been considered to be a signature of cortical "activation" or "arousal" (Moruzzi and Magoun 1949). This activity pattern stands in contrast to regular, synchronous activity typically observed in less alert states. These and many other findings have led researchers to propose that the amygdala plays a key role in

Figure 2.2
Attention response. Stimulation of the amygdala with mild electrical pulses elicits an "attention response" in cat. Before stimulation (left) versus during stimulation (center, right). Reproduced with permission from Pessoa 2010b. Originally adapted from Ursin and Kaada 1960. Illustration by Gatis Cirulis.

arousal and attention (e.g., Kaada 1972; Pribram and McGuinness 1975). Indeed, Karl Pribram and Diane McGuinness (1975) suggested that the amygdala is a core structure in a system involved in a "What is it?" rather than a "What's to be done?" form of processing (Germana 1969).[2] Contrary to Pribram and McGuinness, however, considerable evidence links the amygdala to both forms of processing, as discussed below.

Modern studies have picked up on some of these early themes. For example, Bruce Kapp, William Supple, and Paul Whalen (1994) reported that, in rabbits, electrical stimulation of the central amygdala suppresses low-frequency activity, producing desynchronization (see also Stock et al. 1981). Because of the well-established contribution of the cholinergic system in cortical activation and arousal (Sarter and Bruno 2000), Kapp, Supple, and Whalen investigated the potential effects of a cholinergic antagonist during amygdala stimulation. Desynchronization was markedly *attenuated* by such antagonists, compatible with a role for acetylcholine in cortical activation. Whalen (1998) suggested that engagement of the central amygdala by fear-related cues is part of a more widespread participation of this brain area in modulating the vigilance level of the organism (see below for discussion of mechanisms). Based on these findings as well as those from human neuroimaging, Whalen (1998, 177) proposed that the amygdala is a "*continuous vigilance system*, one that is preferentially invoked in ambiguous learning situations of biological relevance" (italics added).

The idea that the amygdala is attuned to stimuli of biological significance is compatible with a host of neuroimaging findings, including results of amygdala engagement during biological motion (Bonda et al. 1996) and in response to stimuli of social relevance (Adolphs 2010). Neutral-expression faces robustly engage the amygdala, as revealed by neuroimaging studies in humans (Pessoa, McKenna, et al. 2002; Wright and Liu 2006), as well as single-neuron recordings in both monkeys (Gothard et al. 2007; Rolls 2005) and humans (Kreiman, Koch, and Fried 2000a,b; Mormann et al. 2008; Rutishauser et al. 2011). Novelty also modulates amygdala responses. For example, in one study (Blackford et al. 2010), both novel "common" and novel "uncommon" images without explicit emotional content engaged the amygdala more strongly than familiar images (see also Balderston, Schultz, and Helmstetter 2011). Furthermore, salient target faces evoked stronger amygdala responses. This was observed for both emotional and nonemotional face stimuli, and the effect was of the same magnitude for all target faces, providing evidence for a general face salience signal in the human amygdala (Santos et al. 2011; see also Ishai et al. 2004). Along similar lines, C. Hindi Attar and colleagues (2010) reported that motion-defined targets with increased salience further modulated activity in the amygdala when presented together with a biologically relevant face stimulus, supporting a broader role of the amygdala in the detection of behavioral relevance. A particularly compelling example of this was given by Rachel Ludmer, Yadin Dudai, and Nava Rubin (2011), who reported that activity in the amygdala during the moment of "perceptual insight" (making sense of a highly degraded real-world picture) was the best predictor of subsequent memory of the picture (see also Ousdal et al. 2008, 2012).

Findings of how temporal unpredictability influences amygdala responses (Herry et al. 2007) further refine our understanding of its function. In an interesting cross-species study, both mouse and human subjects were exposed to a simple repeating tone, which in one condition was part of a predictable sequence (e.g., stimuli presented every 200 ms) and in another was part of an unpredictable sequence (e.g., stimuli presented with a variable interval with 200 ms mean). In humans, functional magnetic resonance imaging (MRI) results revealed increased responses during the unpredictable versus predictable condition. In addition, evoked responses were observed throughout the entire stimulation period: the human subjects did not habituate, as commonly observed (Buchel et al. 1998; LaBar et al. 1998). In mice, enhanced amygdala engagement was indicated by increased expression of "c-fos" (an "immediate-early" gene) in the basolateral amygdala. Interestingly, the unpredictable sound sequence had a clear behavioral impact on the mouse subjects, preventing rapid habituation of neuronal activity in the same region. In humans, spatial attention was enhanced toward emotional faces when the unpredictable auditory sequence was played, whereas, in mice, based on a stress-inducing maze test, the unpredictable sequence led to both avoidance and anxiety-related behavior.

Taken together, the above findings document an important role for the amygdala in a range of related functions, including arousal, vigilance, novelty detection, and salience detection. They also tend to support the proposal by

Sander, Grafman, and Zalla (2003) that the amygdala constitutes an "evolved system" for *relevance detection*. Based on a careful review of the available evidence, they rejected the notion that amygdala responses are strictly tied to fear-related processing or, more generally, to the processing of unpleasant stimuli (since the amygdala responds to many positive stimuli). Instead, they inferred a "common computational profile" that revolved around the detection of relevance: "An event is relevant for an organism if it can significantly influence (positively or negatively) the attainment of his or her goals, the satisfaction of his or her needs, the maintenance of his or her own well-being, and the well-being of his or her species" (Sander, Grafman, and Zalla 2003, 311). Whalen (1998; see below) has also suggested that the amygdala is more readily engaged when stimuli have more than one interpretation, that is, in situations where there is ambiguity. In such situations, the gathering of additional environmental information is beneficial and possibly critical. Thus the amygdala would be a key component of an *information gathering system* (Whalen 1998).

Another relevant conclusion that can be garnered by this brief review is that the amygdala is engaged by stimuli of biological relevance, even though they need not *strongly* influence emotional state (Whalen 1998). This is the case, for instance, for pictures of affectively neutral faces and of faces expressing emotions such as fear. It is also noteworthy that in early stimulation studies, whereas mild amygdala stimulation produced alerting-type responses, stronger stimulation generated reactions more closely related to fear or anger (Kaada 1960).

Amygdala Responses and Context Dependence

How flexible is the participation of the amygdala in relevance detection? To test the idea that the amygdala is at least partly flexible in its responses, Jay Van Bavel, Dominic Packer, and William Cunningham (2008) manipulated their subjects' "goals" during a task that involved rating famous people. In the "attitude" condition, subjects evaluated each name on a scale from strongly negative to strongly positive to indicate their overall attitude toward each name. In the other two conditions, subjects provided ratings in only one direction: they rated only the positive aspects of each name (scale from none to very good) in the positive condition, and they rated only the negative aspects (scale from none to very bad) in the negative condition. Therefore both positive and negative aspects of each name were relevant to the processing goal in the attitude condition, whereas only positive or only negative aspects of each name were relevant in the positive and negative conditions, respectively. Activity in the amygdala tracked the *fit* between experimentally manipulated evaluative goals and the valence of target stimuli. Thus the amygdala was most active when subjects evaluated positive stimuli (e.g., "Mother Teresa") in the positive condition, or negative stimuli (e.g., "Adolf Hitler") in the negative condition, or both positive and negative stimuli in the attitude condition. The authors' results suggest that the amygdala plays a relatively flexible role in evaluation processes in accordance with current goals.

Another example of the considerable context dependence of amygdala responses was reported by Alumit Ishai and colleagues (2004). Subjects were initially presented with a target face and asked to detect it in a stream of several faces, each displayed for 2 seconds. The target face could be neutral or fearful, albeit facial expression was task irrelevant. Among the stream of faces, distractor faces could also be neutral or fearful. As expected, when first presented, fearful faces evoked stronger responses than neutral ones. Likewise, when the detected target face was fearful, amygdala responses were stronger than when it was neutral. Notably, however, distractor faces, regardless of their valence, evoked only negligible activation. Among the stream of face stimuli, the target stimulus was presented a total of three times; likewise, one of the distractors was also presented three times. The objective of having stimulus repetition was to probe "repetition suppression," namely, decreased evoked responses to repeated stimuli. Other investigators had reported that repetition suppression was stimulus specific, as well as size and location invariant, and was observed under anesthesia (Miller, Gochin, and Gross 1991), indicating that the phenomenon relied on a fairly automatic, intrinsic response of neurons (Desimone 1996; Wiggs and Martin 1998). Ishai and colleagues (2004) showed, however, that repetition suppression depended on context (whether a stimulus was a target or a distractor) and stimulus valence. For instance, repetition of fearful targets was associated with *stronger* suppression (i.e., greater decreases) than repetition of neutral ones. In fact, by the third presentation of the target, responses to fearful targets were not even discernibly different from zero.

In sum, the findings of the studies above demonstrate that amygdala responses are strongly context dependent. For example, in the study by Ishai and colleagues (2004), the same face evoked robust responses when it was a target stimulus, but essentially very weak (or no) responses when it was a distractor or a repeated target within a stream of visual items. Hindi Attar and colleagues (2010) suggest that the amygdala may flexibly integrate stimulus valence with current goals, motivations, and contextual demands.

Attention

Attention is a complex, heterogeneous concept corresponding to disparate but conceptually related mental processes. Although associated with arousal and vigilance and linked to diverse mechanisms (e.g., "bottom-up competition," "top-down selection"), it is often thought to assume a more specific role. Generally speaking, we can think of a central function of attention as highlighting some stimuli (or task components) for further processing, while at the same time de-emphasizing or discarding other stimuli (or task components). When viewed through this wide lens, many amygdala functions can be profitably seen as attentional.

Nonhuman Animals

Although conditioning naturally involves learning, it is subject to several attentional modulations and, more broadly, can be viewed as a simplified system for studying *selective information processing*. In this context, it is worth considering the following problem: "How can a limited-capacity information processing system that receives a constant stream of diverse inputs be designed to selectively process those inputs that are most significant to the objectives of the system?" (Grossberg and Levine 1987, 5015). Indeed, selection of information for further analysis is a key problem that needs to be solved for effective learning (Kruschke 2003), among many other processes.

Selective aspects of learning are investigated in several behavioral paradigms by studying the rate of learning a conditioned stimulus (CS), also known as "associability." Associability determines how much processing is devoted to a conditioned stimulus, and thus the extent to which new information can be learned about it. Several theoretical frameworks of Pavlovian conditioning (e.g., Pearce and Hall 1980) propose that in a situation where a conditioned stimulus is completely predictive of an unconditioned stimulus (US), the conditioned stimulus is actually not worth learning about. In other words, learning about a stimulus should be reserved for stimuli whose consequences are less well established. In this sense, associability can be greatly increased by surprising events. Notably, increased learning when a predictive relation is altered to a less predictive one depends on the integrity of the central amygdala, including its projections to cholinergic neurons in the basal forebrain (Cardinal et al. 2002). For example, control rats learned the relationship between a light and a tone faster when this relationship was less predictive than when it was completely consistent. In contrast, rats with central amygdala damage showed the opposite pattern (Holland and Gallagher 1993).

The importance of the central amygdala for selective attentional processing is further highlighted in an experiment by Peter Holland, Jung-Soo Han, and Michela Gallagher (2000). On each trial after a ready signal, rats were required to poke their noses into one of three ports, guided by the brief illumination of one of them. Rats with central amygdala lesions were slower to learn the task than control animals. Additional attentional challenges, such as reducing the duration of the illumination and varying the duration of the ready signal, had greater impact on the performance of lesioned than control rats. Interestingly, the results were similar to those observed after damage to the basal forebrain system, compatible with the idea that the observed effects of central amygdala lesion are mediated by this system, as discussed below.

The role of expectations (hence surprise) in modulating amygdala responses has also been documented in the basolateral amygdala in a trace conditioning experiment with monkeys (Belova et al. 2007).[3] During the task, neural responses were stronger when monkeys *incorrectly* predicted reward or punishment. For instance, basolateral amygdala neurons driven by reward (water) responded more strongly to the reward when it occurred unexpectedly (i.e., when punishment was expected). Likewise, many neurons driven by punishment (air puff) responded more strongly to unexpected than to expected air puffs. Interestingly, this expectation-dependent modulation of responses resembles "prediction error" signals that are central to formal models of learning (Pearce and Hall 1980; Rescorla and Wagner 1972; Sutton and Barto 1988). Whereas, in some models, prediction errors drive learning directly, in others they do so indirectly, determining the amount of "event processing," which then influences the rate of learning—a process that can be viewed as an attentional function. In this regard, a connection between prediction errors and attention in learning has been revealed for the rat basolateral amygdala (Roesch et al. 2010). Thus, although the central amygdala has been linked to attention (Holland and Gallagher 1999), recent studies indicate that the basolateral amygdala is involved in some forms of attention, too.

In all, the amygdala participates prominently in several selective information processing functions that are encountered during learning—one of the multiple forms of *affective attention* discussed in this chapter. In particular, "surprise," which is directly linked to prediction errors, is an important variable in determining the engagement of amygdala nuclei (see also McNally, Johansen, and Blair 2011 for a discussion of prediction in the context of fear).

Spatial Attention The attentional functions discussed above are *non*spatial in that they do not depend on the location of an object in the visual field. Does the amygdala contribute to spatial attention as well? To test this possibility, Christopher Peck, Brian Lau, and Dan Salzman (2013) trained monkeys to perform a task in which reward-predictive visual cues were presented in different spatial configurations. They found that during task performance, individual amygdala neurons (mostly in the basolateral amygdala) encoded the motivational significance of visual stimuli as well as their spatial location. For

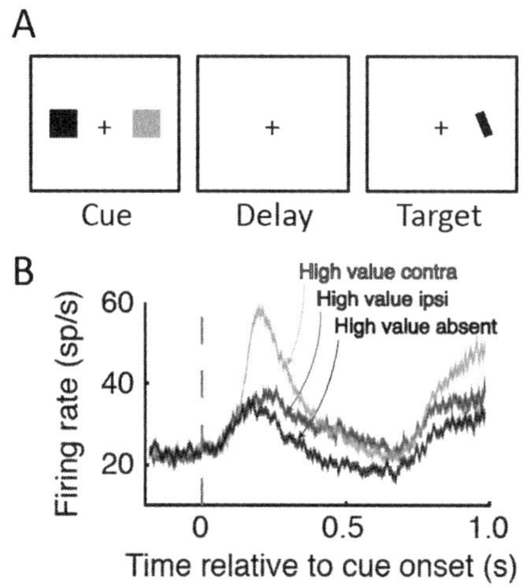

Figure 2.3
Amygdala responses to value and space. (A) Sequence of events in the two attention tasks. After monkeys achieved central fixation (not shown), two cues appeared at either side of the fixation point for 300 ms. The cues were followed by a delay in which no peripheral stimuli were present. The brief appearance of a near threshold-oriented patch (50 ms) at one of the two locations served as the target (here shown as a bar but a grating in the actual experiment), and the monkey correctly detected it by saccading to its location. (B) For this cell, responses to high-value cues were stronger when they were presented in the contralateral relative to ipsilateral visual field. Sp/s, spikes per second. Panel B reproduced with permission from Peck, Lau, and Salzman 2013 and kindly provided by Christopher Peck.

example, some neurons responded more strongly when a high-value cue was presented contralaterally (figure 2.3). More generally, individual amygdala neurons selectively combined information about space *and* value and signaled the location of reward-predictive stimuli with both negative and positive excursions in firing rate. Furthermore, neuronal activity was correlated with the trial-by-trial allocation of attention, suggesting that the representation of value and space in the amygdala influences the direction of spatial attention toward motivationally relevant stimuli. Finally, when a monkey directed attention, a signal representing space and value was actually sustained in the amygdala throughout the trial, long after the visual cues disappeared. Although the authors only tested for spatial selectivity at the level of the visual hemifield, the results suggest that the amygdala may at least be involved in shifting attention to the left or right visual field on the basis of stimulus value.

Mechanisms of Vigilance and Attention

How is the amygdala involved in regulating the nonspatial attentional functions discussed above? The central amygdala has significant projections to several basal forebrain structures, subcortical nuclei found in the ventral part of the forebrain. These include the magnocellular basal nucleus, which contains the basal nucleus of Meynert and other cell groups (Semba 2000).[4] The basal forebrain originates an "ascending" (i.e., toward cortex) cholinergic (and GABAergic) projection that innervates extensively throughout the cortical mantle. It is therefore in a favorable position both to influence sites across the brain, such as sensory regions, which play a central role in responding to environmental stimuli, and to affect information processing based on the content of the information available to the central amygdala. These topographically widespread effects result in increased vigilance, alertness, or attention (Sarter and Bruno 1999, 2000; Sarter, Bruno, and Turchi 1999; see chapter 6 for further discussion in the context of motivation).

Acetylcholine released onto cortical neurons facilitates their response (Everitt and Robbins 1997). There is evidence that these basal forebrain cholinergic effects operate at a time scale commensurate with fast attentional processes (Herrero et al. 2008; Noudoost and Moore 2011a; Parikh et al. 2007; Parikh and Sarter 2008)—and not simply at a slower time scale more consistent with "cortical arousal states," as often assumed. Furthermore, direct stimulation of the basal forebrain was shown to enhance the cortical coding of natural scenes in visual cortex (Goard and Dan 2009) by improving the reliability of visual responses; specifically, stimulation reduced the trial-by-trial variability of responses when a natural movie stimulus was presented. Reduced response variability was also observed following the pairing of acoustic stimuli with activation of the basal forebrain (Froemke et al. 2012), which improved detection of near-threshold stimuli in behaving animals. Finally, other studies suggest that applying acetylcholine and stimulating the basal forebrain can increase the signal-to-noise ratio of neuronal responses and sharpen tuning curves (e.g., Sillito and Kemp 1983).

Thus, one mechanism by which activity in the central amygdala influences cortical processing is by engaging basal forebrain neurons, whose terminals release acetylcholine onto sensory cortex (figure 2.4).[5] In line with the notion that some of the attentional functions of the amygdala depend on its close link with the basal forebrain, lesions of the latter impair performance on attentional tasks that require both sustained attention and selective mechanisms such as the filtering of irrelevant information (Sarter and Bruno 1999, 2000; Sarter, Bruno, and Turchi 1999).

Humans

Attention helps select specific items that shape perception. Emotional content guides information processing by prioritizing this class of stimuli relative to

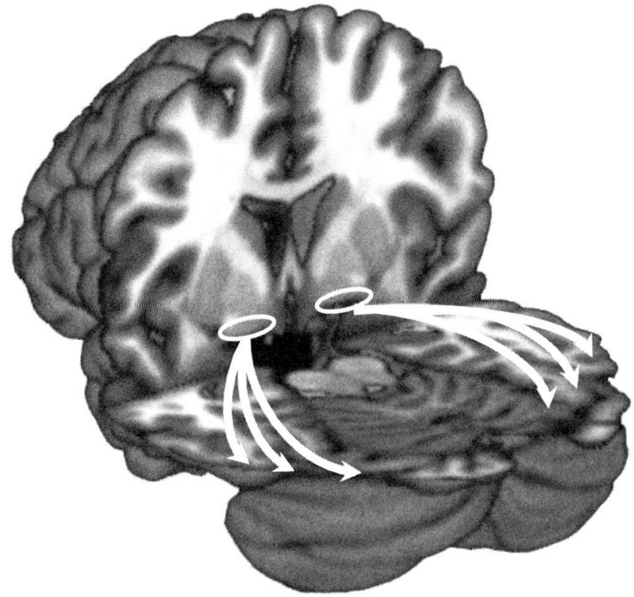

Figure 2.4
Basal forebrain modulation of sensory processing. Represented only schematically here, the basal forebrain (ellipses) is situated atop the amygdala and has widespread projections to the cortical mantle, including to sensory cortex, where they are suggested to influence how sensory information is registered and processed. (For the precise localization of the magnocellular cell groups in humans, see figure 8 of Zaborszky et al. 2008.)

neutral items. A host of experimental paradigms have documented the ways in which emotion-laden items are privileged, most notably, *detection*, *search*, *interference*, *masking*, and the *attentional blink* (Pourtois, Schettino, and Vuilleumier 2012). The last is a particularly rich paradigm to study capacity limitations. Subjects are asked to report the occurrence of two targets, T1 and T2, among a rapid stream of visual stimuli. When T2 follows T1 by a brief delay (200–500 ms), subjects are more likely to miss it, as if they had blinked (hence the name). The attentional blink is believed to reflect capacity-limited mechanisms associated with the processing of the first target (Raymond, Shapiro, and Arnell 1992). The strength of the attentional blink is influenced by the emotional content of the stimuli in question; subjects are better at detecting the second target when it is emotion laden. For instance, in one study (Anderson 2005), subjects were better during T2 words such as "blood" compared to words such as "zipper." Although not discussed further here, emotion-laden T1 stimuli also influence the attentional blink effect—they increase the strength of the blink relative to neutral stimuli (e.g., Schwabe et al. 2011).

In one study, Seung-Lark Lim, Srikanth Padmala, and I reasoned that if the amygdala is involved in shaping perceptual experience when affectively significant visual items are encountered, responses in this structure should be correlated with both visual cortical responses and behavioral reports (Lim, Padmala, and Pessoa 2009). To test this hypothesis, we investigated how affective significance influences visual perception during an attentional blink task that was combined with aversive conditioning (figure 2.5A). During the initial phase of the experimental session, houses or buildings (counterbalanced) were

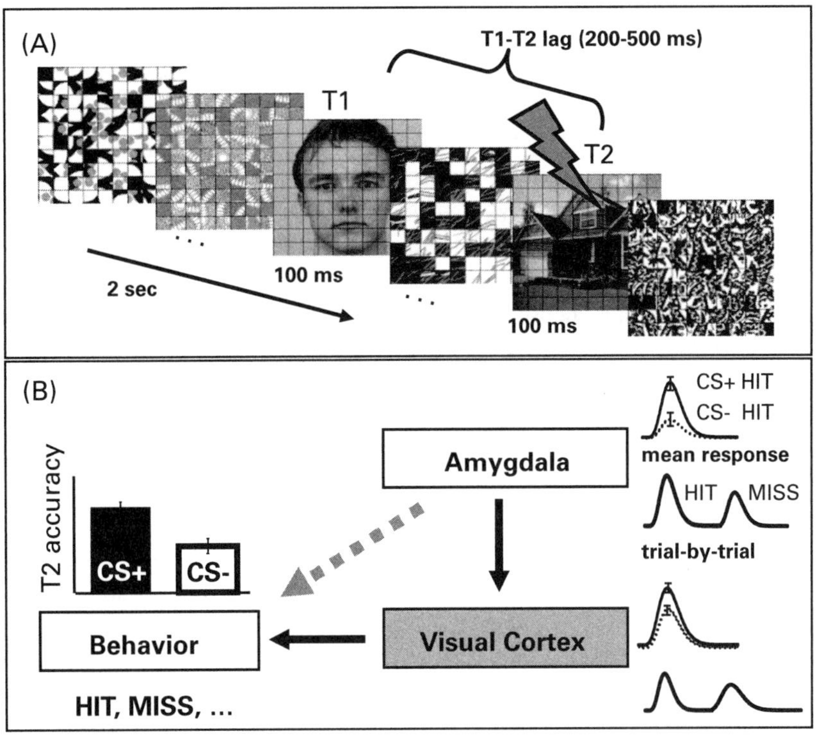

Figure 2.5
Attentional blink paradigm. (A) Subjects were asked to report on the face stimulus (T1) and on whether the stream contained a house, a building, or no scene (T2). Houses or buildings (other than houses) were paired with mild electrical stimulation during an initial learning phase. Thus, when houses were paired with stimulation, buildings were never paired with mild shock (and vice versa). During the main experimental phase, only trials with no stimulation were analyzed. (B) Link between responses evoked in the amygdala and behavior (T2 detection) was hypothesized to be mediated by responses in visual cortex—specifically, in the parahippocampal gyrus given its involvement in the processing of scenes and spatial layouts. My colleagues and I predicted that this relationship would be reflected in mean responses across subjects and in moment-to-moment fluctuations of both brain responses and behavior. Reproduced with permission from Lim, Padmala, and Pessoa 2009.

paired with mild shock. Behaviorally, following aversive learning, affectively significant scenes (CS+; houses or buildings, but not both) were better detected than unpaired (CS–) ones (72% vs. 62% correct, respectively) during the blink period. In terms of brain activity, both amygdala and visual cortex responses were stronger during CS+ than during CS– trials. Increased responses in these regions were associated with improved behavioral performance. Even though amygdala responses were predictive of behavioral performance, once responses in visual cortex were taken into account, the initial relationship was no longer statistically significant (figure 2.5B). Formally, the influence of the amygdala on behavior was statistically mediated via visual cortex (Baron and Kenny 1986). Mechanistically, the link between amygdala and visual cortex may itself have been via the basal forebrain, as discussed above (see also chapter 7 for a discussion of mechanisms).

We (Lim, Padmala, and Pessoa 2009) further hypothesized that if fluctuations in evoked brain responses determine the accuracy of the detection of the second target, *trial-by-trial* variability in response strength should predict behavioral reports. At the trial level, to evaluate this prediction, we performed logistic regression and modeled the probability of a hit trial (i.e., correctly reporting "house" or "building") as a function of trial response amplitude. In visual cortex, the mean logistic regression slopes, which represented the strength of the predictive effect, were significant for both CS+ and CS– trials, indicating that trial-by-trial fluctuations in functional MRI signals reliably predicted perceptual T2 decisions (figure 2.6A and B). Importantly, the predictive power of the logistic regression fit was stronger during CS+ than during CS– scenes. A similar trial-by-trial analysis was performed for the amygdala (figure 2.6C and D). The mean logistic regression slope was significant for CS+ trials, but not for CS– trials, showing that signal variability in the amygdala contributed to perceptual T2 decisions more robustly when these stimuli were affectively significant. In addition, the region interactions summarized in figure 2.5B were observed in a trial-by-trial analysis, too, revealing that the specific pattern of trial-based variability in brain responses in both the amygdala and visual cortex was closely related to fluctuations in behavioral performance.

Finally, we anticipated that if amygdala signals help shape perception, the strength of the predictive effect between visual cortex and behavior should depend on the strength of those signals. In other words, because we could not directly influence amygdala or visual cortex responses, we investigated how the relationship between evoked responses in visual cortex and behavior itself varied as a function of response strength in the amygdala. We found that when amygdala responses were relatively weak, the relationship between visual cortex and behavior was also relatively weak; when they were strong, the

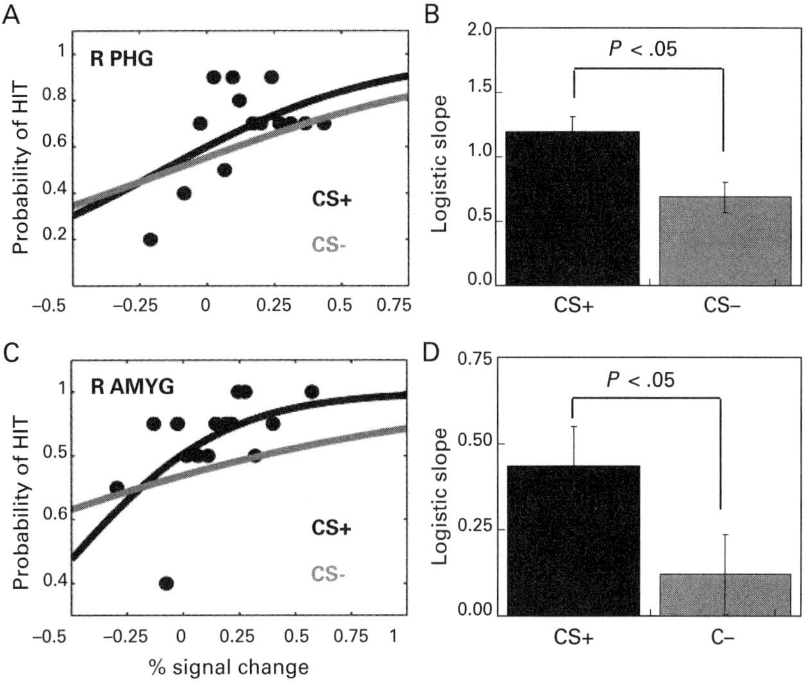

Figure 2.6
Trial-by-trial analysis. (A) Logistic regression of responses in the right parahippocampal gyrus (R PHG) as a function of affective significance (CS+ and CS–) for a sample subject (dichotomous variable: hits vs. misses). The slope of the fit indicates the strength of the predictive effect. For clarity, only binned data for the CS+ condition are shown. (B) Mean logistic slopes across subjects for the parahippocampal gyrus. (C) The same analysis as in panel A but for the right amygdala (R AMYG). (D) Mean logistic slopes across subjects for the amygdala. Reproduced with permission from Lim, Padmala, and Pessoa 2009.

association between visual cortex and behavior was stronger (figure 2.7). Thus the relationship between visual cortex and behavior varied according to the response strength of the amygdala. These results are consistent with the notion that trial-by-trial fluctuations in visual cortex, which are strongly tied to behavior, depend on amygdala responses.

Together with the results from a growing body of attentional blink studies, as well as from studies using related paradigms (e.g., Most, Chun, et al. 2005), our results suggest that affective significance determines the fate of a visual item during competitive interactions by enhancing sensory processing. In such manner, the amygdala helps separate the significant from the mundane. One way to interpret our results is in terms of an attentional function of the amygdala. For example, in studies of attention, fluctuation of responses in visual

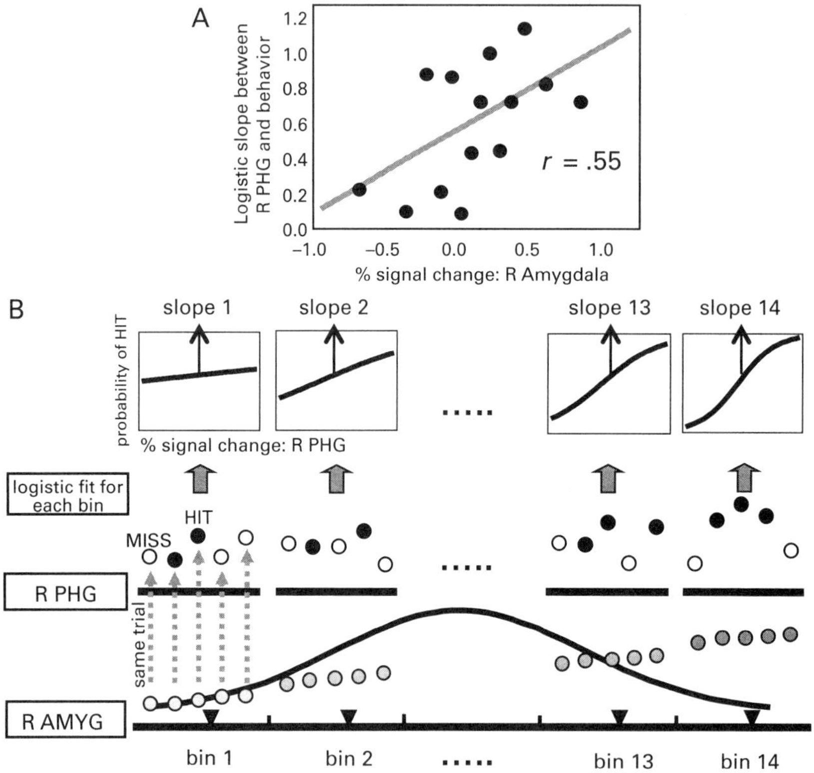

Figure 2.7

Trial-by-trial fluctuations in amygdala responses and magnitude of visual-cortex-to-behavior relationship. (A) The strength of the visual cortex-to-behavior relationship (as indexed by the slope of the logistic fit) was correlated with the magnitude of responses in the amygdala: the stronger the response in the amygdala, the tighter the relationship between visual responses and behavior. (B) Main steps in assessing this relationship. Single-trial amygdala responses were grouped from low to high in fourteen bins, and corresponding responses in visual cortex were determined. Filled circles represent the magnitude of right amygdala (R AMYG, bottom row) and right parahippocampal gyrus (R PHG, middle row) responses, as indicated by the height from the horizontal bars. Idealized logistic regression fits (top row) show an increase in the slope value as a function of the magnitude of evoked responses in the amygdala. Dotted arrows indicate the positions of the median response for each amygdala bin, which were correlated with the slope of the logistic fit for that same bin. Thus the scatter data in panel A were obtained by taking the fourteen pairs of median amygdala response and corresponding logistic fit slope. Reproduced with permission from Lim, Padmala, and Pessoa 2009.

cortex is often conceptualized as dependent on *source* regions in parietal and frontal cortex (Corbetta and Shulman 2002; Kastner and Ungerleider 2000), and these mechanisms are typically viewed as linked to how the processing of attended objects is prioritized. In our attentional blink study (Lim, Padmala, and Pessoa 2009), we observed something quite similar. We found the amygdala behaved much like an "attentional device" would be expected to, namely, as a device that helps to prioritize the processing of certain stimuli over others.

Another attentional contribution of the amygdala is highlighted by studies that have investigated how eye movements are directed toward salient regions of the face. To clarify the relationship between amygdala activation and gaze-orienting behavior, one study (Gamer and Buchel 2009) acquired both functional MRI data and eye movements at the same time during the evaluation of facial expressions. The subject's initial fixation was manipulated by unpredictably shifting the faces downward or upward on each trial, such that the eyes or the mouth were presented at fixation. Amygdala activity was specifically enhanced for fearful faces with the mouth aligned to fixation, and this differential activation predicted gazing behavior preferentially targeting the eye region. These results indicate an amygdala role in reflexive gaze initiation toward fearful, widened eyes. A more direct causal involvement for the amygdala in these findings was obtained by studying patients with amygdala lesions. Ralph Adolphs and colleagues (2005) showed that a subject with bilateral amygdala damage exhibited reduced gaze to the eyes. In a subsequent study, Daniel Kennedy and Adolphs (2010) established that the subject's reduced eye contact was particularly pronounced in the first fixation to the face, and less pronounced in subsequent fixations (figure 2.8). It therefore appears that the salience of the eyes attract one's gaze in an amygdala-dependent fashion.

Amygdala Output Projections and Affective Attention

The attentional blink study reviewed above (Lim, Padmala, and Pessoa 2009) suggests that the amygdala modulates cortical responses in visual cortex. As discussed previously (see figure 2.4), this modulation might rely on basal forebrain mechanisms. A second, more direct pathway for modulatory signals relies on amygdala output projections, as discussed next.

The pattern of connectivity between the amygdala and visual cortex is well characterized in monkeys (Freese and Amaral 2005). Not only does the amygdala receive highly processed inputs from anterior portions of inferior temporal cortex, but, remarkably, its efferent projections reach nearly all levels of the ventral visual stream, even primary visual cortex (area V1) as well. These "feedback" connections, originating in the basolateral amygdala, are distinct from those discussed previously in the context of the central amygdala and

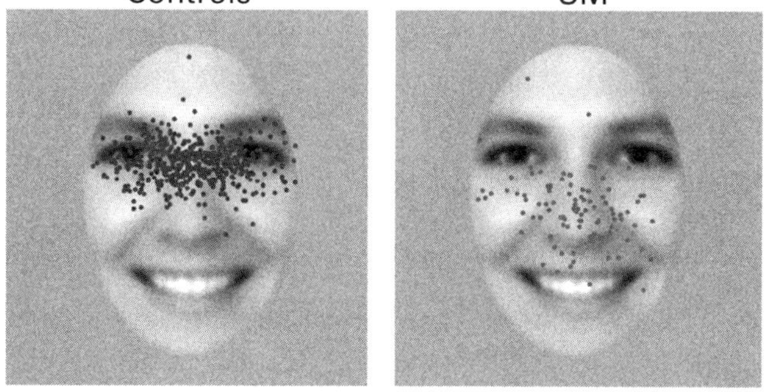

Figure 2.8
Locations of first face fixations for control subjects and patient SM. Each dot represents a single fixation (control fixations, n = 500; SM fixations, n = 100). Reproduced with permission from Kennedy and Adolphs 2010.

have led researchers to propose that they may exert a modulatory influence on visual responses according to the affective significance of the item in question, a proposal consistent with the well-documented finding that emotion-laden stimuli evoke stronger responses in visual cortex than neutral ones. For instance, stronger responses to fearful faces than to neutral ones are observed in the fusiform gyrus (Pessoa, McKenna, et al. 2002; Vuilleumier et al. 2001), and stronger responses to a fear-conditioned grating stimulus are observed in early, retinotopically organized visual cortex, including visual areas V1 and V2 (Padmala and Pessoa 2008). Projections from the amygdala to visual cortex may provide the substrate for these effects.

Evidence supporting this notion comes from a study that compared visual cortex responses to faces in patients with a compromised amygdala and in controls (Vuilleumier et al. 2004). Whereas fearful faces evoked stronger responses than neutral faces in the control group, no significant differences were detected in the patient group (see also Benuzzi et al. 2004). Hence output projections from the amygdala appear to be an important mechanism by which it influences visual cortex responses according to affective significance, and thus also competition taking place in visual cortex (Pessoa, Kastner, and Ungerleider 2002).[6] In the study by C. Hindi Attar and colleagues (2010) discussed above, subjects were instructed to attend to moving dots and to detect short intervals of coherent motion, while ignoring task-irrelevant faces shown at the same spatial location. Trials containing fearful faces evoked stronger responses in the amygdala than neutral ones, but they evoked substantially *weaker* responses in area MT/V5. Because this area responds robustly to moving stimuli, the authors implicated it in the processing of task-*relevant* motion targets. The interference effect was also observed in the behavioral data, where subjects exhibited slower responses to motion when the task-*irrelevant* faces were emotional. These results indicate that emotion-laden faces compete more strongly for resources in visual cortex—as do neutral stimuli that are more potent (e.g., more salient; Desimone and Duncan 1995; Pashler 1998).

Is the Amygdala Necessary for the Competitive Advantage of Affective Stimuli?

As discussed, the strength of the attentional blink is influenced by emotional content: subjects are better at detecting the second target when it is emotion laden. In other words, emotional content counteracts the attentional blink. In one study (Anderson and Phelps 2001), such counteracting was *not* observed in patients with lesions of the amygdala. These findings have been generally interpreted to mean that the amygdala is *necessary* for the advantage of emotion-laden stimuli in perception—not only in the attentional blink, but also in other paradigms with affective visual stimuli.

This conclusion has been challenged by several recent studies. Dominik Bach and colleagues (2011) reported an advantage for emotion-laden stimuli in the attentional blink in two patients with bilateral amygdala damage. When T2 was an emotion-laden word, a reduction of the attentional blink *was* observed even though the patients did not have a functioning amygdala (the reduction was comparable to that observed in control subjects). Studies employing other tasks have been investigated as well. One case study demonstrated that a patient with bilateral amygdala damage still exhibited a behavioral advantage for fearful faces during tasks involving rapid detection or manipulations of awareness (Tsuchiya et al. 2009). Another study revealed that enhanced search for fearful faces persisted after amygdala lesions (Piech et al. 2010). Finally, when patients with unilateral amygdala damage were asked to perform a task where emotional distractors impaired the perception of subsequent targets in control subjects, they exhibited the same effects observed in the controls (Piech et al. 2011). In all, recent findings from lesion studies indicate that the amygdala is *not* necessary for the advantage of emotion-laden information in some of the common paradigms employed, as typically assumed.

Determining the precise role of the *intact* amygdala in the processing of emotion-laden stimuli represents an important project for future research. Although the amygdala does not appear to be necessary, it is likely to play one or more primary roles during normal brain function, along with other structures such as orbitofrontal cortex, or so the lesion data described above suggest (see chapter 7).[7]

Value Representation

Linking stimuli (or, more generally, object representations) to estimates of biological *value* is another important dimension of amygdala function, pertinent when considering studies of negative manipulations as well as those of appetitive manipulations and reward. Thus a whole gamut of tasks such as *reinforcer revaluation* demonstrate the role of the basolateral amygdala in the representation of value in animal learning studies. For example, both rats and monkeys with basolateral lesions fail to adjust their responding (e.g., orienting and food cup approach in rats) to a conditioned stimulus after an unconditioned food stimulus is devalued (e.g., through satiety). In one study (Malkova, Gaffan, and Murray 1997), monkeys first learned to discriminate pairs of objects, which were either "positive" (paired with food) or "negative" (unpaired). Two different food rewards (peanuts, fruit snacks) were employed, one of which was presented on a given trial. Visual discrimination learning proceeded until monkeys performed this task rather well—that is, until they could discriminate between objects paired with food and objects that were not paired with food. Prior to the subsequent critical sessions, the animals were then satiated with one of the foods by having ample access to this food type (but not the other). During critical trials, the monkeys were required to choose between two objects, each of which had been paired with one of the food items—unlike in the initial discrimination learning phase, both items were "positive." After being satiated with one of the foods, intact monkeys avoided displacing objects that covered this food type in favor of objects that covered the food that had not been recently consumed. Monkeys with amygdala lesions did not exhibit this behavior, however, and instead displaced a similar number of objects that had been associated with the two different foods. These results indicate that the amygdala is necessary for learning the association between stimuli and the *value* of particular rewards (see also Cardinal et al. 2002; Holland and Gallagher 1999; Murray 2007).

A paradigm that illustrates this issue is second-order conditioning. In one experiment (Hatfield et al. 1996), rats first experienced light-food pairings, which were readily learned as evidenced by appropriate food cup responding. Both animals with intact amygdala and those with basolateral damage learned this task equally well. When the rats then experienced tone-light pairings, intact rats acquired conditioned food cup responses to the tone but rats with lesions did not. Thus, although amygdala damage had no effect on the acquisition of first-order conditioned responses to the light, it appeared to prevent second-order conditioning to the tone. These findings suggest that, in rats with basolateral lesions, the light failed to acquire reinforcing value based on its first-order pairings with food (see also Everitt and Robbins 1992).

Another study reveals that amygdala neurons may help track value moment to moment (Belova, Paton, and Salzman 2008). During a trace conditioning task, monkeys learned the association between visual stimuli (CSs) and either positive (water delivery) or negative (air puff to the face) outcomes (USs). During a trial, presentation of different visual stimuli elicited approach or defensive behaviors. Stimuli included unconditioned stimuli, learned reinforcement-predictive visual stimuli, and a familiar stimulus long associated with reinforcement, the fixation point (which had a positive value to monkeys because they chose to foveate it to initiate trials). Different populations of neurons in both the basolateral and central amygdala were observed that tracked the positive or negative value of the current state. Positive value–coding neurons increased their firing during the fixation interval and fired more strongly after positive conditioned stimuli and rewards than after negative conditioned stimuli and punishments. Negative value–coding neurons did the opposite, decreasing their firing during the fixation interval and firing more strongly after negative CSs and punishments than after positive CSs and rewards. Hence the overall pattern of responses in the amygdala indicated that neurons tracked moment-to-moment value.

Furthermore, amygdala responses appeared to provide a *graded* representation of value that spanned positive and negative valences. For example, in a positive value–coding cell, responses elicited by the cue were high on large-reward trials, low on punishment trials, and intermediate on small-reward trials. Similarly, in a negative value–coding cell, responses to the cue were strong on punishment trials, weak on large-reward trials, and again intermediate on small-reward trials (figure 2.9, plate 1). This pattern of response implies that amygdala neurons did not simply represent the association of the conditioned

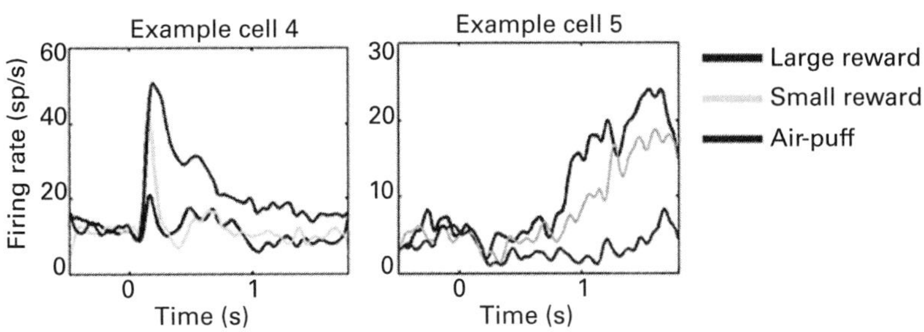

Figure 2.9 (plate 1)
Graded representation of value. (Left) Cell responses during presentation of the visual conditioned stimulus from a cell that encoded negative value. (Right) Cell responses during the "trace interval" (interval between the visual stimulus offset and outcome delivery) from a cell that encoded positive value. Reproduced with permission from Belova, Paton, and Salzman 2008.

stimulus with the sensory properties of a preferred unconditioned one (water or air puff). Instead, the responses to the CSs reflected an *integration* of information about multiple reinforcers with different sensory properties—combining somatosensory, gustatory, and auditory pathways. These response properties suggest that amygdala neurons build associations between learned cues and the value of the corresponding unconditioned stimulus, which can differ in sensory modality, valence, and magnitude.

Consistent with the idea that the basolateral amygdala is involved in value representation, this brain area also exhibits signals linked to reward expectancy. Yasuko Sugase-Miyamoto and Barry Richmond (2005) investigated a task that contained one or more phases of color discrimination. Specific cues indicated whether reward could be obtained immediately or whether additional phases were needed for obtaining a reward, in which case more effort was necessary to obtain it. Amygdala neurons signaled whether trials would require one, two, or three phases. Presumably, these signals could then be used to adjust the motivational level of the animal while it attempted to obtain the reward.

Results from a study that recorded neuronal signals intracranially in humans further suggest amygdala involvement in value representation (Jenison et al. 2011). On individual trials, subjects were given $3 and asked to make a purchase decision regarding a food item (sweet and salty junk food)—that is, to bid between $0 and $3 for the item. They were allowed to keep all unspent money. Single neurons, located primarily in the basolateral amygdala, responded in a manner compatible with the computation of stimulus value at the time of decision making.

In summary, the above findings illustrate that the amygdala plays an important role in encoding value, which is observed not only at the level of the stimulus, but also at the level of "representations." For instance, conditioned responses to a tone that originally was paired with food are reduced if the food is devalued by subsequent injection of a toxin, even though this occurs in the absence of the tone itself (Cardinal et al. 2002; Holland and Gallagher 1999). Evidence for participation in value representations is particularly well established for the basolateral amygdala, as revealed by a rich set of learning paradigms; nonetheless, some forms of value encoding are observed in the central amygdala as well (Belova, Paton, and Salzman 2008).

Decision Making: Nonhuman Animal Data

A central component of decision making requires the evaluation of the costs associated with different candidate actions relative to their potential rewards. In some experimental studies, animals are required to choose between options that yield smaller, easily obtainable rewards and larger, more costly rewards. Examples of "costs" include delaying delivery of the reward and requiring increased physical effort to obtain it. Imposing these costs leads to a "discounting" (i.e., devaluation) of larger rewards, thus increasing preference for smaller, lower-cost rewards (Floresco et al. 2008). A growing literature shows that the basolateral amygdala is of critical importance in these forms of decision making. Lesions or inactivations of the basolateral complex reduce preference for larger, delayed rewards (Winstanley et al. 2004), as well as for larger rewards that require increased effort (Floresco and Ghods-Sharifi 2007; Ghods-Sharifi, St. Onge, and Floresco 2009), in addition to reducing tolerance for uncertainty (Ghods-Sharifi, St. Onge, and Floresco 2009).

During a delay-based decision making task, an animal is typically confronted with a simple choice: immediate delivery of a small reward versus delayed delivery of a larger reward. Within certain parameters, animals, like adult humans, will choose delayed larger rewards over immediate smaller ones. In another paradigm, a rat must choose between two levers, one of which provides a small reward, the other a large reward; in a study by Catharine Winstanley and colleagues (2004), because the trial length was kept constant across conditions, the optimal strategy was to choose the lever linked to the larger reward (since it maximized the total reward provided). Lesions of the basolateral amygdala increased impulsive choices (i.e., no-delay choices), indicating that the intact amygdala *reduced* the discounting associated with the delay, which allowed the animal to wait longer. Hence maintaining a representation of the reward value online was compromised by basolateral damage.

Another way to impose a choice between two options is to contrast them based on their associations with different amounts of reward and effort. For example, the animal may be asked to choose a food well located to the left or right of a decision location. But, to acquire the high reward, the animal is required to exert greater effort (e.g., climb over a scalable barrier) than it is for a low reward. Intact rats typically choose the high- over the low-reward option (given that they are typically hungry in these experimental manipulations). In contrast, rats with inactivations or lesions of the basolateral amygdala choose the low-reward option (Floresco and Ghods-Sharifi 2007; Ghods-Sharifi, St. Onge, and Floresco 2009). In other words, a compromised basolateral complex reduces an animal's preference to work harder to obtain rewards of larger magnitudes. These findings indicate that the basolateral amygdala is involved in encoding the expected magnitude of rewards linked to different choices.

Yet another strategy to evaluate decision making is to manipulate the probability of reward delivery. For instance, an animal may be faced with choosing between a certain smaller reward and an uncertain larger one. In a study by Sarvin Ghods-Sharifi, Jennifer St. Onge, and Stan Floresco (2009), rats chose between a certain-reward lever and a "risky" lever that was linked to 50, 25, or 12.5 percent probability of receiving the larger reward (where the probability was fixed during a block of trials). Rats with lesions of the basolateral amygdala exhibited less tolerance for the risky choice, revealing a form of risk aversion.

A study by Grabenhorst, Hernádi, and Schultz (2012) investigated decision making in rhesus monkeys. The animals performed in a free-choice economic task and chose between "saving" a liquid reward with interest for future trials and "spending" the already accumulated reward immediately. Importantly, the design dissociated the internal process of choice (save vs. spend) from action selection. Behaviorally, the results suggested that monkeys kept track of accumulated rewards over successive save-trials and based their choices on this information. In terms of brain responses, activity of individual basolateral amygdala neurons predicted the behavioral choice independently of specific behavioral actions (figure 2.10). This study provides some of the most compelling evidence that the amygdala *codes* economic choices, thus revealing a more direct role of this structure in economic decisions—beyond a role in valuation itself.

Thus, as these studies show, the basolateral amygdala plays an important role in decision making (see also Seymour and Dolan 2008), a role that necessitates integration of many types of information, including response cost, incentive magnitude, incentive valence, motivational state, and prior learning history, to gauge potential rewards in choosing between distinct courses of action. On the whole, the basolateral amygdala can lead to choices that generate greater long-term payoffs.[8] Conversely, damage to this region reveals behaviors that are at times impulsive, lazy, or risk averse (Floresco et al. 2008; for related findings not discussed here, see the work by Geoffrey Schoenbaum, Charles Pickens, Thomas Stalnaker, and colleagues, in particular, Pickens et al. 2003; Stalnaker et al. 2007).

Of course, the basolateral amygdala does not operate in a vacuum in shaping decision making. The value representations in the basolateral complex and how they guide choice behaviors are parts of circuits that include the frontal lobe, the ventral striatum, and related dopaminergic structures (Cardinal et al. 2002; Floresco et al. 2008; Haber and Knutson 2010; Hampton et al. 2007; Robbins and Everitt 1996).

Decision Making: Human Data

Impairments in decision making as experienced during the Iowa Gambling Task are well known for patients with ventral-medial prefrontal cortex lesions, as studied by Antoine Bechara, Hanna Damasio, and Antonio Damasio (2000).

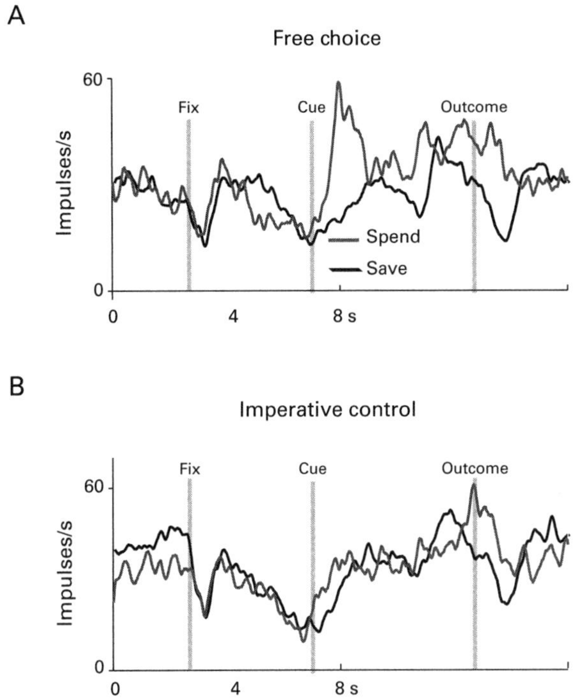

Figure 2.10
Choice coding in the basolateral amygdala. (A) Choice-predictive responses in the free choice task. Response time courses for save (black) and spend (gray) choices. The differential response did not reflect cue differences, as both cues appeared in all trials. (B) Activity in the forced choice control task (where the required choice was indicated to the subject) did not distinguish between save–spend trials. Reproduced with permission from Grabenhorst, Hernadi, and Schultz 2012 and kindly provided by Fabian Grabenhorst.

Intriguingly, patients with bilateral amygdala lesions exhibit a similar deficit (Bechara et al. 1999). Like ventral-medial patients, they fail to choose consistently from the advantageous versus disadvantageous decks of cards. However, Bechara and colleagues also obtained evidence that the contributions of the amygdala and ventral-medial prefrontal cortex during the gambling task are partly dissociable. For example, ventral-medial patients generated skin conductance responses when told that they had won or lost play money, whereas amygdala patients did not.

The impairments in decision making during the Iowa Gambling Task agree with observations that amygdala patients exhibit poor judgment and decision making in real-life situations (Tranel and Hyman 1990). For instance, one patient exhibited "mild improprieties and irrationalities in her social behavior" (Tranel and Hyman 1990, 352). More recent investigations of patients with a compromised amygdala have extended our understanding of potential impairments in decision making. In addition to tasks involving uncertainty, such as the Iowa Gambling Task, tasks involving risk (i.e., probabilistic gains and losses) have also been investigated. For instance, in one study (Weller et al. 2007), amygdala patients were impaired during scenarios involving potential gains (see also Talmi et al. 2010), but not when potential losses were involved. Specifically, they showed elevated risk taking and insensitivity to differences in expected value between choice options in the gain domain, while exhibiting a pattern of decision making that was very similar to that of healthy individuals in the loss domain—namely, taking more risks when it was advantageous to do so and fewer risks when it was disadvantageous to do so (as opposed to ventral-medial patients, whose performance was compromised in both situations). A second study involving patients with amygdala lesions (Brand et al. 2007) also documented impairments in risky decision making. Interestingly, the deficit appeared to be more intense in those patients who also demonstrated deficits in executive function (e.g., a modified card sorting task).

A recent study (van Honk et al. 2013) investigated decision making in three subjects with selective bilateral lesion of the basolateral amygdala (with central amygdala sparing), an extremely rare condition. In an economic "trust" game, the amygdala lesion subjects invested nearly 100 percent more money in unfamiliar persons than healthy controls. This generosity was not attributable to risk-taking behaviors in nonsocial contexts. The findings suggest that selfish behaviors (of the kind displayed by control subjects) are learned from interactions with the social environment that require the basolateral amygdala.

The results of neuroimaging studies also support the concept that the amygdala encodes value in particular and is involved in decision making more generally. In an appetitive conditioning task, amygdala responses evoked by a stimulus predictive of a positive item decreased after devaluation (Gottfried, O'Doherty, and Dolan 2003). When a food changed from appetitive (given hunger) to disagreeable (given satiety), the responses evoked by the predictive cue were attenuated. At the same time, amygdala responses were maintained to stimuli predictive of other palatable (i.e., nondevalued) stimuli. These findings suggest that the amygdala encodes predictive reward value, consistent with animal learning paradigms.

Humans are quite sensitive to the way choices are framed (Kahneman and Tversky 1979). That the amygdala's responses are sensitive to this framing effect suggests it participates in biasing decision making. In one study (De Martino et al. 2006), amygdala responses were driven by a combination of a subject's decision and the frame (gain vs. loss) in which it took place, rather than the valence of the frame per se. These results accord with the notion that frame-related valence information, as provided by amygdala responses, is incorporated into the relative assessment of options, thus affecting individual decisions (see also Coricelli et al. 2005). However, inasmuch as humans with amygdala damage still exhibited an intact framing effect, though choice preference was altered (Talmi et al. 2010), the amygdala does not appear to be causally related to framing. Amygdala responses are also sensitive to the experience of regret. In one experiment (Coricelli et al. 2005), subjects became increasingly regret aversive during the session, a progression reflected by changes in responses in the amygdala (among other regions).

To investigate the contribution of the amygdala to reward-related signals in prefrontal cortex, two patients with bilateral amygdala lesions were scanned with functional MRI during a reversal learning task where they learned which of two choices was the more rewarding and then flexibly switched their choice when contingencies changed (Hampton et al. 2007). Behaviorally, both patients exhibited markedly different performance from the controls: both were significantly more likely to switch stimulus choice than the controls, overall or following receipt of reward. In terms of brain responses, choice-related signals (switch vs. stay) in anterior cingulate cortex and in the anterior insula (abutting lateral orbitofrontal cortex) were significantly attenuated in the amygdala patients, as were expected reward signals during the choice period in orbital and medial prefrontal cortex. Combined, these findings reveal that signals in orbital, medial, and lateral prefrontal cortex linked to computing an expected reward and determining behavioral choice based on that reward depend on signals from the amygdala for their integrity. They also demonstrate the importance of amygdala–prefrontal cortex interactions to the computation of expected reward value in humans (see also Park et al. 2011), supporting a model of decision making where reward-related signals are used to guide behavioral decisions.

During decision making, some situations are naturally ambiguous (when information is missing), whereas others involve risk (when chance is involved and even when the probabilities are available). Some studies have reported amygdala activation during decisions involving ambiguity (e.g., Hsu et al. 2005); others have questioned this finding (Bach et al. 2011; Bach, Seymour, and Dolan 2009). During risky decision making, when high-risk trials are contrasted to low-risk ones, the amygdala is not routinely engaged, although the anterior insula and dorsal-medial prefrontal cortex are (Mohr, Biele, and Heekeren 2010).[9]

Value Representation and Decision Making: Humans versus Animals

Although the studies reviewed above document a significant role for the amygdala both in value representation and in decision making, we need to consider an apparent contradiction when we compare findings from humans and nonhuman animals. Whereas amygdala lesions in humans seem to lead to risky decisions, such lesions in rats render the animals risk averse (as well as less willing to exert effort or tolerate uncertainty in order to gain larger rewards). It is possible to reconcile these contradictory findings by noting that in both cases, human and animal, amygdala lesions lead to suboptimal behavioral choices.[10] Thus lesioned humans in the Iowa Gambling Task picked from the disadvantageous decks of cards (Bechara et al. 1999), and lesioned rats chose the disadvantageous, certain reward condition, which yielded fewer pellets (on average) than the riskier conditions (Ghods-Sharifi, St. Onge, and Floresco 2009). Although damage to the amygdala may deprive the organism of circuits that establish the relative values of different rewards, generating suboptimal choices, true species differences may also be partly responsible for the discrepant behavioral outcomes. Finally, it is also conceivable that in the nonhuman animal studies reviewed here, the involvement of the amygdala was more closely related to other computations than those suggested since, in these complex tasks, multiple processes may have contributed to the underlying behavioral changes. Although the data reviewed in the previous three sections indicate that the amygdala makes important contributions to the representation of value and decision making, more data are clearly needed for both humans and nonhuman animals since many issues remain unresolved.

Amygdala Evolution

Readers skeptical of the participation of the amygdala in the broad array of mental processes described here might object that the amygdala is too "primitive." Not only is it clearly found in all mammals; it is also discernible in reptiles, amphibians, and even fishes. In a comprehensive review, Fernando Martínez-García, Amparo Novejarque, and Enrique Lanuza (2007, 309) propose that "topological and functional data suggest that ray-finned fishes and land vertebrates probably share an ancestor that already possessed an amygdala." If so, an early version of the amygdala dates back more than 400 million years. Thus, skeptics might contend, the amygdala may be better viewed in terms of fear-related mechanisms associated with an effective "alarm system"—one that has been evolutionarily conserved for good reasons.

That said, this putative alarm system has had plenty of time for modifications. For instance, the *basolateral* amygdala is an amniote innovation (amniotes include reptiles, birds, and mammals; Martínez-García, Novejarque, and Lanuza 2007). For its part, the primate amygdala may have evolved into a *less* specialized system so as to cope with new environmental problems (Sander, Grafman, and Zalla 2003).[11] This may have occurred through expansion of connections with a wider range of cortical and subcortical territories, such as direct connections between the amygdala and lateral prefrontal cortex (Ghashghaei, Hilgetag, and Barbas 2007), between the amygdala and the reticular nucleus of the thalamus (Zikopoulos and Barbas 2012), and between the amygdala and visual cortex (Tigges and Tigges 1985), all of which appear to be present only in primates.[12] Indirect connections via highly connected cortical regions such as anterior cingulate and orbitofrontal cortex may also have been involved. Altered and enhanced connectivity may allow a system to expand the repertoire of functions it participates in. Despite the constraints on brain evolution (see Striedter 2005), dramatic changes in connectivity have been documented—for example, those involving somatosensory cortex and the thalamus in several groups of mammals (Kaas 2004; Krubitzer 2009). Degrees of connectivity should also be taken into account: whereas mice have about 10 cortical fields, and macaque monkeys more than 50 fields, humans may have more than 100 (Krubitzer 2009)—possibly as many as 150–200 (Glasser and Van Essen 2011).[13] The combinatorial nature of connectivity is such that, in humans, a highly interconnected region like the amygdala (see chapters 8 and 9) is in a position to participate in an impressive array of cognitive-emotional behaviors.

More broadly, to understand the function of a highly conserved element in a new context, it is not sufficient to presuppose that it has undergone little change. It is crucial to understand the structural and functional *embedding* of the element in its new context. For instance, a specific molecule in a new context may be part of a much richer cascade of interactions, possibly involving multiple molecular circuits. Likewise, a brain region with new inputs and outputs may well be involved in new mechanisms. Consider a hypothetical new output from the central amygdala to lateral prefrontal cortex. The existence of such a pathway would enable outputs that, say, typically affected the descending circuits controlling bodily functions to influence brain regions crucial to a host of cognitive capacities. Affective signals would be directly connected to cognitive ones, and this exchange could then generate novel types of signal interaction. The same argument can be made in reverse, namely, by considering a hypothetical new connection from lateral prefrontal cortex to the central amygdala. The key point here is that a few changes in brain architecture have the potential to significantly change the functional repertoire of a region. And even though this example is hypothetical, it is broadly concordant with anatomical data in primates.[14] In the present context, it is also interesting to consider the superior colliculus, a sensorimotor structure found in the midbrain of the earliest vertebrates (Wurtz and Albano 1980; see chapter 3). According to some, it has been retained in a conserved form and is present in the dorsal midbrain of modern mammals with relatively minor modifications (Stein 1981). Although the superior colliculus predates the evolutionary expansion of the cerebral cortex by several hundred million years (Gaither and Stein 1979), in mammals, it participates in many circuits involving cortex. For example, it receives inputs from visual, parietal, and anterior cingulate cortex (see Comoli et al. 2012 for references), which may be associated with initiating or modulating approach and defense behaviors in rats (Comoli et al. 2012; see chapter 3). In primates, dense connections originate from the frontal eye field in frontal cortex (Preuss 2007). These newer circuits via cortex clearly change the functional capabilities of the superior colliculus.

Another intriguing example concerns orbital prefrontal cortex, parts of which are specific to primates. These parts provide the earliest sites for convergence of *visual* information with gustatory, olfactory, somatosensory, and visceral inputs. Critically, most of the visual inputs to orbital PFC arrive in sectors (in Brodmann areas 11 and 13) that are specific to primates (Passingham and Wise 2012). Again, these changes in connectivity provide new avenues for signal integration.

In general terms, an element's new structural embedding generates a novel functional landscape. Given nonlinear relationships, the new gamut of functions may involve processes that are significantly different from the earlier ones. Obviously, this does not negate the possibility that, in some cases, conservation of structure is indeed linked to conservation of function (consider, for example, the family of Hox genes that determines the overall body plan; Carroll 1995).

Conclusions

In the past two decades, our understanding of the amygdala has greatly expanded. We have learned a great deal not only about the amygdala's contribution in fear-related processes, but also about its many other functions. As the examples in this chapter illustrate, the amygdala's role is indeed quite broad and connected to functions described as cognitive, such as attention (Holland and Gallagher 1999) and decision making (Seymour and Dolan, 2008).

As reviewed, both animal and human studies have documented several ways in which the amygdala operates like attention. Indeed, some forms of *affective attention* are very similar to "regular" attention to neutral items, including competitive interactions (Hindi Attar et al. 2010) and, possibly, increases in response gain. Preliminary findings even suggest that the amygdala participates in visual *spatial* attention, at least in a coarse manner (Peck, Lau, and

Salzman 2011)—for example, enhanced responses were observed when valuable cues appeared in the contralateral hemifield. As further developed in chapter 7, the contributions of the amygdala to attention originate from its participation in several circuits.

Another form of attention deserves to be mentioned here, one that engages the body. As noted above, the central amygdala is at times viewed as a "controller of the brainstem" (Cardinal et al. 2002) and uses its widespread projections to the hypothalamus and brainstem nuclei to coordinate behavioral, autonomic, and neuroendocrine responses.[14] Given the effects of these structures on bodily states and the regulation of the internal milieu, a more direct link with emotional states, notably fear, is established (figure 2.11). Because this type of bodily mobilization, which relies on "descending" circuits, co-occurs with the recruitment of cortical regions involved in salience processing, including dorsal-medial prefrontal cortex and the anterior insula (Hermans et al. 2011), it should be viewed as mobilizing not only the body but also the brain.

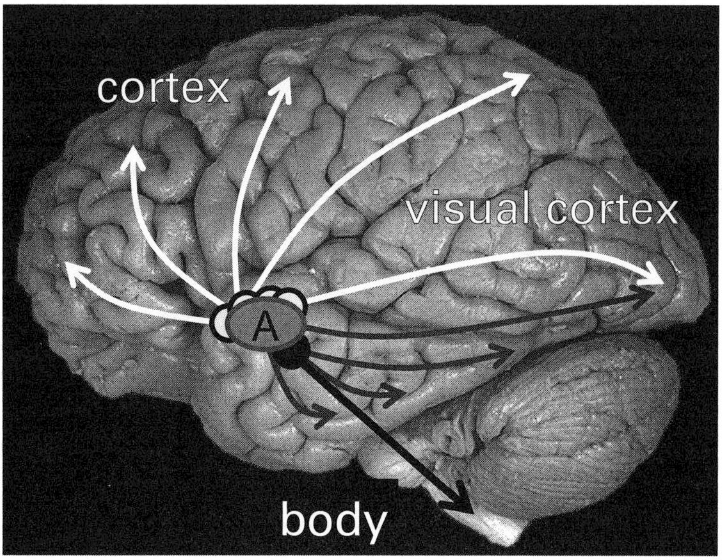

Figure 2.11
Affective attention. Affective information shapes information processing in the brain through several mechanisms. The mechanisms of affective attention highlighted here depend on the amygdala (ellipse with "A"). Diffuse projections from the basal forebrain (white arrows) modulate responses throughout the cortical mantle, including sensory cortex. Efferent projections from several amygdala nuclei (gray arrows) reach multiple stages of visual cortex, including both "late" and "early" ones. The central amygdala also originates descending projections (black arrow) via the hypothalamus and other brainstem nuclei that mobilize bodily resources as a function of affective significance. Reproduced with permission from Pessoa 2010a.

Decision making is an even broader concept than attention. It is often studied by employing complex tasks that rely on multiple processes. Although it is hard to isolate the precise contributions of the amygdala to these tasks, it is also certain that decision making is altered in both animals and humans when the amygdala is compromised. One way by which the amygdala may affect decision making is by biasing the representation of value. Insofar as decisions must take into account both costs and benefits, an impaired computation of stimulus or action value will lead to altered behaviors. Furthermore, as several examples have showed, *without* the amygdala, animals can behave in an impulsive, lazy, or risk-averse fashion. In such cases, "emotion" appears to contribute to outcomes frequently linked with reasoning and cognition—for instance, less impulsive, effortful outcomes.

One aspect of "What is it?" processes concerns "registering the input." As advanced by Pribram and McGuinness (1975), the amygdala plays an important role in this regard. But, given its contributions to both attention and decision making, the amygdala's operation far exceeds simply responding to input stimuli. As I have proposed, it participates in deciding what needs to be done when the stimulus is encountered. Moreover, "What is it?" and "What's to be done?" can also be viewed as basic building blocks fundamental to the construction of emotion (see also Kober et al. 2008). Answering "What is it?" and "What's to be done?" requires that the organism mobilize resources as it seeks out additional information from the environment (figure 2.11). And, when answering these questions redirects mental and bodily resources more forcefully, especially by engaging the hypothalamus and brainstem structures (via the central amygdala), an emotional experience will ensue (figure 2.11, black arrow).

Given that the amygdala is richly interconnected with both cortical and subcortical structures (Petrovich, Canteras, and Swanson 2001; Young et al. 1994), it is part of many intersecting networks of regions (Pessoa 2008). That it is increasingly recognized as performing important functions in cognitive, emotional, and social processes reflects the amygdala's important roles in integrating and coordinating signals associated with these domains.

Notes

1. Newer neuroimaging methods of higher resolution show promise in studying both amygdala structure (Saygin et al. 2011; Solano-Castiella et al. 2010) and function (Gamer, Zurowski, and Buchel 2010).

2. Pavlov (1927) originally described the "orienting response" as the "What is it?" reflex.

3. In *trace conditioning*, the initial conditioned stimulus cue is separated from the unconditioned stimulus by a temporal gap, unlike in delay conditioning, where the two stimuli coterminate.

4. The term "substantia innominata" also denotes the nucleus of Meynert but is falling into disuse as more of the "unnamed material" is being characterized.

5. Although cholinergic mechanisms are commonly emphasized, other *non*cholinergic systems may be involved as well (Lin and Nicolelis 2008).

6. Of course, these lesion findings do not rule out potential mechanisms involving the amygdala and other indirect circuits. For further discussion, see chapter 7.

7. The experiments reviewed here do not address the induction and experience of fear per se, for which the amygdala appears to play a pivotal role (LeDoux 1996; see also Feinstein et al. 2011).

8. The amygdala also biases decisions toward larger, uncertain rewards, which may rely on connections between the basolateral amygdala and the accumbens (St. Onge et al. 2012).

9. Interestingly, during gambling tasks, when subjects find out the outcome of the gamble, larger responses in the amygdala are observed for win versus lose trials—that is, for positive trials (Padmala, MacPherson, Lejuez, and Pessoa, unpublished results).

10. I thank Stan Floresco for discussions on this point.

11. When the term "primate" is not further specified, it applies to both human and nonhuman primates.

12. Rats appear to lack an analogue of dorsal-lateral PFC areas of primates; indeed, it is questionable whether they have a *lateral* component of prefrontal cortex analogous to that in primates (Preuss 1995), although this issue is still debated (for an excellent discussion, see Passingham and Wise 2012).

13. A "cortical field" is a purported basic "processing unit," which can be defined based on multiple criteria, such as set of cortical and subcortical connections, neural response properties, distinct appearance in histologically processed tissue, and molecular signature (Krubitzer 2009).

14. Although the central amygdala does not appear to be connected with lateral prefrontal cortex, the basolateral amygdala is, though not strongly so (Ghashghaei, Hilgetag, and Barbas 2007), and the two amygdala subdivisions are also interconnected (Amaral et, al. 1992).

15. The projections of the central amygdala affect several structures, including the periaqueductal gray, the reticular formation, and the pituitary gland.

3 Affective Visual Perception

In discussing the amygdala's involvement in multiple functions such as attention, value representation, and decision making, chapter 2 rejected a narrower characterization of this structure as limited to fear processing or the handling of negative information. The present chapter will evaluate a specific aspect of purported amygdala function, one suggested to involve a subcortical-only pathway that rapidly conveys affective information. The evidence related to the subcortical pathway is discussed in detail, given the functional implications of the associated framework.

How does an animal assign biological *value* to a stimulus? Which stimuli are good and which are bad? Which should be approached and which should be avoided? The early work of Heinrich Klüver and Paul Bucy (1939) implicated temporal lobe structures in the evaluation of affective significance. Indeed, temporal lesions led to what they termed "psychic blindness": "The ability to recognize and detect the meaning of objects on the basis of visual criteria alone is either lost or seriously disturbed. . . . Certain properties of the objects, their being 'dangerous,' 'inedible,' or 'indifferent,' have suddenly become ineffective in determining visually guided reactions" (Klüver and Bucy 1939, 609, 612).

Work by Lawrence Weiskrantz (1956) involving more circumscribed lesions implicated the amygdala in the behavioral changes originally reported by Klüver and Bucy (1939).[1] Subsequent findings not only pointed to the amygdala in the evaluation of affective significance but also led to a framework that runs as follows: emotional stimuli are processed initially by a dedicated, modular system that operates rapidly, automatically, and largely independently of conscious awareness (Tamietto and de Gelder 2010). This framework, termed the "standard hypothesis" (Pessoa and Adolphs 2010), has two central and interrelated components (Öhman 2005; Tamietto and de Gelder 2010): (1) the purported role of the amygdala in the rapid, automatic, and nonconscious processing of emotional stimuli; and (2) a specific subcortical route of information

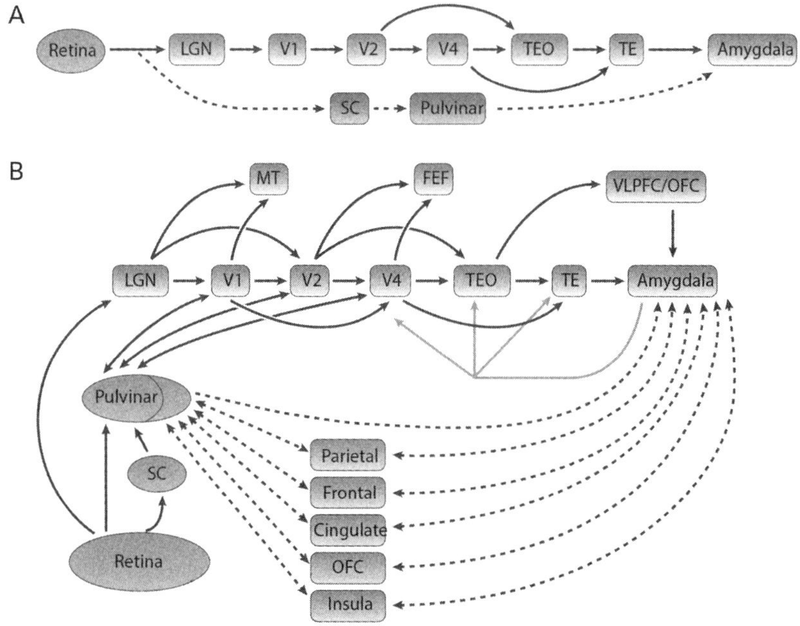

Figure 3.1 (plate 2)
Visual pathways. (A) Traditional flowchart of visual processing typically emphasizes the lateral geniculate nucleus–V1–V2–V4–TEO–TE pathway, although the scheme is not strictly hierarchical. According to the "standard hypothesis," a subcortical pathway involving the superior colliculus and the pulvinar nucleus of the thalamus provides fast and automatic access to the amygdala. (B) Alternative flowchart of visual signals via multiple pathways, including "shortcuts." The "multiple waves" of activation initiate and refine cell responses. For simplicity, feedback pathways, which are extensive, have been omitted. The existence of such feedback pathways dictates, however, that a complex ebb and flow of activation sculpts the neuronal profile both of activation throughout visual cortex and of amygdala responses. Some of the connections between the pulvinar and visual cortex and between the pulvinar and "associational" areas are also indicated. The curved line in the pulvinar schematically separates the medial pulvinar (to the right of the line) from the rest of the structure. FEF, frontal eye field; LGN, lateral geniculate nucleus; MT, medial temporal area (also known as "area V5"); OFC, orbitofrontal cortex; SC, superior colliculus; TE, TEO, inferior temporal areas TE, TEO; VLPFC, ventral-lateral prefrontal cortex; V1, V2, V4, visual areas 1, 2, 4. Reproduced with permission from Pessoa and Adolphs 2010.

processing—the "low road" (LeDoux 1996)—that bypasses the presumably slower, resource-dependent cortex and that culminates in the amygdala by way of the *superior colliculus* and the *pulvinar* nucleus of the thalamus (figure 3.1A; plate 2). The fact that this pathway bypasses cortex is thought to endow the processing of emotion-laden visual stimuli with the properties set forth in the framework's first component.

Broadly speaking, the concept of a modular subcortical pathway is strongly aligned with one of the most entrenched themes in cognitive neuroscience and cognitive psychology: a dual scheme distinguishing automatic from controlled processing (Shiffrin and Schneider 1977). It is also aligned with the idea that emotion and cognition constitute separate mental and neurobiological domains. The standard hypothesis has shaped both basic and clinical research. For example, defects in the modular brain system it proposes are suggested to underlie phobias, mood disorders, and posttraumatic stress syndrome. This chapter provides a critical reexamination of the hypothesis with an emphasis on revising the roles of the brain areas involved.

Standard Hypothesis

The data and theory that underpin the standard hypothesis are not typically articulated in detail, and the central tenets of the hypothesis are often expressed vaguely. The main argument is that, insofar as affectively laden information has survival value, it has driven adaptations in information processing that are reflected in a structurally and functionally modular system (Öhman and Mineka 2001). The standard hypothesis is based not only on empirical data but also on theoretical considerations. In a nutshell, it claims that a system specialized for rapid detection of predators makes sense from an evolutionary perspective.

The purported modularity of the system entails automaticity (Dolan and Vuilleumier 2003): owing to the potency of affective information, this information is processed independently of attention and awareness (see chapter 4 for further discussion). For example, threat-expressing faces are reported to be processed preattentively in visual search paradigms (Öhman, Lundqvist, and Esteves 2001), and fearful facial expressions break into consciousness more quickly than happy ones during continuous flash suppression (a technique to render visual stimuli nonconscious; Yang, Zald, and Blake 2007). Moreover, hemodynamic responses in the amygdala are proposed to occur in response to backward-masked fearful faces (putatively rendering the faces invisible; Morris, Öhman, and Dolan 1998; Whalen et al. 1998) and even in response to unmasked fearful faces in patients with blindsight (Morris et al. 2001; Pegna et al. 2005; see "Blindsight" under "Is Affective Visual Processing Independent of Attention and Awareness?" below).

The standard hypothesis assumes as well that the anatomical components of the system enable emotion processing to occur entirely subcortically (Morris, Öhman, and Dolan 1999). This assumption has its roots in rodent studies demonstrating the existence of a subcortical pathway through the auditory thalamus to the amygdala that is sufficient for some forms of auditory Pavlovian fear conditioning (LeDoux 1996). A similar subcortical route is assumed to exist for visual information in primates, including humans. The notion of such a subcortical pathway is appealing because it is assumed to be faster than a cortical one, and processing of affective stimuli is thought to be adaptive in part because it is fast. For instance, judgments of threat can be made from facial stimuli that are displayed as briefly as 39 ms (backward masked; Bar and Neta 2006). Because the route is assumed to be subcortical, processing of visual information along this pathway is assumed to be coarse. Thus coarse (i.e., low–spatial frequency) information from affective stimuli is believed to engage subcortical visual processing, consistent with findings that the amygdala is activated more strongly by emotional faces presented with low than high spatial frequency (Vuilleumier et al. 2003).

This chapter will discuss several shortcomings of the standard hypothesis and will describe an alternative model. The focus will be on visual processing in the primate brain, particularly in the pulvinar nucleus of the thalamus, given its critical connective role in the purported subcortical pathway. The chapter will review the functional properties of affective vision central to the standard hypothesis and highlight the issues of speed and coarseness. It will then briefly discuss other important notions linked to the hypothesis, including the role of attention and the modularity of the brain (these will be covered in detail in chapters 4 and 8, respectively; see also Adolphs 2008; Duncan and Barrett 2007; Lewis 2005; Pessoa 2005, 2008). Finally, it will outline an alternative model—the "multiple waves model"—that assigns a larger function to cortical processing of affective visual information. I suggest that this model can satisfactorily explain the findings used to support the existence of the subcortical route proposed by the standard hypothesis.

Before evaluating the standard hypothesis in greater detail, two general considerations are in order. First, to dissect the flow of visual information in the primate brain is extremely difficult, chiefly because its visual processing is highly distributed, both temporally and spatially (Bullier 2001). Although the majority of retinal ganglion cells project to the lateral geniculate nucleus (LGN) in the thalamus, there are additional projections to the superior colliculus, to the pulvinar, and to several other subcortical nuclei (see figure 3.1B; plate 2). In total, at least ten pathways from the retina have been established (Cowey 2004); thus neurons in visual cortex can receive information via multiple channels. Moreover, the actual visual "drive" into a region (e.g., ascending retinal projections to the LGN) constitutes a small fraction of the total synaptic inputs to that region, most of which reflect extensive intrinsic processing and feedback projections (Douglas and Martin 2004). These facts greatly complicate the accounts of even standard (i.e., nonemotional) visual neuroscience (Masland and Martin 2007; Nassi and Callaway 2009).

Second, it is often also difficult to determine precisely all the anatomical and physiological components of visual processing linked to a given brain structure because the information available is derived from different species (e.g., rats and mice, monkeys, humans) and from different methods (e.g., electrophysiology, functional MRI, and lesion studies).

General Functional Issues

Speed of Visual Processing

The electrophysiological responses evoked by visual stimuli can be modulated by emotional content, and this modulation has been reported to occur at short latencies, in some human studies, at around 100 ms after stimulus onset (Pizzagalli, Regard, and Lehmann 1999; Halgren et al. 2000). In addition, the "N170" component of the electroencephalographic (EEG) signal, or the "M170" component in magnetoencephalography (MEG) studies, which is linked to face processing, is modulated by facial expression in some cases (Eger et al. 2003; Japee et al. 2009). In contrast, numerous investigations have shown such effects only at longer latencies, ranging from 200 to 400 ms (e.g., Krolak-Salmon et al. 2004). And some studies, including intracranial ones, have observed emotional modulation only at latencies from 600 to 800 ms and longer (e.g., Brazdil et al. 2009). It is unclear why the timing of emotional modulation varies so widely (from less than 100 ms to more than 800 ms), but one possibility is that the effect is highly context dependent (see below)—which would run counter to some of the main tenets of the standard hypothesis.

Both EEG and MEG are excellent techniques to study the temporal evolution of brain signals, but because signals measured at the level of sensors do not uniquely constrain the neural sources—the "inverse problem"—localizing responses is problematic with either technique. Subcortical neural sources pose additional problems, given their deep origin and noncortical structure. Thus, although signals detected in EEG/MEG studies reveal fast emotional modulation and are used as evidence in favor of fast subcortical processing, because their origin is unclear, the signals might not arise in the proposed subcortical pathway.

Affective modulation of brain responses certainly can be fast. But how does its timing compare to the speed of visual processing in general? One way to assess this is to measure and contrast response latencies across brain areas (Bullier 2001). For example, do responses in the proposed subcortical pathway occur earlier than those in cortical sites? Single-neuron recordings in monkeys show that responses in cortex (even in frontal cortex) are detected with latencies that are within the range of those observed in subcortical areas. Figure 3.2

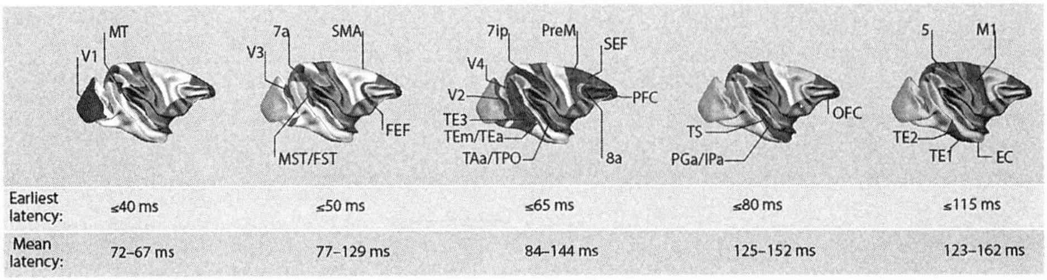

Figure 3.2 (plate 3)
Response latencies to visual stimulation in macaque cortex. The earliest latencies are remarkably short, and even mean response latencies reveal very fast cortical processing. Areas that became active at the given latency after visual stimulation are shown in red, those activated earlier in yellow and those not yet activated in white. Areas for which no information was available are shown in dark gray. EC, entorhinal cortex; FEF, frontal eye field; FST, fundus of superior temporal cortex; 5, 7a, 7ip, 8a, areas 5, 7a, 7 intraparietal, 8a; IPa, superior temporal area IPa; M1, primary motor cortex; MST, medial superior temporal cortex; MT, medial temporal area (also known as "V5"); OFC, orbitofrontal cortex; PFC, prefrontal cortex; PGa, superior temporal area PGa; PreM, premotor cortex; SEF, supplementary eye field; TAa, anterior subregion of superior temporal area TA; TE1–TE3, inferior temporal areas TE1–TE3; Tem/TEa, medial and anterior subregions of inferior temporal area TE; TPO, superior temporal area TPO; TS, superior temporal sulcus; V1–V4, visual areas 1–4. Reproduced with permission from Pessoa and Adolphs 2010 and adapted from Lamme and Roelfsema 2000.

(plate 3) illustrates that in macaque cerebral cortex, the earliest latencies are remarkably short, and even mean response latencies demonstrate extremely fast cortical processing (Lamme and Roelfsema 2000). Visual response latencies in the pulvinar are between 60 and 80 ms and overlap with latencies observed in early visual cortex areas V1 and V2 (Ouellette and Casanova 2006). In inferior temporal cortex (i.e., "late" visual cortex), latencies can be as short as 60–85 ms (Lamme and Roelfsema 2000) and, strikingly, in some frontal sites such as the frontal eye field, as short as 40–70 ms. These latencies again overlap with those in area V1 (Nowak and Bullier 1997; Schmolesky et al. 1998). Thus, although mean response latencies increase gradually from posterior to anterior visual cortex areas, there is considerable overlap of response times across the brain, as Jean Bullier (2001) has demonstrated (figure 3.3). In the context of the standard hypothesis, it therefore seems that pulvinar responses are not particularly fast. That said, visual response latencies in the superior colliculus are somewhat faster than those observed in the pulvinar, showing an early, transient response around 40–70 ms that may support rapid eye movements during visual orienting (Boehnke and Munoz 2008; note that these response times overlap with frontal eye field responses, which are involved also in eye movements and attention).

Before discussing responses in the amygdala, it is worth describing a recent study that reported responses to face stimuli in the pulvinar (Nguyen et al.

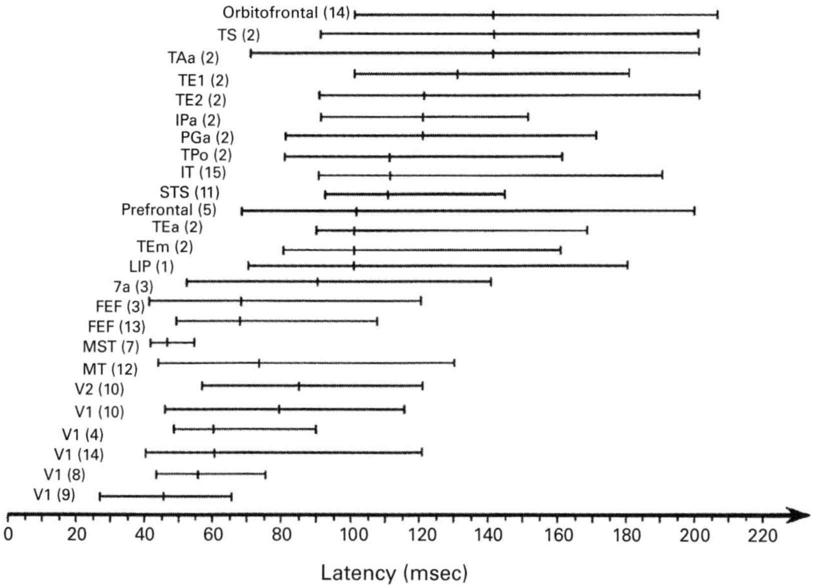

Figure 3.3
Latencies of visual responses of neurons in different cortical areas. For each area, the central tick marks the median latency and the extreme ticks the 10 and 90 percentiles. Numbers in parentheses refer to bibliographic references given in Bullier 2003. Considerable overlap in response timing is observed. LIP, lateral intraparietal area; STS, superior temporal sulcus; TPo, temporal-parietal-occipital region. (See caption to figure 3.2 for key to other abbreviations.) Annotated and reproduced with permission from Bullier 2001.

2013). A subset of the cells studied exhibited short latencies; on average approximately 60 ms after stimulus onset. Faster responses (30–50 ms) were also observed, although these were found when schematic, "face"-like patterns were employed. But because such schematic faces included stimuli with a "mouth" above the "eyes," the critical stimulus feature that engaged the neurons is unclear (if indeed they were responding to "faceness").

What are the response latencies of neurons in the amygdala? In the monkey amygdala, latencies to visual stimuli range from 100 to 200 ms (Gothard et al. 2007; Inagaki and Fujita 2011; Kuraoka and Nakamura 2007; Leonard et al. 1985; Nakamura, Mikami, and Kubota 1992), although shorter response latencies to unspecific stimuli (e.g., fixation spots) have been reported (Gothard et al. 2007). Differences in evoked responses between threatening and neutral or appeasing facial expressions range from 120 to 250 ms (Gothard et al. 2007). Intracranial studies in humans have reported that amygdala responses to visual stimuli (Mormann et al. 2008; Oya et al. 2002) and modulation of amydala responses by affective content (Krolak-Salmon et al. 2004; see also

Oya et al. 2002) both start at around 200 ms, although, intriguingly, response latency varies widely (200–800 ms) across experiments.

In summary, subcortical visual processing is not qualitatively faster than cortical processing. Moreover, the crucial variable is not when the initial stimulus response occurs but when reliable differences between affective and nonaffective stimuli can be detected. According to some proposals (e.g., Tovee and Rolls 1995), most of the information encoded by visual neurons may be available in 100-ms segments of activity (i.e., spiking data within a 100 ms block); a fair amount of information is available in segments of 50 ms, and possibly even in segments of 20–30 ms (note that these segments are taken after the normal latency for neurons to start firing in response to a stimulus). Although these considerations demonstrate the remarkable speed of neuronal computation (at least under some conditions), they add tens of ms to the time that is required to, for example, discriminate between stimuli. In other words, if the response latency in a given region is, say, 120 ms, the time required for differentiating affective and neutral stimuli would be expected to be at least 140–170 ms, and probably longer. A final point is that responses in humans appear to be slower than in monkeys. For example, according to one study (Yoshor et al. 2007), the fastest recording sites in human subjects had response latencies of just under 60 ms and were probably located in area V1 (or possibly area V2), whereas the fastest responses observed in monkey area V1 have latencies shorter than 40 ms (Lamme and Roelfsema 2000). These differences should be taken into account when comparing the speed of processing in monkeys and humans.

A complementary perspective on the speed of visual processing comes from behavioral and electrophysiological studies of scene perception in humans. The evidence suggests that visual processing in general (i.e., including nonaffective information) can be surprisingly fast. Based on a number of tasks involving classification of natural scenes, a substantial amount of information can be gathered from even a single glance at the stimulus. Michelle Greene and Aude Oliva (2009) reported that subjects typically required 19–67 ms to attain 75 percent correct performance on several tasks, including determining a scene's global property (e.g., "natural scene") and basic level categorization (e.g., "forest"); performance reached asymptote at around 100 ms of image exposure. These findings build upon earlier work on scene perception (Potter and Levy 1969), as well as "ultrarapid" visual perception. One investigation showed that EEG responses linked to categorizing images displayed for 20 ms developed approximately 150 ms after stimulus presentation (Thorpe, Fize and Marlot 1996; see also Fabre-Thorpe 2011).

Advances in our understanding of rapid visual perception also stem from computational modeling work inspired by the organization of the visual system, which has revealed that feedforward architectures can account for the performance of humans in rapid categorization tasks (Serre et al. 2007; Serre, Oliva, and Poggio 2007). Several properties of the computer models match neuronal responses in monkey visual cortex (Hung et al. 2005) and responses in human temporal cortex measured intracranially. In one case, single-trial neuronal data from human visual cortex could predict the category of the object presented with data from the first 100 ms after stimulus onset (Liu et al. 2009)—and prediction performance was robust to depth rotation and scale changes. Together, these findings suggest that the feedforward machinery of the visual system is capable of performing complex computations in very short periods of time (such as generating viewpoint-invariant representations).

Another relevant aspect of the work on rapid visual perception concerns attention. There is evidence that visual perception of nonaffective stimuli is not only fast, but also may require less attention than often thought. For instance, rapid visual categorization of novel natural scenes requires little focal attention (Li et al. 2002), indicating that perception outside the focus of attention may extend beyond simple and salient stimuli (see also Peelen, Fei-Fei, and Kastner 2009). These findings bear on the standard hypothesis because they show that many types of complex visual processing can take place even when resources are scarce (but see Evans and Treisman 2005).

Thus visual cortex processing is both efficient and fast (see also Roland 2010). From a behavioral standpoint, sophisticated object perception takes place with very brief presentations. From a physiological standpoint, although mean response latencies increase gradually from posterior to anterior visual cortex, considerable response-timing overlap is observed. As argued below, the data are consistent with a "multiple waves" model of visual cortex processing in which visual information propagates rapidly via several alternative routes. These considerations suggest that the speed of cortical processing has been considerably underestimated in the literature on emotional perception. Consequently, the argument that cortical mechanisms are too slow and in need of a separate subcortical system to account for the properties of affective perception loses much of its forcc.

Is Affective Visual Processing Independent of Attention and Awareness?

A central component of the standard hypothesis, automaticity entails that processing of emotion-laden items take place without need for attention and outside of awareness. To anticipate chapter 4, which will review the literature on attention in greater depth, though the issue is complex, mounting evidence of the effect of attention on emotional processing poses a considerable challenge to the idea of "strong automaticity" (Pessoa 2005). Like regular, nonemotional perception, visual attention is an important factor in determining the behavioral impact and neural fate of an affective visual stimulus.

Blindsight A growing literature also suggests that awareness is a significant factor during emotional perception (Adams et al. 2011; Gray et al. 2013; Hsu, Hetrick, and Pessoa 2008; Pessoa 2005; Straube et al. 2010; see also Kang, Blake, and Woodman 2011). Nevertheless, findings concerning emotional "blindsight" are frequently used to argue in favor of nonconscious subcortical processing. Blindsight is a phenomenon in which patients with lesions in primary visual cortex, though lacking conscious visual experience, have residual visual abilities (Poppel, Held, and Frost 1973; Weiskrantz 1986). More precisely, these patients exhibit above-chance performance on detection and discrimination tasks in the absence of phenomenal visual experience. Several case studies of patients with striate cortex lesions have reported residual visual abilities when presented with emotional facial expressions (de Gelder et al. 1999) or other emotional stimuli (Hamm et al. 2003). In some reports, even with the lack of visual awareness, amygdala activation has been observed (Morris et al. 2001; Pegna et al. 2005). Based on these findings, several authors have argued that emotional visual stimuli activate the amygdala via the subcortical pathway—hence the absence of phenomenal visual experience.

Although the findings of emotional blindsight are noteworthy, the inferences that can be drawn from them are limited. First, the studies do not demonstrate that the amygdala is more important for nonconscious than for conscious processing. For example, in one study involving brief presentations and masking (Pessoa et al. 2006), the pattern of amygdala responses arose when the subjects *reported* seeing something fearful or not, rather than when the fearful face was actually presented. Indeed, amygdala responses during "false alarm" trials (when subjects incorrectly reported seeing a fearful face not presented) were *greater* than those during "correct reject" trials (when subjects correctly reported *not* seeing a fearful face not presented). Second, the studies do not demonstrate that the amygdala shows largely normal activation in the absence of cortical input. Third, and critically, these reports do not take into account the many additional alternative routes that bypass early visual cortex (discussed at length below). Thus, in principle, signals can reach the amygdala in several ways that bypass area V1. In sum, to infer the existence of a subcortical pathway from cases of emotional blindsight is unwarranted.

Objective and Subjective Measures of Awareness Although, as stated above, unaware emotional perception has been challenged, reports supporting it continue to be published. A factor that may help explain the discrepancy of published results is the use of different criteria to determine whether or not a subject is aware of perceiving a stimulus. According to *subjective* criteria, unaware perception occurs when subjects *report* not having seen target stimuli or report being unable to perform the task better than chance (independent of their actual performance). Subjective criteria hold that only the subjects themselves have access to their inner states and that their introspection is a reliable source of information about conscious experiences (Merikle, Smilek, and Eastwood 2001). According to *objective* criteria, unaware conditions occur when a subject's performance in a yes/no or forced-choice task is at chance (e.g., d-prime is zero), such as when subjects fail to detect alternative stimulus states (presence vs. absence of targets). Under such conditions, behavioral effects of unaware stimuli (e.g., faster reaction time for undetected fearful faces), as well as associated physiological or neuroimaging signals, constitute correlates of unaware perception.

In my studies, *objective* criteria were used in studies reporting that awareness is necessary for the processing of emotional faces (e.g., Pessoa, Japee, & Ungerleider, 2005; Pessoa, Japee, Sturman, & Ungerleider, 2006; Japee et al. 2009). Unfortunately, no consensus exists regarding the "best" approach to measure and characterize awareness. In fact, there is a mounting tension between objective and subjective threshold approaches (e.g., Merikle et al., 2001; Seth et al. 2008; Snodgrass, Bernat, and Shevrin 2004). Historically, an important concern with subjective procedures is that they can be quite sensitive to response bias (Eriksen 1960). For instance, subjects may be reluctant to indicate having seen a stimulus when the available evidence is very weak. Such concern came into sharp focus with the development of signal detection theory (Green and Swets 1966). At the same time, the importance of subjective measures of awareness is that they resonate with the intuitively appealing idea that awareness should be based on introspective reports of individuals' inner states (James 1890). In fact, from the subjective awareness perspective, at times, the utilization of objective measures may logically preclude the existence of unconscious perception (Bowers 1984).

I have argued that it is important to investigate both objective and subjective measures of perception. Thus, Remigiusz Szczepanowski and I investigated the perception of minimally visible stimuli by utilizing both types of measures (Szczepanowski and Pessoa 2007) within a single task. Thus subjects performed a single detection task for which they were required to make two evaluations: an evaluation of whether or not a fearful face was present and an evaluation of their response confidence. The first evaluation corresponds to a yes/no choice during a standard detection task. The second evaluation can be considered to be a discrimination between correct and incorrect responses (by providing "low" and "high" confidence ratings). Whereas the detection task explicitly probed subjects' perception of fear, confidence ratings provided an indirect assessment of the accessibility of information about fear. Our experiment thus allowed us to test for potential dissociations between these two measures of fear perception, for instance, whether the successful detection of fearful faces could be accompanied by *random* discrimination of response correctness. All but one subject exhibited above-chance detection of fearful faces for all stimulus durations (including 17- and 25-ms targets). Sensitivity for briefly presented stimuli was demonstrated for subjective perception as well (most often for durations of 25 ms or higher). Therefore, our results showed that sensitivity for briefly presented fearful faces is not limited to detection tasks but that it can be demonstrated for subjective perception as well.

Importantly, our findings revealed a dissociation between the two measures of fear perception such that, for some subjects, the successful detection of fearful faces was accompanied by random discrimination of response correctness. Figure 3.4 illustrates the potential relationship between the critical durations linked with objective and subjective measures. The shorter duration represents the critical duration for reliably *detecting* a fearful-face target, and the longer duration represents the critical duration for reliably *discriminating* between correct and incorrect performance. The shaded area indicates a "dis-

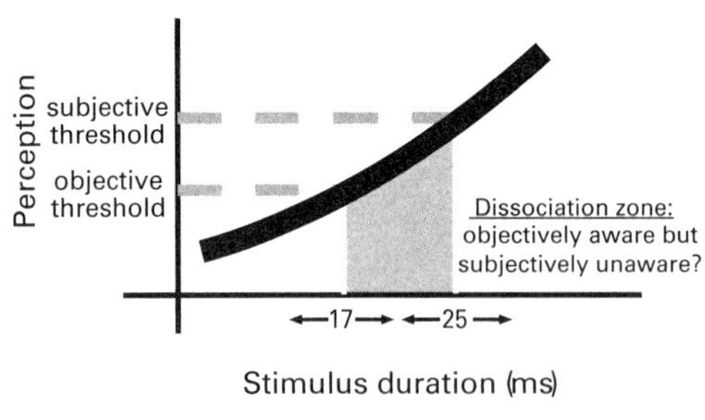

Figure 3.4
Dissociation between objective and subjective perception measures. The dissociation zone refers to stimulus durations for which the subject would be able to reliably detect a fearful-face target but for which the subject would not reliably discriminate between correct and incorrect responses. The arrows around the duration values suggest that, when a dissociation occurs, the values will vary for different subjects. A possible interpretation of this pattern of results is that a subject would be subjectively *unaware* but objectively *aware* of the stimulus, though other interpretations are possible. Adapted with permission from Szczepanowski and Pessoa 2007.

sociation zone" in which the subject is above the objective threshold but below the subjective one (for further discussion, see Szczepanowski and Pessoa 2007; see also Sweeny et al. 2013 and Winkielman and Schooler 2011).

Amygdala as Necessary At the foundation of the standard hypothesis is the notion that the amygdala is *necessary* for critical properties of affective processing, including automaticity and nonconscious mechanisms. In this respect, a well-known study by Adam Anderson and Elizabeth Phelps (2001) compared performance on the attentional blink task in patients with amygdala lesions and in controls. Their findings supported a necessary role for the amygdala: the counteracting of the attentional blink by words with emotional content was not observed in patients with lesions of the left amygdala. Recent studies reviewed in chapter 2, however, have questioned the necessary status of the amygdala for rapid detection, search, or nonconscious processing of affective visual stimuli. The results of these studies strengthen the idea that the amygdala is *not* a necessary component of several of the functions traditionally attributed to the "low road" and that other mechanisms must therefore be invoked to explain the results, at least when the amygdala is damaged.

Processing of Affective Visual Stimuli Involves Both Coarse and Fine Information

According to the standard hypothesis, the subcortical pathway is particularly effective at carrying coarse information, mainly because the superior colliculus and pulvinar are assumed to convey little detailed information. This general notion was inspired by findings in rodents that simple (coarse) auditory conditioning does not require cortex, whereas conditioning that demands more complex stimulus discriminations does (LeDoux 1996; see also below).

Results from human neuroimaging studies appear to be compatible with this notion. Coarse and fine visual information are referred to as "low– and high–spatial frequency" (low– and high–SF) information, respectively, when understood in terms of frequency analysis (also called "Fourier analysis"). For example, in one study (Vuilleumier et al. 2003), amygdala responses were stronger when subjects viewed low– versus high–SF fearful faces, and when they viewed low–SF fearful versus neutral faces. Furthermore, activations in brain areas consistent with the locations of the superior colliculus and pulvinar were greater in response to low–SF fearful faces than to low–SF neutral faces. Although findings of this kind have sometimes been interpreted to indicate that the structure is relatively blind to high–spatial frequency information, the amygdala receives major projections from the anterior inferior temporal cortex

(Amaral et al. 1992) that convey highly processed object information. In fact, the amygdala receives highly processed cortical input from *all* sensory modalities except olfaction (Amaral et al. 1992). For example, in monkeys, functional MRI has shown that electrical microstimulation of a patch of temporal cortex strongly responsive to faces activates the lateral nucleus of the amygdala (Moeller, Freiwald, and Tsao 2008). Notably, electrophysiological studies have revealed that the monkey amygdala contains neurons tuned to the identity of specific faces (Gothard et al. 2007; Rolls 2005)—a property that requires high–SF information (figure 3.5). The human amygdala also displays category-specific responses, including responses specific to faces (Kreiman, Koch, and Fried 2000a; Mormann et al. 2008). Intriguingly, the right amygdala responds selectively to animals as well (Mormann et al. 2011).

Regarding the responses of the human amygdala to face stimuli, one study (Rutishauser et al. 2011) found that more than 50 percent of all amygdala neurons responded to normal images of faces, whereas only 10 percent responded to digitally scrambled images. A substantial proportion of the neurons showed responses selective for whole faces versus parts of faces, indicating that amygdala neurons encode holistic information about faces, rather than just about their constituent features (neurons also responded to facial features, though less vigorously). Also of relevance, most of the neurons that generated their strongest response to whole faces did not distinguish between fearful and happy expressions and may therefore be signaling something closer to stimulus relevance.

In a monkey electrophysiology study, Mikio Inagaki and Ichiro Fujita (2011) explicitly investigated spatial-frequency tuning by amygdala neurons. Neurons in anterior inferior temporal cortex showed distinct tuning characteristics compared to those in the amygdala—they were less dependent on the physical size of the stimulus, hence more invariant to changes. Although this finding is compatible with the existence of distinct routes to these two structures, and possibly with the existence of a subcortical pathway to the amygdala (as suggested by the authors), overall, amygdala neurons exhibited a broad range of response characteristics that overlapped considerably with those in anterior visual cortex.

Behavioral studies also clarify the type of visual information needed during emotional perception. They have shown that the discrimination of facial expressions relies on both low– and high–spatial frequency information (e.g., Smith et al. 2005). The perception of fear is particularly reliant on high–spatial frequency information (Smith and Schyns 2009). Indeed, in their study of a patient with bilateral amygdala lesions, Ralph Adolphs and colleagues (2005) found that the patient's impaired recognition of facial expressions of fear was

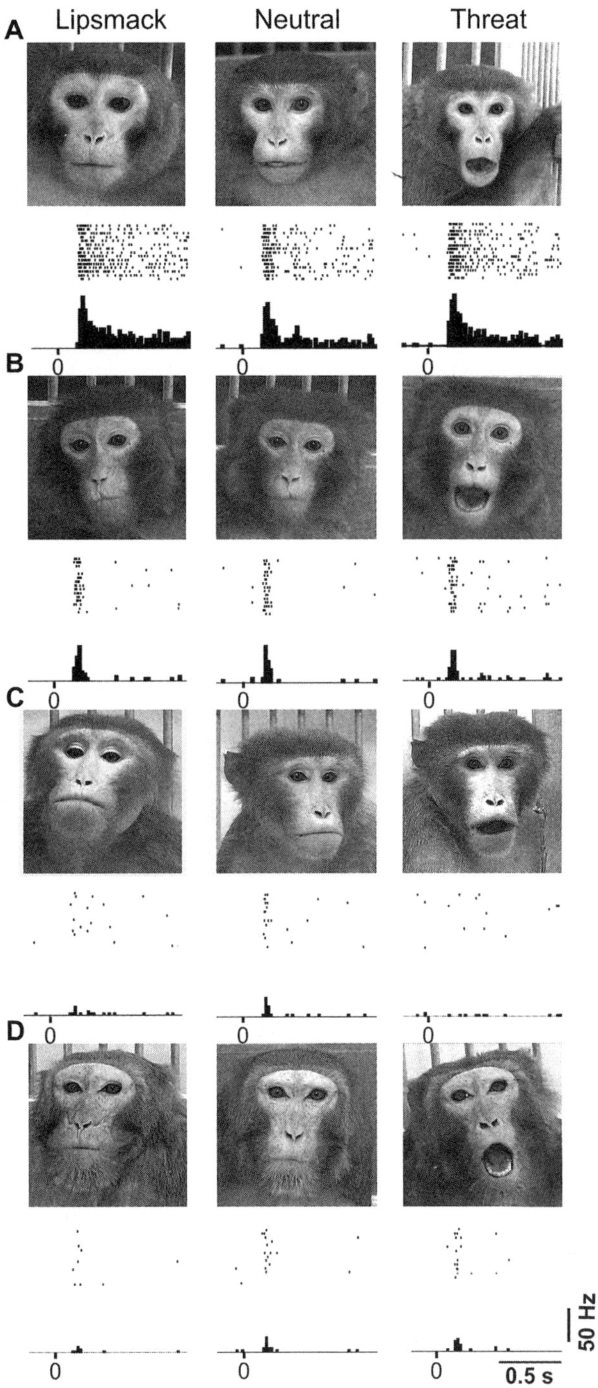

Figure 3.5
Face-identity responses in the monkey amygdala. Each row of images contains three facial expressions displayed by the same monkey. Below each image is the peristimulus time histogram (in 20 ms bins) and single-trial spike rasters of a neuron that responded with a tenfold increase in firing rate to the faces of the two monkeys in rows A and B. Note that the responses were relatively unchanged across facial expression. The "lipsmack" expression is one of appeasement. Reproduced with permission from Gothard et al. 2007.

due to impaired processing of the eye region of faces, and especially to impaired processing of high–SF information about the eyes (see figure 2.8). These results demonstrate the importance of high–spatial frequency information in fear recognition and indicate that the amygdala is required for this type of visual processing.

Although some of the findings reviewed above are consistent with the notion that subcortical areas process coarse visual information, when taken all together, they clearly establish that the perception of emotional expressions involves both coarse and fine information and that the amygdala not only receives but also uses both kinds of information to facilitate recognition of facial expressions.

Physiological and Anatomical Issues

The pulvinar is the key "link element" in the purported subcortical route (figure 3.1A; plate 2). In reviewing physiological and anatomical data on the pulvinar pertinent to the standard hypothesis, this section will first discuss data that bear on whether this structure is better characterized as a relatively *passive* way station or as a *dynamic* element of brain circuitry, then review evidence regarding the existence of a subcortical pathway via the pulvinar, and finally touch on properties of auditory and visual processing, given that one of the motivations for proposing the subcortical pathway is based on auditory conditioning data.

Pulvinar

Pulvinar Input The pulvinar complex, as this set of related nuclei is sometimes called, is the largest nuclear mass in the primate thalamus and thought to have expanded in size as it evolved in primates (Chalfin et al. 2007; Grieve, Acuña, and Cudeiro 2000). In terms of connectivity explicitly relevant to visual processing, the pulvinar receives direct visual input from the retina, indirect visual input via the superficial layers of the superior colliculus, and massive input from striate and extrastriate visual cortex (figure 3.6; plate 4). All of these projections terminate in the inferior pulvinar (see "Pulvinar Anatomy" below for a more detailed description).

Before proceeding to the pulvinar itself, we need to clarify the organization of the superior colliculus (SC), a laminated midbrain structure that acts as one of the centers organizing eye-gaze movements (May 2006; Wurtz and Albano 1980). Primarily visual sensory in nature, the superficial layers of the SC receive direct retinal input and project to the deeper layers, which are both

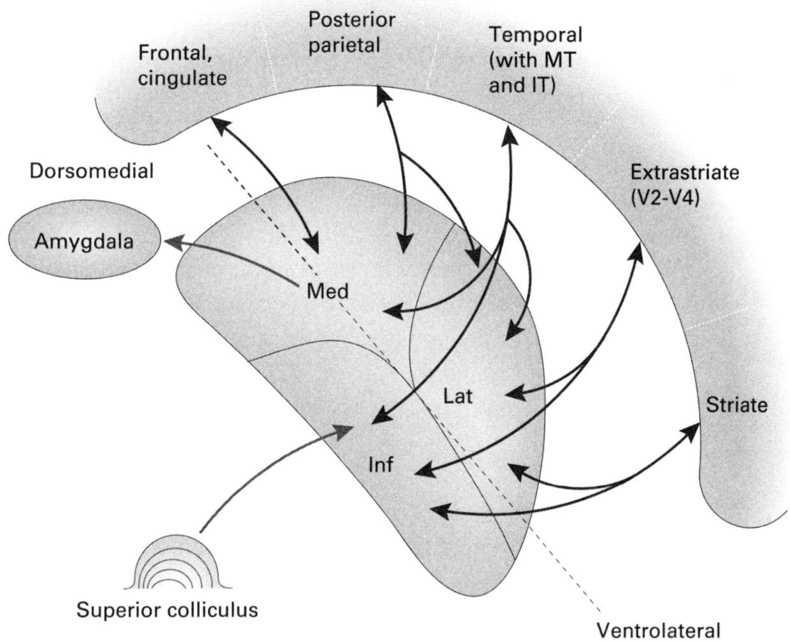

Figure 3.6 (plate 4)
Schematic diagram of pulvinar connectivity. Most pulvinar nuclei and subnuclei (including those not shown here) are involved in thalamo-cortical loops that target different cortical territories. The inferior nucleus is reciprocally connected to striate and extrastriate cortex, the lateral nucleus is connected to association cortex in temporal and parietal lobes (as well as extrastriate cortex), and the medial nucleus is connected to higher-order association cortex in parietal, frontal, orbital (not shown), cingulate and insular regions (not shown), in addition to the amygdala. Thus the medial nucleus is not only connected with the amygdala but is also part of multiple thalamo-cortical loops (note, however, that the connection to the amygdala does not seem to be bidirectional). The superior colliculus is a layered structure whose superficial layers are visual and project to the inferior nucleus. Its intermediate and deeper layers are multimodal and involved in motor preparation, including for eye movements, and project to the medial nucleus (not shown). Inf, inferior; IT, inferior temporal cortex; Lat, lateral; Med, medial; MT, medial temporal area (also known as "V5"). Reproduced with permission from Pessoa and Adolphs 2010, adapted from Stepniewska 2004.

multimodal and motor, receiving input from somatosensory and auditory sources as well as from the basal ganglia and cerebellum. Sensory, association, and motor areas of cerebral cortex provide another major source of collicular input, particularly in more encephalized species, where visual sensory cortex projects to superficial layers while the frontal eye field targets the deeper layers. The deeper layers themselves project to brainstem structures containing gaze-related burst neurons, as well as to the spinal cord and medullary reticular formation regions that produce head turning (see "Role of the Superior Colliculus" under "Multiple Waves Model" below).

Intriguingly, visual response properties of pulvinar cells do not reflect those of cells in the superior colliculus, and the precise contribution that input from the SC makes to pulvinar responses remains uncertain (Stepniewska 2004). Collicular lesions have little effect on electrophysiological responses of pulvinar cells, whereas striate cortex lesions *abolish* responses in the inferior pulvinar (Bender 1983). Likewise, collicular and pulvinar lesions result in different behavioral impairments (Robinson and Cowie 1997; see also the blindsight study Tamietto et al. 2010 for evidence of a dissociation of superior colliculus and pulvinar roles in visual processing). These findings support the idea that the pulvinar may be better characterized as participating in cortical networks than as relaying visual information from the superior colliculus, as does the finding that, unlike "driving inputs" to the lateral geniculate nucleus, those to the pulvinar originate in the cortex, whereas subcortical inputs are typically modulatory (Guillery 1995; Sherman and Guillery 1996).[2]

Pulvinar Function Studies in monkeys and humans with pulvinar lesions suggest that the pulvinar is involved in determining what is salient in a visual scene (Ungerleider and Christensen 1979; Zihl and von Cramon 1979). Consistent with this, the response of pulvinar neurons to visual stimuli is increased when attention is paid to the stimulus or when it has behavioral relevance. For instance, primate pulvinar neurons respond more vigorously to behaviorally relevant targets than to unattended items (Robinson and Cowie 1997). In one monkey study, as many as 92 percent of cells exhibited attenuated responses when stimuli were task irrelevant (passively viewed) versus task relevant (Benevento and Port 1995). Furthermore, the impact of attention on evoked responses in the monkey pulvinar is spatially specific, such that a pulvinar neuron only increases activity when the animal attends to a stimulus that falls within the cell's receptive field (Petersen, Robinson, and Keys 1985). Finally, the pulvinar seems to be critical (as shown by pharmacological inactivation) when a distractor stimulus needs to be "filtered out" (Desimone et al. 1990). Thus it has been proposed that the pulvinar is involved in attention and distractor filtering (Desimone et al. 1990).

Both neuroimaging and lesion studies in humans corroborate the notion that the pulvinar participates in attention. Although early positron emission tomography (PET) studies are compatible with a role of the pulvinar in visual attention (LaBerge and Buchsbaum 1990; Corbetta et al. 1991)—including attentional filtering in the presence of distractors and selective attention to stimulus features, such as shape and color—the low spatial resolution of PET precludes anatomically clear conclusions. More decisive evidence stems from studies at higher resolution using functional MRI. For example, responses in the pulvinar were only observed when the stimulus was attended, but not when it was unattended (Kastner et al. 2004). In addition, lesion studies have described deficits of attention in the visual field contralateral to the pulvinar lesion (Arend et al. 2008).

The pulvinar is important for visual awareness, too. Thus pulvinar lesion studies have uncovered feature-binding deficits (Karnath, Himmelbach, and Rorden 2002; Ward et al. 2002; Zihl and von Cramon 1979); the pulvinar on the right hemisphere was identified as a subcortical node associated with spatial neglect in humans (Karnath, Himmelbach, and Rorden 2002; Mesulam 1981).[3] Monkey physiology and human functional MRI studies have also revealed contributions of the pulvinar to visual awareness. During a visual illusion that induced the intermittent perceptual suppression of a bright luminance patch (Wilke, Mueller, and Leopold 2009), monkey pulvinar neurons showed changes in spiking rate in response to trial-by-trial stimulus visibility, suggesting that they reflected visual awareness. Similarly, a functional MRI study found that the human pulvinar responded trial by trial not to the affective significance of visual stimuli (positive vs. negative conditioned stimulus) per se, but to affective stimuli that were consciously perceived (Padmala, Lim, and Pessoa 2010; figure 3.7A). In another human functional MRI study (Pessoa

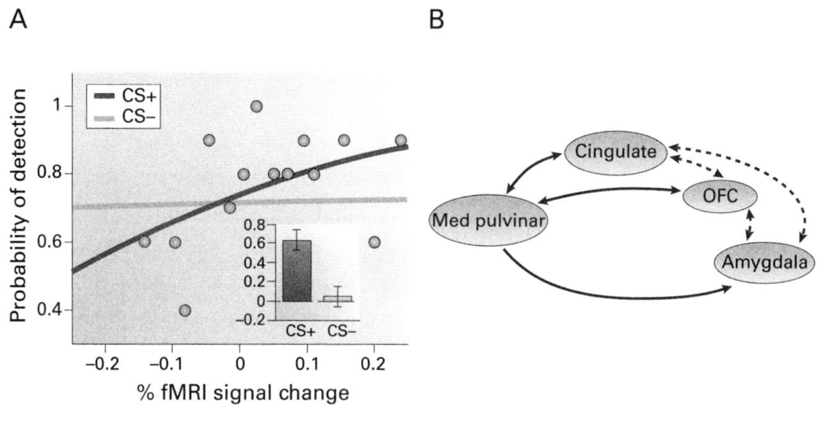

Figure 3.7
Pulvinar and amygdala during processing of affective stimuli. (A) Logistic regression analysis of evoked responses in the left pulvinar as a function of affective significance for a sample subject during an attentional blink task (see figure 2.5A). The slope of the logistic fit indicates the strength of the predictive effect. For clarity, only binned data for the conditioned stimulus (CS+) condition are included. The inset shows mean logistic fit slopes across subjects. (B) The medial pulvinar is proposed to amplify evoked responses of behaviorally relevant stimuli via circuits involving cingulate cortex, orbitofrontal cortex (OFC), and the amygdala, all regions important for stimulus valuation. Reproduced with permission from Pessoa and Adolphs 2010 and adapted from Padmala, Lim, and Pessoa 2010.

and Ungerleider 2004a), pulvinar responses were observed during "false alarm" trials (i.e., where a stimulus change was reported but did not actually occur) but not during "miss" trials (i.e., where a stimulus change occurred but went unnoticed by the subject). Together, the above results do not support the notion that the pulvinar is principally involved during nonconscious processing, and are therefore inconsistent with the *passive relay* role of the standard hypothesis. As stated before, "driving inputs" (as opposed to "modulatory inputs") to the pulvinar appear to originate in cortex (Guillery 1995; Sherman and Guillery 1996; see also below). Thus pulvinar responses may be closely aligned with awareness because of the contributions from cortex, which is thought to be important for conscious perception (e.g., Kouider and Dehaene 2007; Alkire, Hudetz, and Tononi 2008).

Pulvinar Anatomy An implicit assumption of the standard hypothesis is that the pulvinar, whatever its functions are, is basically a single structure—or at least that the part of it that receives collicular inputs is the same as the one that projects to the amygdala. But is the pulvinar in fact organized as a simple relay station that conveys signals from the superior colliculus to the amygdala? A review of pulvinar anatomy shows it is not. (Readers less interested in the details of the neuroanatomy may want to skip ahead to this section's summary paragraph.)

In primates, the pulvinar is a set of thalamic nuclei that accounts for a quarter of the total thalamic mass (Grieve, Acuña, and Cudeiro 2000). Originally partitioned into three subdivisions, namely, inferior, lateral, and medial (Walker 1938), it is now typically divided into four, with anterior pulvinar representing the fourth subdivision. Broadly speaking, the *inferior* pulvinar is reciprocally connected to striate and extrastriate cortex; the *lateral* pulvinar is connected to association cortex in temporal and parietal lobes (though parts also receive extrastriate cortical inputs); and the *medial* pulvinar is connected to parietal, frontal, orbital, cingulate, and insular cortex, in addition to the amygdala (Grieve, Acuña, and Cudeiro 2000; Shipp 2003; figure 3.6; plate 4).

Several anatomical features highlight the extensive *bidirectional* connectivity between the pulvinar and cortex. For example, all twenty to thirty known visual areas connect with the pulvinar, sometimes in a relatively topographic fashion (Shipp 2003; Stepniewska 2004), and, as stated, parietal, frontal, orbital, cingulate, and insular cortex are all connected with the pulvinar as well. At a gross level, it is as if the entire convoluted cortex were "shrink-wrapped" around the pulvinar (Shipp 2003). Based on connectivity data, it has been suggested that the pulvinar may contain two "domains" (Grieve, Acuña, and Cudeiro 2000; Shipp 2003; see figure 3.6; plate 4). Densely connected with visual cortex, including the V1–V4 and MT (middle temporal) areas (Shipp 2003), the "*ventral* domain" therefore has a strong visual component—indeed, it could be called the "visual pulvinar"—and its projections to the dorsal visual stream may mediate some of the visual abilities in people with blindsight (Lyon, Nassi, and Callaway 2010; Berman and Wurtz 2010). The "*dorsal* domain" has connections with cross-modal association cortex, including temporal and parietal areas, such as area 7A and the lateral intraparietal area, that participate in attention (Shipp 2003). The dorsal domain receives highly processed visual input from anterior parts of ventral visual cortex (Shipp 2003). And because it is also connected with cingulate cortex, frontal cortex (including orbitofrontal cortex), insula, and amygdala (Shipp 2003; see below), it has remarkable potential to integrate information from very diverse brain regions (figure 3.6; plate 4). Whereas the connectivity of the pulvinar's ventral domain is restricted to the occipito-temporal cortex, sites in the dorsal domain may be connected with relatively distal parts of the brain, such as parietal and frontal cortex (Shipp 2003). Indeed, many extensive fronto-parietal cortical connections are mirrored by overlapping fields in the dorsal domain (Romanski et al. 1997; Barbas, Henion, and Dermon 1991): where regions in frontal and parietal cortex are interconnected in the cortex, their projection sites in the pulvinar typically coincide (and the connections are bidirectional between pulvinar and cortex)—an organization that further exemplifies the integration ability of the pulvinar's dorsal domain.

In all, the ventral and dorsal domains of the pulvinar have very distinctive connectivity patterns: the ventral domain is strongly *visual*, whereas the dorsal domain is *associational*. Of relevance to the standard hypothesis, visual signals from the retina are conveyed to the superficial layers of the superior colliculus and then to the inferior pulvinar, which belongs to the ventral domain. Connections from the pulvinar to the amygdala, however, originate in the medial pulvinar, which is part of the dorsal domain. All of which serves to challenge the notion that the pulvinar links the superior colliculus and the amygdala in a straightforward way.

Because an indirect pathway might still convey signals to the amygdala—although via a longer (hence slower) route, we need to consider connectivity *within* the pulvinar. There is no good evidence to support the existence of connections from the inferior to the medial pulvinar, which would be required to provide a contiguous pathway from the superior colliculus to the amygdala. Like other thalamic nuclei, the primate pulvinar does not appear to have long-range intrinsic connections, either excitatory or inhibitory (Imura and Rockland 2006), although inhibitory interneurons and local processing may exist within the medial pulvinar (see also Ma et al. 1998).

As a final complication, connections also seem to exist from the intermediate/deep (nonretinal) layers of the superior colliculus to the medial pulvinar (Romanski et al. 1997). Could the intermediate or deep layers be sending direct input to the amygdala via the medial pulvinar? Such a scenario is problematic for the following reasons. Although, for the sake of simplicity, we have treated the medial pulvinar as a unit, it consists of central/lateral and medial subunits with substantially different connectional patterns (see figure 10 of Romanski et al. 1997). Whereas the medial subunit projects to the amygdala, it is the central/lateral subunit that receives significant input from the nonretinal superior colliculus. Thus, again, it seems unlikely that a simple colliculo-pulvino-amygdalar pathway exists.[4] Even more critically, because signals from the intermediate and deep layers of the superior colliculus are multimodal and possibly linked to saccadic eye movements, they would not easily fit the role commonly assigned them in the subcortical pathway of the standard hypothesis.

Next to nothing is known about the connectivity and electrophysiology of the pulvinar in humans, although a diffusion tensor imaging study has reported results consistent with connectivity to ipsilateral superior colliculus as well as temporal visual cortex (Leh, Chakravarty, and Ptito 2008; see also Tamietto et al. 2012 and the conclusion to this book for further discussion).[5] A study in epileptic patients examined evoked responses to electrical stimulation to map the *functional connectivity* of the medial pulvinar and suggested functional connections between medial pulvinar and visual cortex (including sites in occipital and temporal cortex) and between medial pulvinar and the amygdaloid region (Rosenberg et al. 2009).[6] Given the limited resolution and other characteristics of both diffusion tensor imaging and electrical stimulation, we cannot infer the existence of *anatomical* connectivity with any certainty, although it is generally assumed that the human and monkey pulvinar have much the same connectivity.

To summarize, the pulvinar has significant visual and integrative properties. Studies have characterized several ways in which the pulvinar is modulated by attention and awareness; indeed, the pulvinar is likely to be an important "control site" for attentional mechanisms more broadly (Shipp 2004). Anatomically, the pulvinar is a heterogeneous structure that may be viewed as containing at least two "domains" with very distinct connectivity patterns. Furthermore, the connectivity of the pulvinar is such that it is regarded as a higher-order thalamic structure that may be involved in cortico-thalamo-cortical communication (see below), not a simple first-order relay (Sherman 2007). Collectively, these observed characteristics of the pulvinar run contrary to the standard hypothesis, which assumes that largely automatic processing is mediated by a rapid subcortical pathway relying on a passive pulvinar.

Does the Subcortical Pathway Exist in Primates?

The early inspiration and impetus for the standard hypothesis came from work with rodents that first gave rise to the idea of both subcortical ("low road") and cortical ("high road") pathways for processing fear-relevant information (LeDoux 1996, 2000). Work on fear conditioning showed that there are direct projections to the amygdala from the *auditory* thalamus (i.e., medial geniculate thalamus) in the rat (Campeau and Davis 1995; Romanski and LeDoux 1992). In the rat, there is some evidence that the lateral-posterior nucleus of the thalamus—whose properties are related to those of the pulvinar in primates (see Chalfin et al. 2007 for evolutionary considerations)—conveys *visual* information to the amygdala (Shi and Davis 2001; for related findings, see also Linke et al. 1999). Based on a series of lesion manipulations, Changjun Shi and Michael Davis (2001) have argued, however, that this pathway is unlikely to be functional in intact animals. Instead, a pathway connecting the lateral-posterior thalamus to the amygdala *via anterior temporal cortex* appears to be critical for transmitting emotion-laden visual information (figure 3.8).

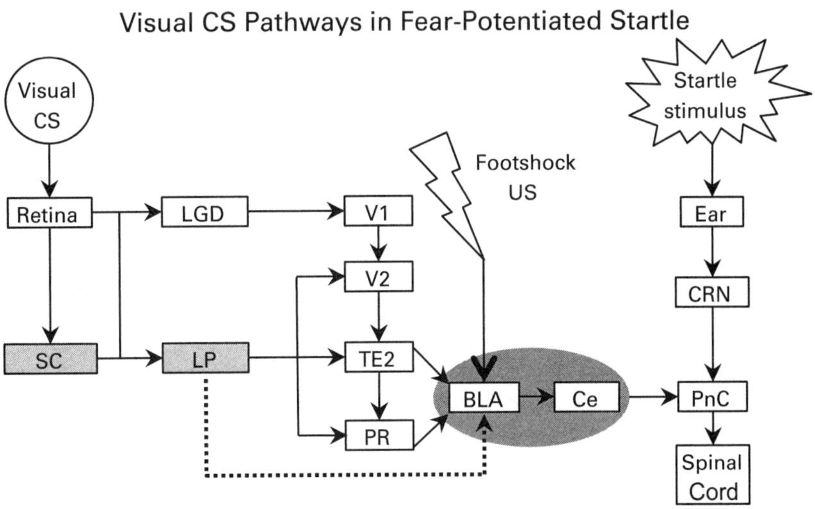

Figure 3.8
Visual pathways in the rat. Schematic diagram of thalamo-cortico-amygdala and thalamo-amygdala visual pathways involved in fear-potentiated startle in the rat. The pathway indicated by the dashed line may not be critical in normal visual fear conditioning. The lateral posterior nucleus of the thalamus in rats is thought to be related to the pulvinar nucleus in primates. BLA, basolateral nucleus of the amygdala; Ce, central nucleus of the amygdala; CRN, cochlea root neurons; LGD, dorsal-lateral geniculate nucleus; LP, lateral-posterior nucleus of the thalamus; PR, perirhinal cortex; PnC, pontine reticular nucleus, caudal part; SC, superior colliculus; TE2, temporal area TE2; V1–V2, visual areas 1–2. Reproduced with permission from Shi and Davis 2001.

More generally, the work of Shi and Davis (2001) highlights the need to assess whether specific brain structures are operational during *normal* vision, namely, in intact animals. In other words, a lesion may identify a pathway that is *capable* of mediating a behavior (e.g., fear conditioning) but one that might *not* be *functional* in intact animals: when the normal route is damaged by lesions, other pathways not typically engaged may take over, supporting the behavior in question.[7]

Another study involving "rewiring" suggests that visual and auditory pathways are also organized differently in rodents (Newton et al. 2004). Mice acquired a conditioned fear response rapidly (in terms of the number of trials) to auditory cues but slowly to visual cues. The authors proposed that the difference in behavior was due to distinct connectivity patterns in the visual and auditory modalities—direct projections to the lateral amygdala from the auditory thalamus but indirect ones from the visual thalamus. To test this hypothesis, the authors induced the growth of *retinal* projections to the medial geniculate nucleus (figure 3.9), which participates in *auditory* processing in normal animals. Acquisition of *visually* cued conditioned fear was accelerated in the rewired mice, where visual stimuli induced activity in the "auditory" thalamus and the lateral amygdala, much as auditory stimuli did in control mice. Together, their data suggest that the rewired pathway conveyed visual information and mediated "rapid" activity-dependent plasticity in structures influencing learned behavior.

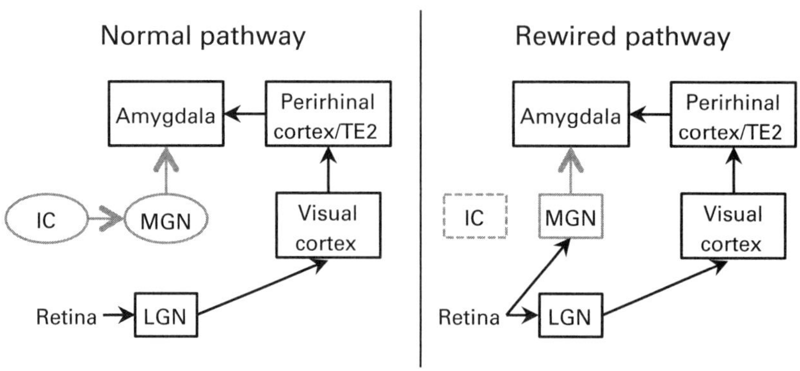

Figure 3.9
Simplified fear conditioning pathways in normal and rewired mice. (Left) Schematic of the principal visual (black) and auditory (gray) conditioned fear pathways in normal mice. (Right) Schematic of the rewired visual (black) cued conditioned pathway. The inferior colliculus (IC) was lesioned bilaterally (dashed box) in neonatal mice to induce retinal projections to the medial geniculate nucleus. LGN, lateral geniculate nucleus; MGN, medial geniculate nucleus; TE2, temporal area TE2. Reproduced with permission from Newton et al. 2004.

What about evidence of subcortical connectivity in monkeys? Here, as we have seen, anatomical studies have shown connections between the superficial superior colliculus and the inferior pulvinar, both of which can be considered "visual" structures (Grieve, Acuña, and Cudeiro 2000; Stepniewska 2004), and between the pulvinar and the amygdala (Jones and Burton 1976; Romanski et al. 1997). But, whereas the inferior pulvinar is extensively interconnected with visual cortex (consistent with visual functions), the pulvinar's projection to the amygdala originates in the *medial* pulvinar (Jones and Burton 1976; Romanski et al. 1997; see also Aggleton, Burton, and Passingham 1980). And, again as we have seen, because there is no good evidence of connectivity between the inferior and the medial pulvinar, a subcortical pathway from superior colliculus to pulvinar to amygdala seems unlikely.

Working with the tree shrew, Martha Bickford, Jonathan Day-Brown, and colleagues have documented a potential subcortical pathway that is in line with the standard hypothesis (Day-Brown et al. 2010). Placed in its own order (Scandentia), separate from insectivores and primates, this squirrel-like mammal is nonetheless considered a "basal primate" (Butler and Hodos 2005). Bickford and colleagues reported projections from the superior colliculus to the "Pd" nucleus of the pulvinar (part of the medial pulvinar), which in turn connects with the lateral amygdala (Day-Brown et al. 2010). Although this study provides perhaps the most compelling evidence for a subcortical visual pathway, the diffuse nature of the observed superior colliculus–to-pulvinar connectivity makes it unlikely that such a pathway could effectively convey visual *form* information. In related work, Ranida Chomsung, Heywood Petry, and Martha Bickford (2008) speculated that the primary signal conveyed from the superior colliculus to the pulvinar in the tree shrew may encode motion (see also Luksch, Khanbabaie, and Wessel 2004 for discussion of how collicular responses mediate sensitivity to motion independently of stimulus details). It is also worth noting that the tree shrew has surprisingly robust visual capabilities in the complete absence of primary visual cortex (Diamond and Hall 1969; Snyder, Killackey, and Diamond 1969). Therefore, even though subcortical pathways involving the superior colliculus may support several aspects of tree shrew vision, the anatomical findings of these studies would hardly seem to apply to humans, who experience *blindness* with lesions in primary cortex.

To summarize, except for the report on the tree shrew, there is scant evidence for a direct subcortical pathway conveying visual information to the amygdala in primates. It is thus unclear how findings from auditory fear conditioning studies in rodents can be applied to visual processing of affective stimuli in primates (see also next section). At the same time, work with rodents suggests that basic forms of vision-based fear conditioning may be mediated by a route linking the thalamus to anterior temporal cortex. This type of cortical bypass connection is proposed below to rapidly convey affective information across the brain.

Subcortical Auditory Processing in Rats Is Qualitatively Different from Subcortical Visual Processing in Primates

The standard hypothesis derived a considerable portion of its historical motivation from the organization of the auditory system in rodents. But the auditory and visual systems differ in important ways. The temporal precision of the auditory system is substantially greater than that of the visual system. In contrast to vision, audition is omnidirectional, such that information from all directions in space can be sampled (though at relatively lower spatial resolution). Furthermore, the functional anatomy of the auditory system is very different from that of the visual system. Properties such as sensitivity to sound frequency, duration, amplitude, pitch, and binaural disparity, which are observed in primary auditory cortex (area A1), are already observed at subcortical levels. In fact, primary auditory cortex should not be thought of simply as primary visual cortex transplanted into the auditory modality since it seems to perform high-level functions. Indeed, given that several subcortical stages exist below the level of the primary auditory cortex, the *inferior* colliculus, which is involved in audition, occupies a processing level similar to that of area V1 in vision, and area A1 is more analogous to visual areas in inferior temporal cortex than to V1 (cell responses in inferior temporal cortex are considerably more elaborate than those in V1; King and Nelken 2009).

These considerations suggest that a subcortical pathway for auditory input to the amygdala in rodents would be quite distinct from the purported subcortical visual pathway in primates. Connections from the auditory thalamus to the amygdala, while bypassing cortex, still convey relatively processed information to the amygdala, in contrast to the suggested primate visual counterpart. Visual responses in the superior colliculus are quite rudimentary. Indeed, Peter Schiller and Fritz Koerner (1971, 924) described them as "event" and "jerk" detectors, noting that "none of the units we studied were shape or orientation specific. Similar responses could be elicited using squares, triangles, bars, or any of numerous other configurations."

Multiple Waves Model

The standard hypothesis has influenced both basic and applied research and, at first glance, has intuitive appeal. Emotional reactions can be fast and relatively impervious to top-down effects when task demands are not high. Nevertheless, a host of problems plague the hypothesis in its basic form, chief among them (1) visual processing of complex scenes in general (i.e., nonemotional scenes) is already surprisingly fast and not appreciably slower than affective visual mechanisms; (2) affective processing is not strongly independent of attention and awareness; (3) coarse visual processing (e.g., low–spatial frequency faces) does not map cleanly onto subcortical processing; (4) new findings show that complete lesions of the amygdala in humans spare rapid and nonconscious aspects of affective processing; and (5) the existence of a subcortical visual pathway linking the superior colliculus, pulvinar, and amygdala receives scant support from anatomical data.

Faced with these problems, the standard hypothesis can no longer be said to hold, although several of its themes and original motivation remain pertinent. We turn now to the multiple waves model Ralph Adolphs and I have proposed as an alternative to it (Pessoa and Adolphs 2010).

Multiple Visual Pathways and Coarse Information Processing

One of the primary motivations for the standard hypothesis is the perceived need for rapid processing: fast—though coarse—visual mechanisms are just what an organism needs to survive in a dangerous environment. As reviewed next, visual routes other than a colliculo-pulvino-amygdalar one are proposed to carry out this role (for further discussion of additional pathways in vision, see Catani et al. 2003; Chen et al. 2007; Cowey 2004; and, in particular, Vuilleumier 2005, which anticipates some of the themes elaborated below).

Although visual processing along the ventral stream, which is crucial for object recognition, has historically been described as occurring in a relatively hierarchical fashion, significant "shortcut" connections link areas V1 to V4 (Nakamura et al.1993), V2 to TEO (Nakamura et al. 1993), and V4 to TE (Felleman and Van Essen 1991), providing the means for faster information transmission to anterior temporal cortex (Lamme and Roelfsema 2000; figure 3.1B; plate 2). Direct connections between the lateral geniculate nucleus and extrastriate regions, including areas V2 (Bullier and Kennedy 1983; Yukie and Iwai 1981) and V4 (Yukie and Iwai 1981), have also been reported.

Functional evidence for *bypass* systems has been observed in several studies. For example, combining electrophysiology and functional MRI in lesioned monkeys, Michael Schmid and colleagues (2009) detected robust visual activation in areas V2 and V3 in animals with lesions of area V1, demonstrating that routes bypassing V1 are sufficiently potent to drive extrastriate visual responses. In a second such study (Schmid et al. 2010), they detected widespread extrastriate activation in the absence of area V1 and observed responses in several visual areas, including areas V2, V4, and MT/V5, as well as parietal cortex. Notably, animals with V1 lesions were able to perform a visual detection task when stimulus contrast was high. Further reversible deactivation of the visual thalamus (lateral geniculate nucleus) showed that successful behavior in this task was dependent on this structure. These findings are particularly significant because they demonstrate the importance of the visual thalamus for some types of blindsight, which it can be argued the monkeys exhibited.

In humans, functional evidence for the role of cortical bypass systems was reported in a case study of a patient with developmental agnosia and prosopagnosia (intriguingly, with no discernible macroscopic cortical lesion; Gilaie-Dotan et al. 2009).[8] In this patient, whereas area V1 was robustly activated by visual stimuli (as measured with functional MRI), intermediate areas V2–V4 were not activated, although robust downstream activation was observed in the parahippocampal gyrus and other anterior regions, with spared visual selectivity. As in normal individuals, the patient's parahippocampal gyrus responded robustly to outdoor scenes, including pictures of houses and "places" (Epstein and Kanwisher 1998; see also Boyer, Harrison, and Ro 2005 for evidence of a V1-bypassing pathway in humans).

In the past few years, researchers have demonstrated several other visual pathways that either bypass the lateral geniculate nucleus or involve the LGN but bypass early visual cortex. For example, Robert Wurtz and colleagues (2011) have described multiple LGN–bypassing visual routes in primates, including one from the superior colliculus to the frontal eye field through the medial dorsal thalamus, and two from the superior colliculus to the middle temporal area (area MT, which is strongly sensitive to motion) in visual cortex via the pulvinar. These pathways take part in multiple aspects of "active vision," including saccade- and attention-related mechanisms. Other primate studies have also revealed LGN–to–MT connections that can convey visual signals directly to MT (Sincich et al. 2004; see also Bridge et al. 2008 for related evidence in humans). Indeed, interest in the contributions of multiple routes in generating visual response properties is producing a more nuanced understanding of visual processing (see, for example, Ponce, Lomber, and Born 2008; Ponce et al. 2011; Passarelli et al. 2011). Intriguingly, multiple pathways also might be involved with more sophisticated behaviors that rely on visual processing, such as reading (Richardson et al. 2011).

Figure 3.10 (plate 5) provides one view of the connectivity between visual areas, emphasizing the projections from subcortical regions (including the lateral geniculate nucleus, superior colliculus, and pulvinar) to cortical regions (Capalbo, Postma, and Goebel 2008).[9] Derived by taking structural data into account and determining the "best-fit" connectivity based on cell response timing, the figure's diagram highlights the property that signals from subcorti-

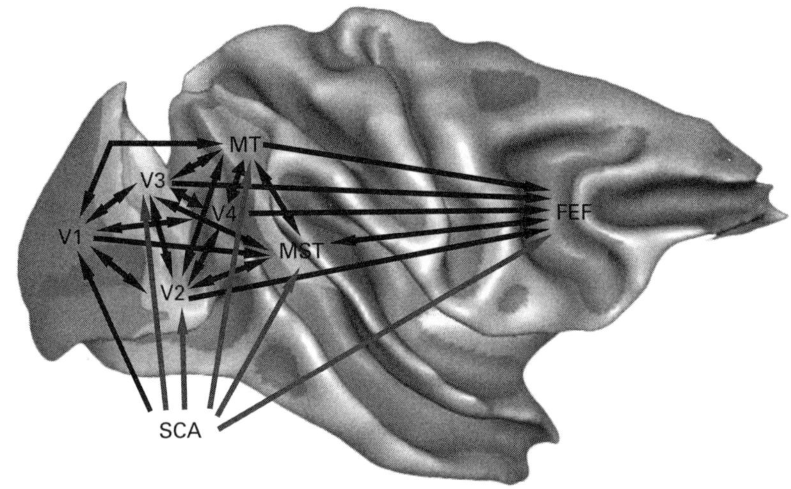

Figure 3.10 (plate 5)
Visual cortex connectivity model. Connectivity was determined by taking structural data into account and estimating the "best-fit" pathways based on cell response timing. Including subcortical regions and their connections (red arrows) provided the best model. FEF, frontal eye field; MST, middle superior temporal area; MT, middle temporal area; SCA, subcortical areas; V1–V4, visual areas 1–4. Reproduced with permission from Capalbo, Postma, and Goebel 2008.

cal areas can be rapidly disseminated across occipital and temporal visual cortex (in addition to frontal cortex). Notably, a model that includes subcortical regions and their connections (red arrows) provides a better account of response timing than do models that contain only cortico-cortical shortcuts, whereas assuming a purely hierarchical structure of the visual system fails to provide a good fit to existing latency data (Capalbo Postma, and Goebel 2008).

Long-range shortcuts also exist, such as projections from visual areas TEO/TE and the superior temporal sulcus that link regions in anterior ventral visual cortex with ventral-lateral and orbital prefrontal cortex (Rempel-Clower and Barbas 2000; Saleem, Kondo, and Price 2008). In orbital PFC, the projections are indeed quite widespread (Saleem, Kondo, and Price 2008). Some bypass connections involve magnocellular projections, known to convey low–spatial frequency and motion information at relatively short latencies, to cortical regions in prefrontal (and middle temporal) cortex.[10] For example, frontal eye field (and MT) cells exhibit latencies 10–40 ms shorter than cells in areas V2 and V4 (Schmolesky et al. 1998). Additional long-range pathways connect regions as early as area V2 with prefrontal cortex (Barbas 1995; Rempel-Clower and Barbas 2000). Jean Bullier (2001) has suggested that low–spatial frequency information may rapidly reach parietal and frontal cortex from early visual cortex, thereby providing coarse information about the gist of a visual scene and supporting object recognition (see also Bar 2003; Fabre-Thorpe 2011; Kveraga, Boshyan, and Bar 2007). For instance, Bullier (2001, 106) proposes "an integrated model that uses asynchronous transfer of information in the geniculo-cortical connection and the rapid activation of the dorsal stream by the M[agnocellular] channel to generate a first-pass analysis of the visual scene." It is conceivable that these "first-pass" distributed volleys of activation are less susceptible to manipulations of attention and awareness (Barrett and Bar 2009; Bullier 2001; see chapter 4 for further discussion).

In summary, there are multiple parallel routes for visual information processing that lead to substantial temporal dispersion of evoked responses (see figure 3.3) and that enable "high-level" regions to respond with surprisingly short latencies (Nowak and Bullier 1997). Each processing stage adds approximately 10 ms to the latency (Nowak and Bullier 1997). The cost of using such bypassing stages may be that, at first, only relatively coarse information is available about a visual item. This is consistent with a coarse-to-fine processing strategy in which the more global content of a stimulus is processed earlier than finer details (Sugase et al. 1999; Sripati and Olson 2009).

Based on the considerations above, I propose that the initial processing of visual information proceeds simultaneously along parallel subcortical and cortical channels, creating "multiple waves" of activation across visual cortex and beyond (see Rudrauf et al. 2008). In this manner, visual stimuli that have affective and motivational significance are capable of engaging multiple brain sites—including the amygdala, orbitofrontal cortex, anterior insula, and anterior cingulate cortex—that can gauge their impact and further redirect resources toward behaviorally relevant items. Thus rapid processing of affective information is possible even in the absence of a specialized subcortical pathway (figure 3.1B; plate 2) or a single specific structure such as the amygdala. It should be emphasized, however, that fast visual processing is still very powerful. For example, information about a visual item is available in short segments of spiking data (30–100 ms; Tovee and Rolls 1995), and stimulus category can be predicted from human intracortical recordings within 100 ms (Liu et al. 2009).

In light of this, I suggest that affective blindsight involves some of the alternate pathways described here. A study of a patient with complete cortical blindness used advanced source modeling to investigate the time course of information processing (Andino et al. 2009). All facial expressions, including neutral ones, evoked relatively short latency responses (70–120 ms) localized to the superior temporal sulcus; emotion-specific responses that were localized to anterior temporal cortex and possibly amygdala occurred considerably later

(120 and 200 ms, respectively). Whereas this study (Andino et al. 2009) suffers from the localization problems alluded to before, its findings are consistent with the proposal that affective significance is computed in parallel along several circuits (see also Rotshtein et al. 2010 for a related proposal). More conclusively, studies like the one by Schmid and colleagues (2010) described above document instances during which "nonstandard" routes of the type described here *do* support blindsight.

Although this section has emphasized the role of multiple pathways during rapid affective perception, they also operate during less challenging situations, such as those involving longer stimulus durations. Findings from a recent study by Danai Dima and colleagues (2011), who investigated the processing of facial affect when subjects viewed angry, fearful, or sad expressions, each presented for 2 seconds, are interesting in this regard. In their analysis, Dima and colleagues employed dynamic causal modeling (see Friston, Harrison, and Penny 2003), a technique that attempts to estimate the strength of directed interactions between neural systems based on functional MRI data. Although results obtained with this technique need to be considered with caution (see Ramsey et al. 2010), their findings indicated that valence signals were communicated simultaneously across parallel channels. Notably, amygdala signals were not sufficient to explain valence-related interactions between visual cortex and frontal cortex, consistent with the existence of separate sources of valence modulation. Given the change of focus from a single specialized subcortical route to a multiple pathways model, it is important to reconsider the roles of both the pulvinar and the amygdala during processing of emotional visual stimuli.

Role of the Pulvinar in Processing Emotional Visual Stimuli

First, it is worth considering the functions of the pulvinar in general. Despite several decades of work, reviews of the pulvinar often note that "surprisingly little is known about its functions" and frequently refer to this structure as "enigmatic" (Grieve, Acuña, and Cudeiro 2000; Stepniewska 2004). Yet a recurring theme is that the pulvinar is involved in attention and in determining behavioral relevance. Based on the data described here, the proposed multiple waves model suggests that the pulvinar helps to coordinate and regulate the flow of multimodal information via a series of thalamo-cortical loops that highlight signals related to behavioral significance (see figure 3.6; plate 4). Notably, the model takes into account that most of the input to the pulvinar comes from cortex.

In the context of emotional processing, the *medial* nucleus of the pulvinar is probably the most relevant one, given that it connects not only with the amygdala, but also with a large array of other brain regions. The model therefore suggests that the medial nucleus is involved in more general functions that impact emotional processes, such as determining the behavioral relevance of a stimulus. For example, it is connected with parietal areas that are engaged in attention, with orbitofrontal and cingulate cortex, which are important for computing an object's biological value, and with the insula, which has a role in emotional feelings. These pathways are all bidirectional, providing ample opportunities for the medial nucleus to modulate and regulate information flow. According to the multiple waves model, the importance of the pulvinar in emotion is due not to its status as a subcortical "labeled line" conveying emotional information to the amygdala but rather to its pattern of connectivity with subcortical and cortical sites that have a role in determining the biological significance of a stimulus.

Studies by Robert Ward and colleagues have investigated the effect of pulvinar lesions on processing affective visual information in humans. A complete unilateral loss of the pulvinar led to a severe deficit in a patient's ability to recognize fearful facial expressions shown in the contralesional visual field (Ward et al. 2007). According to the multiple waves model proposed here, when weak or brief visual stimuli have biological significance, cortico-pulvino-cortical circuits coordinate and amplify signals in a manner that enhances their behavioral impact. This model is also compatible with impairment in recognizing anger (and possibly happiness; Ward et al. 2007). Of note, the essential pulvinar damage was found in the medial pulvinar, the region that in monkeys projects to the amygdala. The proposed model is consistent, as well, with the finding that viewing complex unpleasant images impaired performance in a subsequent simple (neutral) visual task in controls, but not in a patient with pulvinar damage (Ward, Danziger, and Bamford 2005); according to the model, the unpleasant stimulus did not garner additional resources in the patient (which would have interfered with performance, as it did in the controls).

Pulvinar involvement in processing affective information does not seem to reflect emotional content per se, however. In a human functional MRI study, my colleagues and I (Padmala, Lim, and Pessoa 2010) found that, in trials that contrasted affectively significant (CS+) and neutral (CS–) conditions, there was a significant relationship between the magnitude of evoked responses in the pulvinar and the probability of correctly detecting a target on a trial-by-trial basis during the affective but not during the neutral condition (figure 3.7A). These results reveal an emotion-visibility interaction that may characterize the role of the pulvinar more generally. In other words, the pulvinar amplifies responses to stimuli of potential value to the animal (such as those signaling the possibility of shock in the experiment; figure 3.7B).

Cortico-Thalamo-Cortical Communication Despite progress, our current understanding of pulvinar function is largely incomplete. In an interesting proposal by S. Murray Sherman (2007), the pulvinar participates in regulating *cortical* communication, with direct cortical connections between two areas supplemented by an indirect pathway coursing through the pulvinar or other higher-order thalamic nuclei (figure 3.11). Data consistent with this proposal were reported in rat somatosensory cortex, where activity was found to be driven by a cortico-thalamo-cortical pathway (Theyel, Llano, and Sherman 2010). Additional results of a monkey physiology study lend further support to the proposal. By recording simultaneously in the pulvinar and cortical visual areas V4 and TEO (the latter in inferior temporal cortex), Yuri Saalmann and colleagues (2012) obtained evidence that maintaining attention in the absence of visual stimulation (delay period after a cue stimulus disappeared) depended on *pulvino*-cortical interactions. In contrast, direct *cortico*-cortical influences during this delay period were weak (though strong when the cue was shown). It is particularly intriguing that the *relative* contribution of the pulvinar on cortico-cortical interactions was largest during the delay interval. At this juncture in the trial, cortical signals would presumably benefit the most from the support of the pulvinar. To conclude, as stated by Brian Theyel, Daniel Llano, and S. Murray Sherman (2010, 87), "corticothalamocortical information transfer may represent an important addition to, or even replacement of, the current

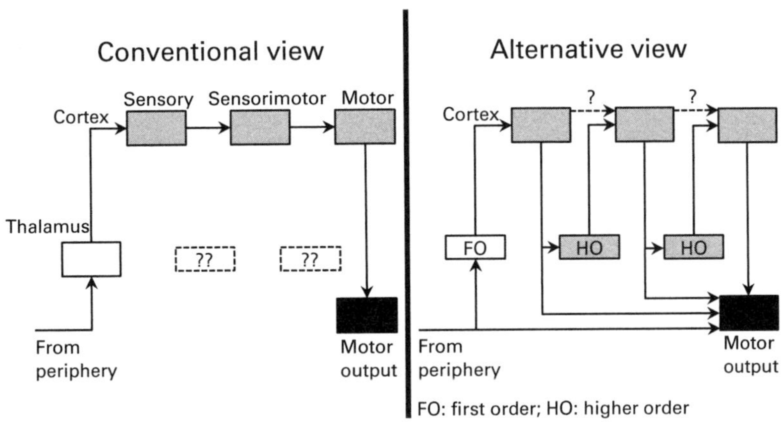

Figure 3.11
Conventional and alternative views of thalamo-cortical circuits. In the conventional view, cortical communication is accomplished via pathways between cortical sites. In the alternative view, as proposed by Sherman and colleagues, higher-order thalamic nuclei play a prominent role in this communication, and direct cortico-cortical pathways may be less important. FO, first order; HO, higher order. Reproduced with permission from Sherman 2007.

dogma that corticocortical transfer of primary information exclusively involves direct corticocortical pathways."

Role of the Amygdala in Processing Emotional Visual Stimuli

Bypass connectivity data in visual cortex suggest that responses in the amygdala should be, at times, quite fast. Although fast responses are possible, the amygdala receives signals from anterior portions of ventral visual cortex, providing it with inputs that have potentially undergone considerable elaboration. In other words, in many cases the amygdala operates on signals that have received extensive processing, leading to responses that are highly selective (e.g., Mormann et al. 2008). Given the multiple pathways that reach the amygdala, it is not surprising that response latencies should span a considerable range, even within a single paradigm. Furthermore, neurons in the amygdala should exhibit a broad range of response characteristics, as observed in the study by Inagaki and Fujita (2011).

What roles does the amygdala play in the processing of affective visual stimuli? The structure's connectivity pattern provides some clues. The predominant source of visual input to the amygdala, specifically to the basolateral amygdala, comes from higher-order visual cortex in the anterior temporal lobe (Amaral et al. 1992). This suggests that the amygdala is a convergence zone for highly processed sensory information. In addition, there are loops between visual cortex and the lateral amygdala, and this feedback is thought to modulate visual responses (Vuilleumier et al. 2004; see chapters 2 and 7). Further integrative functions of the amygdala stem from its extensive connections with much of cortex. In addition to its well-recognized connections with medial and orbital territories of prefrontal cortex, the amygdala is also connected to lateral prefrontal cortex, albeit in a weaker manner (Ghashghaei, Hilgetag, and Barbas 2007).What is more, the architecture of prefrontal cortex is such that, on average, inputs from the amygdala reach approximately 90 percent of PFC after a single connection within frontal cortex (Averbeck and Seo 2008). Finally, the amygdala seems to be part of a "core brain circuit" (Modha and Singh 2010) that is topologically central in terms of global brain connectivity and whose functions probably include aggregation and distribution of information (issues dealt with at greater length in chapter 9).

In light of the foregoing considerations, the amygdala's contribution to processing of affective visual information arises not from a subcortical source of visual input, but rather from the structure's broad connectivity with cortex and with other subcortical structures. Given this connectivity, the impact of the amygdala on behavior can be mediated through many routes, for instance, via both visual and prefrontal cortex, a possibility consistent with findings of our study combining the attentional blink task with fear conditioning (Lim, Padmala, and Pessoa 2009; discussed in chapter 1. For emotion-laden stimuli, trial-by-trial fluctuations in evoked responses in the amygdala predicted whether a target would be detected. Furthermore, the amygdala's influence on behavior was mediated by both visual and prefrontal cortex (as suggested by statistical path analysis). I propose that during the handling of affectively significant items, the amygdala enhances sensory processing through both direct (amygdala-to–visual cortex) and indirect (amygdala-to–prefrontal cortex–to–visual cortex) pathways (see chapter 7 for further discussion).

Prosopagnosia and Capgras Syndrome Some patients suffering from prosopagnosia have a deficit in face perception but exhibit skin conductance responses when they see familiar faces (Bauer 1984; Tranel and Damasio 1985). A prediction of the multiple waves model is that this effect is mediated via routes that bypass regions in temporal cortex whose lesions compromise face perception. Intriguingly, Capgras syndrome (Capgras and Reboul-Lachaux 1923; Ellis and Young 1990) poses a somewhat reverse condition to prosopagnosia. In some instances, patients have altered *familiarity* of persons close to them, such as their parents—these are deemed to be "impostors" who look exactly like the real persons. In one study (Hirstein and Ramachandran 1997), in contrast to normal persons, a patient's skin conductance responses to photographs of familiar people, including his parents, were *not* larger in magnitude than his responses to photographs of unfamiliar people. Capgras syndrome has been suggested to involve damage to the inferior temporal cortex (Capgras patients also have face processing impairments; Young et al. 1993), like other cases of visual hypoemotionality, where patients exhibit deficits of visually evoked emotions with preserved emotional responses to nonvisual stimuli (Bauer 1982; Habib 1986; Sierra et al. 2002). The link between Capgras syndrome and visual cortical damage poses problems for the subcortical pathway of the standard hypothesis. Presumably, a subcortical route would be able to carry signals leading to enhanced skin conductance responses even in Capgras syndrome patients. I am not aware that such responses have ever been observed.

Role of the Superior Colliculus in Processing Emotional Visual Stimuli

Although the focus in this chapter has been on the pulvinar's role as the key "link element" in the purported subcortical pathway of the standard hypothesis, some considerations regarding the superior colliculus are in order here. The superior colliculus is a layered structure at the "roof" of the midbrain. It is called the optic tectum (meaning "roof") in nonmammals and its circuitry and function are believed to be conserved phylogenetically (Butler and Hodos 2005). Indeed, the colliculus is considered to be an ancient visual system that is found throughout vertebrate orders, including ray-finned fishes. It is generally described as having six (or seven) layers, the top two (or three) of which exhibit visual properties and constitute the "visual colliculus." These *superficial* layers are the recipient of direct retinal input, and response latencies to visual stimuli there are quite short (40–70 ms). Response properties change considerably as one moves down to the *deep* layers. They become visual-motor and discharge in close temporal relation to saccadic eye movements (Wurtz and Albano 1980). In addition, responses to auditory and tactile stimulation are observed, and many cells are multimodal (Stein and Meredith 1993).

Not surprisingly, the connectivity of the superficial and deep colliculus is quite different. What is surprising, though, is how different they are. The superficial layers receive inputs from the retina, primary visual cortex and surrounding areas, as well as the frontal eye field. The deep layers receive fibers from these regions, in addition to inputs from frontal, parietal, and temporal cortex, as well as from regions in the basal ganglia.

Here, the *deep* colliculus is of interest insofar as it has been implicated in several defensive behaviors. A series of experiments by Peter Redgrave, Paul Dean, and colleagues in the 1980s led them to suggest that, in rodents, the superior colliculus generates avoidance or escape movements directed away from stimuli that signal emergency or danger (see Dean, Redgrave, and Westby 1989). For example, a rapidly approaching (looming) stimulus can trigger defensive responding that relies on the superior colliculus. The colliculus has a considerable number of descending projections to several structures in the midbrain and pons that are involved in these behaviors, allowing the deep layers to rapidly engender defensive responses.

One particularly interesting midbrain structure is the periaqueductal gray (also called central gray), which surrounds the cerebral aqueduct (a duct containing cerebrospinal fluid). This structure is involved in the integration of behavioral responses to threatening or aversive stimuli. Whereas specific sectors of the periaqueductal gray mediate active coping strategies (such as confrontation or flight), others mediate more passive coping strategies in the face of an inescapable stressful encounter (Bandler and Shipley 1994). Intriguingly, the deep layers of the superior colliculus are adjacent to the periaqueductal gray, leading to the suggestion that they may in fact be the "same" structure that happens to be split into two by coursing fiber tracts (Holstege 1991). Regardless of the exact relationship between the deep collicular layers and the periaqueductal gray, they are bidirectionally connected to one another, affording effective communication between them.

The work discussed in the previous paragraphs was done in the rat (see also Brandão et al. 1994; Schenberg et al. 2005). The circuitry may be similar to that in other mammals and possibly primates—for instance, the descending projections of the superior colliculus appear to be similar in the rat and primate (Huerta and Harting 1984). In primates, cortex is essential for form vision, though basic stimulus features, including looming, may also engage the colliculus and lead to rapid downstream activation of defensive behaviors. Until recently, however, no evidence of the involvement of the primate superior colliculus in defense-like behaviors had been reported and colliculus-related defense mechanisms were assumed to be absent in primates. To the surprise of the authors, in a recent study, activation of the deep collicular layers in monkeys evoked cowering, escape-like responses, high-pitch vocalizations, and attack of objects (DesJardin et al. 2013). These initial findings, if confirmed, would reveal that the deep layers of the superior colliculus participate in defensive behavior in primates, too.

In this section, the discussion thus far has centered on the "nonvisual" deep layers of the colliculus. Interestingly, in rodents, there is evidence that the "visual" *superficial* layers project to the substantia nigra (which is part of the basal ganglia) and are critical for the transmission of short-latency visual information to this structure (Comoli et al., 2003). In rodents, Redgrave (in Smith et al. 2011, 16105) has suggested that "an unpredicted and biologically significant behavioral event causes a short-latency response in the superior colliculus, which is relayed to the basal ganglia." However, the ability of this circuit to support fast *affective* responses remains unknown. But because the superior colliculus supports only very crude form vision, the circuit would not be capable of rapidly distinguishing between different shapes, thus severely limiting the types of responses that it can generate. To conclude, the superior colliculus is a heterogeneous structure with a superficial portion that is mostly visual and a deeper part that receives visual, auditory, and somatosensory signals from cortex, as well as inputs from the frontal eye field. The deeper colliculus is thus substantially multimodal and associational. Here, as in the case of the pulvinar, we find again that the part of the structure that is associational is the one that has the most relevance to emotional processing.

Direct Subcortical Inputs to Visual Cortex

More than twenty-five years ago, Jürgen and Margarete Tigges (1985, 353) reviewed findings regarding subcortical sources of pathways to visual cortex: "The inputs provided by these heterogeneous subcortical projections undoubtedly contribute to the complexity of intracortical processing of visual information. It will be very exciting to learn from future experiments how these many and diversified subcortical projections influence the mode of operation of cortical columns which appear to be such a fundamental aspect of cortical organization." They listed more than twenty structures with direct connections, including the locus coeruleus, raphe nucleus, reticular formation, lateral hypothalamus, basal forebrain, and claustrum, in addition to the amygdala and thalamic nuclei (including the lateral geniculate nucleus and the pulvinar).

Two such pathways to visual cortex, the first from the locus coeruleus (discussed further in chapter 9) and the second from the raphe nucleus, seem to hold special promise. Neurons in the locus coeruleus, which is located in the lower part of the brainstem (pons), respond to a variety of sensory stimuli, including visual ones (Aston-Jones and Cohen 2005; Berridge and Waterhouse 2003). The locus coeruleus may act as an integrative center for external sensory events and internal vegetative requirements. This integration, in turn, can potentially influence responses in visual cortex (Tigges and Tigges 1985). Because the (serotonergic) fibers from the raphe nucleus, also located in the brainstem, form impressive and profuse arborizations in layer IV-C of primary visual cortex, the major target of fibers from the visual thalamus, they are in a position to modulate very early visual processing. After nearly three decades, however, our understanding of the functions carried out by these and other subcortical connections to visual cortex remains rather limited.

Conclusions

The evidence reviewed in this chapter suggests that the idea of a subcortical pathway specialized for the processing of emotional visual stimuli as maintained by the standard hypothesis is much in need of revision. The multiple waves model proposed by Ralph Adolphs and me (Pessoa and Adolphs 2010) effects such a revision, with several implications for the characterization of amygdala function in the processing of emotional visual information, as outlined next. The amygdala plays significant functions in a wide array of networks. Though the precise contribution of the amygdala in these networks is still unknown, it does not map specifically onto emotion but, instead, corresponds to broader and more abstract dimensions of information processing, including salience, ambiguity, unpredictability (Whalen 1998; Sander, Grafman, and Zalla 2003; Adolphs 2008; Hsu et al. 2005), and other aspects of *biological value*.

Indeed, as I argued in chapter 2, the amygdala promotes *selective information processing* and thus plays an important attentional role. It serves to allocate resources to stimuli, at least in part by influencing (through its connectivity) the anatomical components required to prioritize particular features of information processing in a given situation. Such a function would come into play most prominently for affectively significant stimuli. Notably, the amygdala may not be unique in this respect since there are other, largely parallel, networks with architectures that enable related functions, too—such as the network comprising cortex and the pulvinar. The role of the amygdala in the proposed multiple waves model is consistent with a large body of findings and can in fact accommodate several views of amygdala function (Aggleton 1992, 2000; Whalen and Phelps 2009). It is also well understood that the amygdala contributes to several aspects of emotional expression and mobilization of bodily resources (via the central nucleus). Among others, this is an important distinction between the roles of the amygdala and pulvinar during processing of affective visual stimuli.

The multiple waves model also stresses the speed and temporal dispersion of cortical processing, rendering moot the assumed need for a fast subcortical route. Many visual properties can be established very rapidly by the initial wave of cortical response. This implies that there is ample time for feedback to influence ongoing signals. Consequently, unraveling the flow of visual information within cortex and subcortex presents formidable difficulties, difficulties we must overcome if we are to advance knowledge of affective processing. Ultimately, the fate of a biologically relevant stimulus should not be understood in terms of a "low road" versus a "high road," but in terms of the "multiple roads" that lead to the expression of observed behaviors.

There is an enormous amount of literature implicating the amygdala in affective dysfunction in nearly all psychiatric illnesses, most notably, mood disorders. In this respect, the proposed multiple waves model suggests that, rather than focusing on neurons within the amygdala, we should focus on connections within cortex and between cortex and subcortical structures such as the amygdala. In large part, the substrate of brain function is to be found not so much within neurons as within networks—a theme that will be developed in detail in chapter 8.

Notes

1. It is noteworthy that the "circumscribed lesions" in Weiskrantz 1956 were in fact fairly broad and included parts of the temporal pole. Thus "even within the medial temporal region, there is no certainty that damage to the amygdaloid complex, as such, is essential" (Weiskrantz 1956, 389). However, given subsequent work on the amygdala, Weiskrantz's findings have been typically associated with this structure.

2. In neuron-to-neuron communication, researchers at times differentiate between "driving inputs" and "modulatory inputs." As suggested by Sherman and Guillery (1998, 7121): "The former carry the message, defining the essential patterns of activity, whereas the latter can alter the effectiveness of the drive without contributing significantly to the general pattern of the

message." In general, this distinction is not clear cut, however, and often difficult to establish. For further discussion, see also Markov and Kennedy (2013).

3. "Neglect" is a common and disabling condition following brain damage in which patients fail to be aware of items to one side of space. For example, individuals with right-sided brain damage often fail to be aware of objects to their left.

4. I thank Alan Anticevic for discussions of this point.

5. Diffusion tensor imaging is an MRI technique that capitalizes on the diffusion of water molecules along fiber tracks to estimate anatomical connectivity between brain regions.

6. Two regions are functionally connected if their responses are correlated. See chapter 8 for further discussion.

7. Based on these considerations, Shi and Davis (2001) argued that even the *auditory* subcortical pathway to the amygdala might not typically drive affective information in intact animals that have undergone fear conditioning.

8. "Agnosia" refers to the difficulty or inability to recognize visual stimuli, such as animals or man-made objects, whereas "prosopagnosia" refers to the difficulty or inability to recognize faces.

9. Note that areas along the ventral surface were not included in the diagram in figure 3.10 (plate 5) but are also connected with the pulvinar, including areas TE and TEO (Webster, Bachevalier, and Ungerleider 1993).

10. The "magnocellular system" (or "magno channel") is particularly sensitive to moving stimuli and exhibits faster cell responses.

4 Attention and Emotion

The processing of emotion-laden visual stimuli is rapid and occurs under a variety of challenging conditions, which may involve brief durations, crowded displays, and masking, as well as situations in which a stimulus is task irrelevant or its consequences are unintended. Hence emotional processing is generally characterized as "preattentive," "automatic," and "unaware." Chapter 3 described how these properties are frequently associated with subcortical mechanisms and with the standard hypothesis. The present chapter will discuss the interplay between attention and affective vision with an emphasis on the question of automaticity.

Attention, Automaticity, and Resources

Before reviewing the experimental data, it is useful to discuss what is meant by terms such as "preattentive," "automatic," and "unaware." In the 1970s and 1980s, early vision was thought to consist of two sequential stages (Treisman and Gelade 1980): a first, *preattentive* stage, where processing was rapid and occurred in parallel across the entire visual field; and a second, *attentive* stage, where processing was limited in capacity, and thus serial. The existence of a preattentive stage was suggested by psychophysical data, such as the relatively "flat slopes" of responses observed in certain types of visual search procedures. For instance, detecting a red circle among green circles was found to be largely independent of how many green circles were displayed—hence the flat slope as a function of the number of distractors. The preattentive stage was also based on the idea of "built-in analyzers" capable of determining basic stimulus attributes, such as spatial frequency, orientation, color, motion, and disparity. These analyzers were thought to correspond closely to properties of neurons in early visual regions, such as primary visual cortex, and the information gathered at the preattentive stage to be made available to the subsequent attentive stage, which was responsible for assembling basic features into meaningful objects (Treisman and Gelade 1980). Critically, the attentive stage was believed to be resource limited: its processing could not be deployed across the visual field all at the same time. Instead, attention was required.

The preattentive-attentive scheme is an instance of a *bottleneck* model of attention. An alternative scheme involves capacity models of attention (e.g., Kahneman 1973), which associate attention with a source of "energy" that can be flexibly allocated and according to which initially demanding tasks can gradually be performed with less attention as a result of practice. Some researchers have proposed that, in this manner, attentive tasks can become like preattentive ones; they can reach a point where they use no attention at all (for further discussion, see Moors and De Houwer 2006).

Early usage identified processes as "automatic" when they were effortless, nonconscious, or involuntary. As pointed out by Joseph Tzelgov (1997), this definition allowed diverse phenomena involving different psychological mechanisms, such as preattentive processing, well-practiced cognitive or perceptual-motor skills, and even social information processing, to be viewed under a single theoretical umbrella.

Finally, "unaware" processing refers to those conditions in which a subject does not appear to have access to (representations of) the items in question. For instance, even when words are presented in such a manner that subjects appear not to be aware of them, a Stroop effect may be observed (Marcel 1983).

Problems with Traditional Terms

Although it is beyond the scope of this chapter to provide an in-depth evaluation of preattentive, automatic, and unaware processing, some explanatory remarks are in order. Let us first consider "preattentive" processes. The notion that a preattentive stage feeds into a capacity-limited, attentive stage faces several problems (see Di Lollo et al. 2001; Nakayama and Joseph 1998). For one, the list of basic features said to be processed preattentively, which was based on properties of the early visual system (e.g., orientation), has grown far too long. It includes complex visual attributes like those based on shading information (Enns and Rensink 1990; Kleffner and Ramachandran 1992)—disregarding, for the moment, emotion-related properties. For another, more critical problem, features whose processing was once deemed preattentive are subject to capacity limitations when the experimental manipulations are sufficiently strict (Di Lollo et al. 2001; Joseph, Chun, and Nakayama 1997).

In addition to these well-established problems, many recent studies also challenge the usefulness of the term "preattentive." To give but a couple of examples, data from an intracranial study in humans (Ossandon et al. 2012) suggest that "preattentive" search may rely on much the same mechanisms as those required during "attentive" processing. In both a classic pop-out condition requiring the detection of a gray letter among an array of black distractors and a second, effortful search condition, sustained activity was observed in the patients' "dorsal attentional network" (see chapter 7), which includes frontal and parietal areas that guide attention. Furthermore, the dorsal-lateral prefrontal cortex, in particular, though often envisioned as representing top-down factors such as goals, is engaged very early on when a visual target stimulus is defined purely by bottom-up factors—for instance, when a red square must be detected among green square distractors (Katsuki and Constantinidis 2012; Wardak, Vanduffel, and Orban 2010; but see Buschman and Miller 2007).

"Automaticity" encounters problems not only with counterevidence—and plenty exists—but also with usage of the term itself, which has varied across a range of contexts and can be quite confusing to readers unfamiliar with the nuances involved. For instance, some researchers have suggested that automaticity can be controlled (e.g., Tzelgov 1997), even though automaticity and control are typically viewed in opposition. That a process can be automatic *and* still require resources would be a contradiction in terms for some researchers, but perfectly legitimate for others who understand automaticity to refer to processing that is neither part of the task nor beneficial to the task intentionally performed (Tzelgov 1997). Elsewhere (Pessoa 2005), I have characterized automaticity linked to implicit or unintentional processing as "*weak* automaticity," as opposed to "*strong* automaticity," which is linked to processing purportedly independent of resources, that is, capacity *un*limited—much as preattentive, early visual processing was once thought to be. Thus the processing of emotional faces would be considered "weakly automatic" if, for instance, the valence of the faces affected performance (or brain signals) when those faces were irrelevant to the task at hand. For their processing to be deemed "strongly automatic," however, the faces would have had to influence performance (or brain signals) when they were task irrelevant *and* when processing was sufficiently exhausted by the "main" task. For example, suppose a subject had to perform a highly demanding cognitive task with centrally presented stimuli while emotional faces were flashed in the subject's visual periphery. Under these conditions, if effects of the faces were detected, the subject's processing of these faces would be considered "strongly automatic." A final distinction should be drawn here. Processing might be influenced by a top-down factor, such as attention, but not strictly require attention. It might simply be handled more efficiently when attention is devoted to it.

Given these considerations, it is not surprising that Gordon Logan (1988, 583–584) summarized the situation surrounding the term more than twenty years ago as follows: "Empirically, automaticity is reasonably well understood . . . disagreement arises over the interpretation of the facts." Thus, at the very least, a clear operational definition is required if confusion is to be avoided when automaticity is invoked. Yet, when precise usage is employed, *domain-specific* definitions result that likely do not apply to other domains (Tzelgov 1997). When all is said then, it is unclear whether the concept is really useful—in the sense of integrating disparate phenomena—or simply adds more confusion.

Probably as many objections to automaticity have been raised as suggestions that a certain type of processing is automatic. A good example is the intuitively appealing notion that abrupt visual onsets (e.g., a letter appearing in what had been a blank location) capture attention automatically. Initial studies suggested that this was the case for abrupt onsets obtained with differences in stimulus luminance but not in stimulus hue (Jonides and Yantis 1988; Yantis and Jonides 1984). However, in subsequent experiments, the same researchers showed that abrupt onsets are "not strongly automatic" (Yantis and Jonides 1990, 133). This situation also illustrates a fairly common recurring pattern of initially reporting a phenomenon as automatic and later, upon more refined experimental probing, revealing limitations to the effect—recent instances include binocular rivalry (Zhang et al. 2011) and gist/scene perception (Cohen, Alvarez, and Nakayama 2011; Mack and Clarke 2012).

Finally, it is worth considering the relationship between "awareness" and the other terms, in particular, "automaticity" (see also Bargh and Morsella 2008, and the conclusion to this book). Historically, several researchers have equated these two terms—or at least suggested that unaware effects are automatic. Yet their relationship and the link between attention and awareness are far from simple, as both conceptual and experimental work has suggested (Koch and Tsuchiya 2007; Lamme 2003; Most, Scholl, et al. 2005). To complicate matters, awareness can be measured in many ways, and how it is measured has an important effect (Seth et al. 2008).[1] According to *objective* criteria, unaware perception occurs when a subject's performance in a forced-choice task is at chance. Under such conditions, behavioral effects of unaware stimuli (e.g., faster reaction time to undetected fearful faces), as well as the associated brain signals, would constitute correlates of unaware perception. According to *subjective* criteria, unaware perception occurs when subjects verbally *report* that they are unable to perform the task better than by chance (independently of their actual objective performance). There is far from any consensus in the field as to the best (or acceptable) approach to measure awareness (see Seth et al. 2008). In any case, it does *not* follow that an unaware process requires no attention. This is because attentional resources might have to be allocated (to a region of space or task) for unaware effects to be expressed.

In summary, terms like "preattentive," "automatic," and "unaware," which are routinely employed to describe the effects of emotional information, are pregnant with unintended meanings and relationships. In the case of "preattentive," the term should simply be avoided. In the case of "automatic" and "unaware," their use should be made as precise as possible. Subsequent sections will review some of the specific evidence for and against the idea that attention is required for emotional processing before returning to some of the issues touched on above.

Evaluating the Experimental Evidence

Task-Irrelevant Processing and Emotional Perception

Emotional stimuli effectively divert processing resources and interfere with performance even when they are task irrelevant (Pessoa and Ungerleider 2004b; Vuilleumier 2005). For instance, when subjects performed auditory tasks (e.g., word discrimination), their reaction times were slower when they viewed distractor pictures that were unpleasant versus neutral (Bradley, Cuthbert, and Lang 1996; Buodo, Sarlo, and Palomba 2002). Strikingly, interference has been observed even when the primary task is quite basic, such as detecting a simple visual stimulus (Pereira et al. 2006). Interference effects are evident not only in terms of behavioral performance, but also physiologically. For example, specific event-related potential (ERP) components resulting from picture viewing were modulated by emotional content when the main task involved detecting a checkerboard stimulus interspersed with emotional picture presentation (Schupp et al. 2003a). The studies described here are often interpreted to suggest that the processing of emotional stimuli is *obligatory*. Although "obligatory" is seldom defined precisely, its usage frequently suggests a fairly strong reading of the term.

Emotional Perception Does Not Require Attention

A stronger case can be made that emotional processing is obligatory when the spatial focus of attention is explicitly manipulated—in the experiments of the previous paragraph, although task irrelevant, emotional stimuli were fully attended. In the first study of this type (Vuilleumier et al. 2001), the attentional focus was manipulated by having subjects maintain central fixation while comparing either two faces or two houses presented off fixation. Because subjects compared either the faces or the houses to each other on every trial, the focus of subjects' attention was varied by having them attend to the left and the right of fixation (while ignoring top/bottom stimuli) or above and below fixation (while ignoring left/right stimuli). In each case, subjects indicated whether the attended stimuli were the same or different. When conditions involving fearful faces were contrasted to those involving neutral ones, differential responses in the amygdala—which are often considered a signature of emotional processing—and visual cortex were *not* modulated by the focus of attention. The results thus indicate that the processing of emotional items does not require attention. Related findings were observed when overlapping competing stimuli (such as semitransparent overlapping images) were employed. In such cases, object-based attention is said to be manipulated because objects occupy the same position in space; in other words, more than attention to spatial location is needed. Amygdala responses evoked by fearful faces were equivalent whether or not the faces were attended (Anderson et al. 2003).

Emotional Perception Requires Attention

Based on findings such as those summarized in the previous section, it was suggested that emotional stimuli constitute a *privileged* stimulus category that is not only prioritized but also processed in a mandatory manner independent of attention. Because, however, visual processing capacity is known to be generally limited, competition among visual items "selects" the most important information at any given time (Desimone and Duncan 1995; Grossberg 1980).Whenever there is spare capacity, it can be used for dealing with items that are not at the focus of attention (Lavie 1995). And when the relevant task does not demand all of the available attentional capacity, irrelevant stimuli will unintentionally capture resources that enable their processing (Lavie 1995). This line of reasoning, which has been applied to regular, nonemotional stimuli, implies that the automaticity of affective processing can be tested by attentional manipulations that more fully consume resources. Thus a critical variable in understanding the extent of unattended processing is the degree to which the experimental manipulation uses up resources.

Several functional MRI studies have attempted to follow this line of reasoning. For example, centrally presenting subjects fearful versus neutral faces while they performed a highly demanding peripheral perceptual task evoked differential responses in neither the amygdala nor visual cortex (Pessoa, McKenna, et al. 2002). In a follow-up study, when the difficulty of the peripheral task was parametrically manipulated, a valence effect (i.e., responses to fearful faces greater than neutral ones) was observed on tasks of low difficulty (low demand), but not on those of medium or high difficulty (medium or high demand; Pessoa, Padmala, and Morland 2005). Attentional modulation of the emotional valence effect has also been observed for peripherally displayed faces (Silvert et al. 2007), as it has in studies employing object-based manipulations (Mitchell et al. 2007). Posner-type manipulations of attention have revealed that amygdala activation relies on attention, too (Brassen et al. 2010). Furthermore, the need for attention has been observed even when stimuli of increased potency are employed, such as those paired with shock (Lim, Padmala, and Pessoa 2008; but see Cornwell et al. 2011; for related behavioral evidence, see Fox, Yates, and Ashwin 2012; Yates, Ashwin, and Fox 2010).

Event-related potential studies, which, unlike functional MRI studies, offer temporal information on the order of milliseconds, have investigated how emotional perception depends on attention, too. In one case, although the early processing of emotional pictures was intensely attenuated when subjects performed demanding tasks, when they passively viewed the same emotional images, responses to them were stronger than to neutral images (Schupp et al. 2007). Responses to peripheral emotional pictures also rely on the availability of resources. In one study (De Cesarei, Codispoti, and Schupp 2009), emotional pictures in the near periphery modulated brain activity only when subjects viewed them passively, but not when they were engaged in a distractor task (determining whether a rectangular outline contained a gap).

It is worth pointing out that emotional stimuli can be prioritized in many ways, notably, leading to attention capture (Fox et al. 2001), a situation when processing is interpreted to be more automatic. They are also associated with an increased difficulty of disengagement (Fox, Russo, and Dutton 2002). Still the distinction between capture and disengagement is problematic because both types of mechanism are resource dependent and need attention to occur. As previously discussed, even capture linked to abrupt visual onsets can be suppressed when attention is previously focused on a different spatial location (Yantis and Johnston 1990).

Emotional Perception Does Not Require Attention, Again

The findings that emotional perception is automatic and the competing findings that it requires attention can be reconciled by making use of the concept of *attentional demand*. When demand is low, "spillover" capacity will be available to handle task-irrelevant items (Lavie 1995)—processing will appear to be "automatic." As demand is increased, however, fewer resources will be available and, in the limit, interference effects from task-irrelevant items will be eliminated—processing will appear to be capacity limited. Whereas attentional demand can be used to explain a broad set of results, some findings appear to resist this explanation. In one study (Muller, Andersen, and Keil 2008), subjects performed a difficult target detection task while task-irrelevant, emotionally arousing pictures were shown in the background. Despite the difficulty of the task, emotional pictures interfered behaviorally with the main task. Parallel effects were observed in steady-state visual evoked potentials, which were reduced during the presentation of emotional versus neutral background images (such reduction was thought to reflect the withdrawal of processing resources from the main task by the emotional distractors).

A recent magnetoencephalography study found evidence for mandatory processing of fearful faces (Fenker et al. 2010). Both low- and high-demand conditions were investigated in separate experiments. During the high-demand condition, the target was defined by a conjunction of features, such as a red-green (vs. blue-yellow) vertical bar. In the low-demand condition, subjects determined the orientation of an oriented bar (vertical vs. horizontal) presented in a given color. To establish the neural impact of the task-irrelevant faces, the authors probed the "N2pc" response component, which is believed to reflect attentional selection in visual search (Eimer 1996). Notably, they observed that lateralized fearful faces elicited an N2pc response component for approximately 240–400 ms in contralateral visual cortex even during conditions of high demand.

It has been advanced that competition dampens the effect of emotion-laden items in typical individuals, but not in specific populations. Indeed, behavioral and event-related potential results in spider-fearful subjects (though not necessarily phobics) were not altered by standard manipulations of the number of distractor items (Norberg, Peira, and Wiens 2010). In addition, a functional MRI study with spider phobics reported unchanged amygdala responses based on attention (Alpers et al. 2009), although the attentional manipulation in the study was not strong. Brian Cornwell and colleagues (2011) have proposed that, whereas attention is required for the processing of emotion-laden items during a safe context, anxiety elicited by threat broadens attention and enhances sensitivity to potential danger cues, even when perceptual systems are taxed. When their subjects were threatened by shock, greater amygdala responses to fearful versus neutral distractor faces were preserved under high attentional demand.

Finally, in a recent study, Swann Pichon, Beatrice de Gelder, and Julie Grèzes (2012) found that, even though responses in the amygdala are strongly influenced by task demands—that is, they are *not* automatic—other brain areas involved in reflexive defensive behaviors are impervious to these demands. The regions that were *not* modulated by attention included the hypothalamus and the periaqueductal gray (in the brainstem).[2] In a sense, these findings "push down" the level at which automaticity is observed on the neuroaxis, to "more basic" subcortical sites, but not to the amygdala.

Operationalizing Attentional Demand

In the preceding discussion, attentional demand was used to help explain discrepant results. But how should it be operationalized? Nilli Lavie (1995) proposed that this be done through "perceptual load," which is typically manipulated by varying the number of distractor items presented in the visual display, also called the "set size." For example, a subject might be asked to search for a target letter "X" or "N" when presented by itself or among five other letters. During the former, low-load condition, performance is not strongly challenged and other task-irrelevant items in the display interfere with behavior (e.g., a peripheral fearful face might slow down reaction time). In contrast, during the latter, high-load condition, the visual system is more robustly taxed and other task-irrelevant stimuli do *not* interfere with performance as much, if at all.

Although Lavie's perceptual load model has been applied to a wide set of results in attention research, it has been challenged by work on "dilution" effects that demonstrated "reverse load" effects (Tsal and Benoni 2010; Wilson, Muroi, and MacLeod 2011). In these studies, *increased* task-irrelevant processing (greater interference) was observed with a larger number of distractors, even though perceptual load would predict reduced task-irrelevant effects with larger set sizes because resources would be further taxed.

It is also possible to operationalize attentional demand in terms of task difficulty. Although this appears at first to be a reasonable measure, because a task may be difficult due more to sensory limitations (e.g., low-contrast stimuli are used) than to capacity limitations (Hsu and Pessoa 2007; Lavie and de Fockert 2003), more sophisticated approaches are needed. In other words, because a task associated with low accuracy does *not* necessarily deplete processing resources, researchers need to carefully consider the parameters that contribute to performance in attentional tasks. More broadly, multiple factors interact to determine how stimuli *compete* in visual cortex and beyond, including task difficulty, set size, spatial arrangement, and cueing, among others. For example, a spatial cue (central arrow pointing to the target location) was found to eliminate task-irrelevant processing effects even when the target stimulus appeared among a uniform set of distractors (all distractors were the letter "O"), a condition deemed to be of low perceptual load (Johnson, McGrath, and McNeil 2002; see also Yantis and Johnston 1990 for conditions that influence visual selection). Douglas Johnson, Allison McGrath, and Carrie McNeil (2002, 287) summed up the challenge facing those attempting to operationalize attentional demand: "A simple and parsimonious explanation of selective attention is at best elusive."

The above considerations are relevant for processing visual stimuli in general and affective visual stimuli in particular. Inasmuch as tasks in early studies did not greatly tax attentional resources, given the accumulating evidence of capacity limitations during emotional perception, tasks in subsequent studies have challenged attention more vigorously. A common manipulation in these tasks is to vary the number of distractors, although the interpretation of this manipulation is not necessarily straightforward; indeed, as mentioned above, "reverse" effects have been documented. In the coming years, it will be important to investigate a broader range of task parameters and how they impact the processing of task-irrelevant emotional items.

Interim Summary

Results from studies employing behavioral and neuroscience methods in diverse paradigms indicate that emotional processing, though prioritized, depends on resources in many contexts. By and large, attentional demand and competition can account for the discrepancy between results suggesting that emotional perception is automatic and those suggesting that it depends on attention. Thus, to reveal that emotional perception is *not* immune to the effects of attention, processing resources need to be largely consumed; otherwise, performance will appear to be relatively automatic. Yet attentional demand does not explain all the available data. At present, the reasons for this discrepancy remain unclear, suggesting that it would be profitable for future studies to tackle this issue more directly—including the role of the various factors that influence selection, as described in the previous section.

Because specific subject populations, such as those with phobias, may be particularly susceptible to affective stimuli, individual differences are important predictors of sensitivity to emotional stimuli. In particular, studies from the literature on anxiety have revealed that high-anxious subjects display greater interference from threat-related stimuli (MacLeod, Mathews, and Tata 1986). Some studies have investigated the extent to which amygdala responses to threat-related distractors depend on individual anxiety levels. In one study (Bishop et al. 2004), low-anxious subjects exhibited increased amygdala responses only to attended fearful faces, whereas high-anxious subjects exhibited increased responses to both attended and unattended threat-related stimuli. These findings suggest that the threat value of a stimulus varies as a function of a subject's anxiety level, but other results demonstrate that attention is important even for high-anxious subjects (Bishop, Jenkins, and Lawrence 2007; Fox, Russo, and Georgiou 2005).

Temporal Paradigms and Mechanisms of Prioritization

In contrast to the spatial manipulations discussed in the previous section, the attentional blink is used to investigate *temporal* capacity limitations. As described in chapter 2, in this paradigm, subjects are asked to report the occurrence of two targets, T1 and T2, within a stream of rapidly presented visual stimuli. They have difficulty detecting the second of the two targets when it follows the first by a brief delay. This section discusses findings from attentional blink studies and their implications for the understanding of emotional perception.

In our study reviewed in chapter 2 (Lim, Padmala, and Pessoa 2009), my colleagues and I asked subjects to detect a T2 target that was a picture of either a house or a building, and stimuli of one of the categories were paired with mild electrical stimulation during an initial conditioning phase (figure 2.5).[3] In the context of the present chapter, two findings of this study are particularly noteworthy. First, although pairing with mild shock reduced the magnitude of the attentional blink, it did not eliminate it. In fact, performance on T2 detection was only 72 percent correct. In other words, even stronger emotion-laden stimuli are subject to the blink and are not processed automatically—strong automaticity would have entailed *no* attentional blink. In support of our finding, a previous study also reported that fear-conditioned faces (by pairing with a loud tone) were subject to an attentional blink (Milders et al. 2006). Second, the attentional blink allowed us to probe the impact of detected and undetected T2 targets, namely, the impact of stimulus awareness on brain responses. As expected, correctly detected pictures of the scene category previously paired with shock evoked greater responses in the amygdala and visual cortex than neutral targets (i.e., pictures of the scene category that was not paired with shock). In contrast, for trials on which the shock-paired T2 went undetected, the same contrast did not reveal significant differential responses. In other words, affective significance did *not* robustly influence brain responses when subjects were *unaware* of the T2 stimulus, but only when the T2 scene was correctly detected. These findings, again, argue against the automaticity of emotional perception.

In a behavioral study of the attentional blink, Timo Stein and colleagues (2010) used a clever manipulation to influence T1-related demands as a strategy to probe T2 detection performance. The difficulty of T1 detection was manipulated by using "flanker interference." Specifically, during the low-demand condition, the center face was flanked by identical copies of that face; during the high-demand condition, it was flanked by randomly sampled faces.

For the T1 task, subjects were asked to report the gender of the central face; for the T2 task, to detect (present vs. absent) faces that could be either fearful or happy, though their valence was task irrelevant. During the low-demand condition, subjects detected fearful faces better than they did happy faces, replicating previous reports, whereas this advantage disappeared during the high-demand condition, during which they detected fearful and happy faces equally often. The results demonstrate that the privileged access of fearful faces to awareness is not automatic, but instead depends on resources. Of note, a blink was observed across conditions—a property that appears to be shared by *all* affective emotional blink experiments. It was the *advantage* for fearful faces that was eliminated during the high-demand condition. In other words, capacity limitations for the processing of emotional faces were obtained for both levels of demand.

Together, the results from attentional blink experiments suggest that emotion-laden stimuli are subject to the blink, countering the notion of strong automaticity. More broadly, these results indicate that affective perception is under the control of attentional mechanisms not only during spatial competition tasks, but during temporal competition tasks as well. And some of the findings are particularly noteworthy because they involved stimuli that were emotionally potent given their pairing with mild shock. At the same time, in support of the idea that emotional items have a special advantage, affective items show a reduced attentional blink.

A final point deserves discussion here. The interpretation of findings of attentional blink studies is complex because both attention and awareness are intertwined in the paradigm. What is becoming increasingly clear, however, is that attention and awareness, though related, may also be partially dissociated. As Steven Most, Brian Scholl, and colleagues (2005, 224) have pointed out: "Environmental cues can trigger automatic orienting responses, but these reflexive responses by themselves do not directly produce awareness. . . . Visual stimuli that do not become part of a cycle of expectation, exploration, and reinterpretation may never be noticed at all." Indeed, implicit attention capture and awareness can be dissociated (Most, Scholl, et al. 2005; for further discussion on attention and awareness, see Astle, Nobre, and Scerif 2010; Cohen et al. 2012; Koch and Tsuchiya 2007; van Boxtel, Tsuchiya, and Koch 2010; and Van den Bussche et al. 2010).

Timing of Emotion and Attention Effects

Techniques that provide fast temporal information, notably, electroencephalography and magnetoencephalography, have been used to probe the timing of affective processing in humans. Although functional magnetic resonance imaging studies have also investigated this question in depth, rapid effects of emotional items that are independent of attention may have been missed by functional MRI, which gives a temporally "blurred" version of neural events. This section considers two noteworthy studies that used MEG or intracranial recordings to evaluate responses evoked in the amygdala when researchers explicitly manipulated attention and emotion.

In the first study, Qian Luo and colleagues (2010) used MEG to investigate the amygdala responses of subjects viewing task-irrelevant fearful and neutral faces. Although the amygdala is located deep within the brain and therefore challenging to probe with EEG or MEG, advanced source analysis techniques have attempted to measure signals from this structure (Ioannides et al. 1995; Streit et al. 2003). On each trial, the subjects were asked to discriminate the orientation of two peripherally located bars (same vs. different). As in previous studies, attention was manipulated by varying task difficulty. During the low-demand condition, the bar orientation difference was high (90 degrees), making the task very easy. During the high-demand condition, the bar orientation difference was low (15 degrees), making the task relatively harder.

MEG revealed a significant effect of facial expression in the left amygdala, specifically, increased gamma-band activity in response to fearful versus neutral expressions soon after stimulus onset (30 to 60 ms).[4] Consistent with automaticity, attentional demand had *no* effect on the subjects' left amygdala responses in this early temporal window, whereas it did have an effect on the subjects' right amygdala responses to fearful versus neutral faces during a later period (280 to 340 ms). Luo and colleagues thus suggest that "emotional automaticity is a matter of timing" (as reflected in the title of their study) and conclude that functional MRI may simply miss the fast first pass of emotional information, which is presumed to be automatic.

As mentioned in chapter 3, it is not entirely clear whether signals from deep structures in the brain, such as the amygdala, can be localized with current EEG or MEG methods. An approach that bypasses this problem is to record directly from the amygdala in humans (e.g., during presurgical preparation). This was the strategy adopted by Gilles Pourtois and colleagues (2010). In the same paradigm discussed earlier in the chapter, when presented with two houses (e.g., to the left and right of fixation) and two faces (e.g., below and above fixation), the patient was asked to determine whether the horizontal or vertical stimulus pair was identical. Intracranial recordings from face-sensitive sites in the patient's lateral amygdala showed an early and systematic differential neural response between fearful and neutral faces, regardless of attention. Differences were observed from 140 to 290 ms. Furthermore, comparing trials with task-relevant versus -irrelevant faces (regardless of facial expression) revealed a sustained attentional effect in the left amygdala that started at 710 ms after stimulus onset. By using techniques that offer millisecond temporal resolution, these two studies tried to determine the temporal evolution of affective processing and how it is influenced by attention. However, they raise important questions.

Let us first consider the MEG study. It was suggested that responses in the amygdala are modulated by affective content within 30–40 ms after stimulus onset, possibly via a fast subcortical pathway. Yet the timing is puzzling in light of known response latencies in the visual system. For instance, the earliest responses in the lateral geniculate nucleus of the thalamus, which receives direct retinal input, are observed at approximately 30–50 ms (Lamme and Roelfsema 2000).[5] But, as discussed in chapter 3, additional time (at least 20–30 ms and probably quite a bit more) would be needed for signals to start conveying *differential* responses as a function of stimulus type. Notably, response latencies in the amygdala range from 100 to more than 200 ms. Thus the mechanisms by which valence modulation could occur in the time window observed by Luo and colleagues (30–40 ms) remain elusive and appear to be inconsistent with the properties of amygdala responses in primates.

The timing observed in the intracranial study by Pourtois and colleagues (2010) is more in line with existing data: they detected affective influences on responses starting at 140 ms after stimulus onset. A key concern, however, is that the task they employed was not sufficiently challenging. The patient was correct 95 percent of the time during face trials and 97 percent of the time during house trials. Although this resulted in balanced task performance for faces and houses, the task was extremely easy. As discussed earlier, attentional resources will be available when the central task is not taxing, and therefore effects of valence under these conditions are not surprising. And although the effect of valence may be referred to as "automatic" in the weaker sense of implicit processing of task-irrelevant information, the task does not allow for a stricter test of automaticity.

A final issue merits discussion. A common objection to functional MRI is that it is not sensitive to brief events (e.g., Brosch and Wieser 2011). Several examples show that this is not necessarily the case, however. Thus Robert Savoy and colleagues (1995) found the blood oxygen level dependency (BOLD) signal measured by functional MRI to be sensitive to quite brief (as short as 34 ms) and transient events (figure 4.1). Remarkably, even submillisecond stimulation has been shown to evoke detectable hemodynamic responses (Hirano, Stefanovic, and Silva 2011; see also Ogawa et al. 2000). Furthermore, functional MRI responses have been consistently reported for stimuli that are

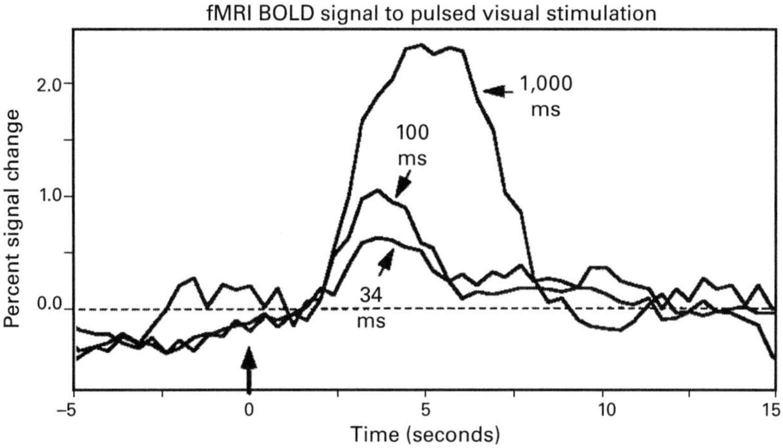

Figure 4.1
Functional MRI BOLD responses to pulsed visual stimuli. Original data from Savoy et al.1995. Reproduced with permission from Rosen, B. R., Buckner, R. L., and Dale, A. M. (1998). Event-related functional MRI: Past, present, and future. *Proceedings of the National Academy of Sciences of the United States of America*, *95*(3), 773–780, Copyright 1998 National Academy of Sciences, U.S.A.

presented very briefly (approx. 30 ms) even when masked (Morris, Öhman, and Dolan 1998; Whalen et al. 1998), a result replicated many times with affective stimuli. Thus, even though it is clearly desirable to obtain millisecond-level data such as provided by techniques like MEG, functional MRI is not blind to brief, transient events, as further illustrated in figure 4.2.

In sum, both studies described here conclude that the effects of emotional content temporally precede those of attention and that prior discrepancies in the literature may stem from the temporal characteristics of the functional MRI signal. The points raised above suggest, however, that this conclusion is unwarranted.

Conceptual Proposal: Moving beyond the Current Impasse?

Although a great deal has been learned about the extent and limits of affective vision, the two camps—*unlimited* versus *limited* processing—have entrenched views, and an impasse is evident. Perhaps this is not surprising. After all, emotional stimuli are sufficiently potent to exhibit a host of properties that show they are "superior" to neutral items. Clearly, they are processed when unattended, at least when the attentional manipulation is not sufficiently severe. At the same time, emotional processing is subject to capacity limitations, as revealed by several experimental manipulations.

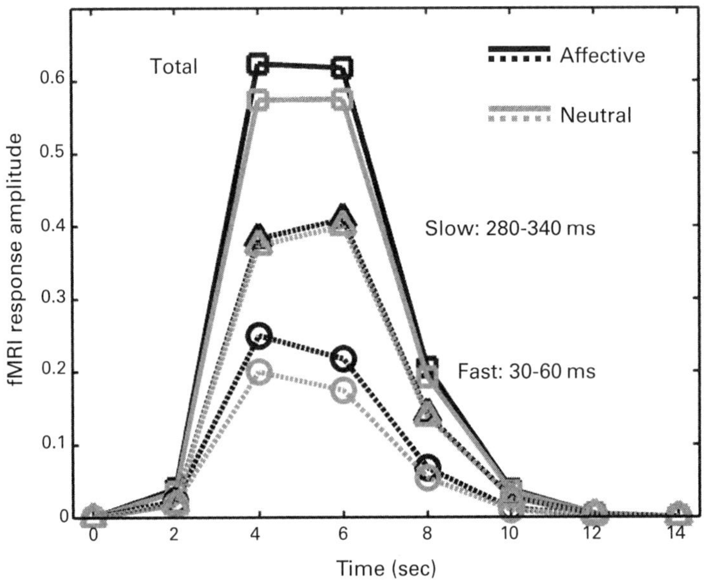

Figure 4.2
Functional MRI responses to simulate MEG results by Luo and colleagues (2010). Simulated responses generated by convolving an "input function" with a canonical hemodynamic response. The input function included "fast" events ("on" between 30 and 60 ms) and "slow" events ("on" between 280 and 340 ms). In the study by Luo and colleagues, during the hard condition, only fast responses varied as a function of valence (attention did not affect early but did influence late responses). To simulate this, the "fast" effect of emotional content was obtained by assuming a higher intensity for affective versus neutral "fast" events; "slow" events were of the same magnitude. The simulation shows that fast responses are not inherently invisible to functional MRI and would be expected to generate differential responses (results labeled "fast"). No differential response is observed for the "slow" component because the simulated input did not vary between conditions (a slight displacement was used for display). An actual functional MRI study would pick up the "total" signal containing the contributions of both fast and slow components and should be sensitive to differences present in only the first time window. Black lines: Responses to affective and neutral conditions are indicated by gray and dashed lines, respectively; fast and slow components by solid lines; total signal is measured in arbitrary units. Adapted with permission from Pessoa 2010c.

Are more empirical data needed to resolve this debate? A recent study by Andrea Shafer and colleagues (2012) suggests that the solution is to say that affective visual processing is *both* automatic *and* modulated by attention. Because they observed that some brain regions influenced by emotional content, including the amygdala, were *not* affected by task difficulty, whereas other regions were, Shafer and colleagues interpreted their findings as supporting both viewpoints. A similar position was taken by Hadas Okon-Singer, Joseph Tzelgov, and Avishai Henik (2007, 147), who concluded that the "processing of emotionally significant stimuli is automatic yet requires attention."

Although this is a possible way out, it limits the use of "automatic" to only *weak* versions of automaticity.[6]

I contend that broader conceptual issues suggest the impasse will *not* disappear based solely on additional experimental results, however valuable they might be. On the one hand, advocates of limited processing can always claim that attentional resources were not sufficiently consumed—if only the manipulation were stronger, the impact of affective items would go away. On the other hand, showing that an emotional effect has disappeared is always subject to the "null problem" (claiming the absence of an effect), which is fraught with considerable difficulties—especially within the framework of conventional frequentist statistical analysis. For instance, my colleagues and I (Pessoa et al. 2002) detected a clear statistical interaction in the right amygdala, which revealed the need for attention during the processing of emotional faces. Yet it could also be argued that the pattern of our results for the left amygdala was consistent with automatic processing (figure 4.3). Thus, even though we detected no statistical differences between unattended fearful and neutral faces in the left amygdala during the challenging bar-orientation task, the case could be made that the difference *would* have been significant with greater statistical power (indeed, with a sufficiently large N, any difference is potentially significant)—we would have observed a valence effect even during the demanding attentional manipulation. Curiously, statistical power comes into play in other situations as well. For instance, when attention by emotion interactions (which indicate that attention is important for emotional perception)

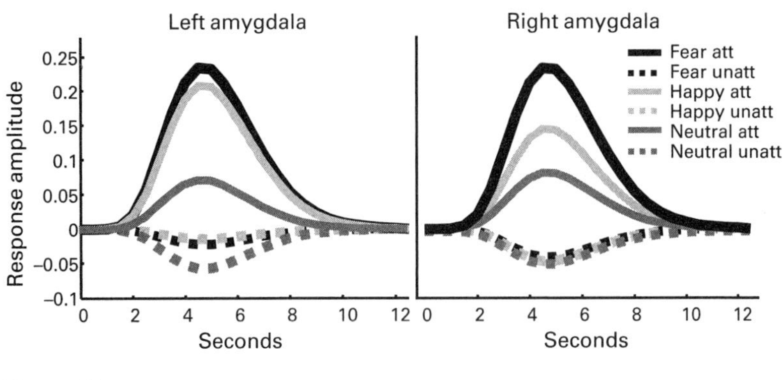

Figure 4.3
Capacity limits and statistical power. Results from Pessoa et al. 2002, describing how attention influences valence effects. Although no differential responses were detected statistically for unattended faces in either the left or right amygdala, it could be argued that a study with greater statistical power might have picked up responses to fearful faces even during the unattended condition in the left amygdala (note that responses to fearful faces were greater—less negative—than neutral ones). Adapted with permission from Pessoa, Gutierrez et al. 2002.

are not detected, results may be interpreted in *favor* of automaticity (Luo et al. 2010). It is noteworthy that attaining adequate statistical power to assess interactions is more difficult than in the case of other simple or main effects (Murphy and Myors 2004).

The preceding problems are similar to those of establishing automatic processing more generally. As quoted earlier, a central problem in the area was agreeing over the "interpretation of the facts" (Logan 1988). One way out of the impasse is provided by Agnes Moors and Jan De Houwer (2006), who assert that every process is uncontrolled, efficient, unconscious, and fast, to *some degree*. Thus, in the context of emotional processing, when it is said that processing is "automatic," the question is, *relative to what*? Stating that processing is automatic relative to a neutral item is a starting point. Establishing finer comparisons would offer a fuller characterization of the phenomenon, for instance, relative to abrupt onsets, a class of undeniably highly potent stimuli (Jonides and Yantis 1988).

I propose that an approach based on the notion of *resources* is useful in the context of affective processing. Given limited mental resources, performance is impaired when demands are greater than available capacity (figure 4.4A). Although the concept of resources has at times been viewed as vague (Navon 1984) and problematic (Logan 1988), it should be possible to refine it (e.g., Lavie et al. 2004; Park et al. 2007; see also Pessoa 2009; Chun et al. 2011; and chapter 7; for earlier conceptualizations, see Kahneman 1973; Norman and Bobrow 1975). Accordingly, instead of focusing on whether processing is preattentive, automatic, or both, studies should aim to understand how a certain process is influenced by the consumption of resources, for instance, through dual-task procedures. In particular, evidence that the processing of emotional stimuli does *not* require attention would be provided by studies that demonstrate the behavioral or neural consequences of emotional items even when resources are depleted—or, in practice, in the "near absence" of attention given that complete absence is difficult, if not impossible, to ascertain. To a certain degree, however, disagreement concerning the relative merits of attentional manipulations should be expected. But at the very least, the notion of resources entails that care be taken to establish that capacity has been effectively taxed (figure 4.4B), for instance, by showing the impact of the task manipulation on a set of "comparison tasks" (see, for example, Li et al. 2002).

To motivate the dynamic framework developed here, consider a processing architecture in which a single boundary between preattentive and attentive processing exists (figure 4.5A; for discussions, see Cave 1999; Cave and Batty 2006). Whereas, in this preattentive/attentive scheme, some visual features are handled preattentively by virtue of being optimally matched to properties of

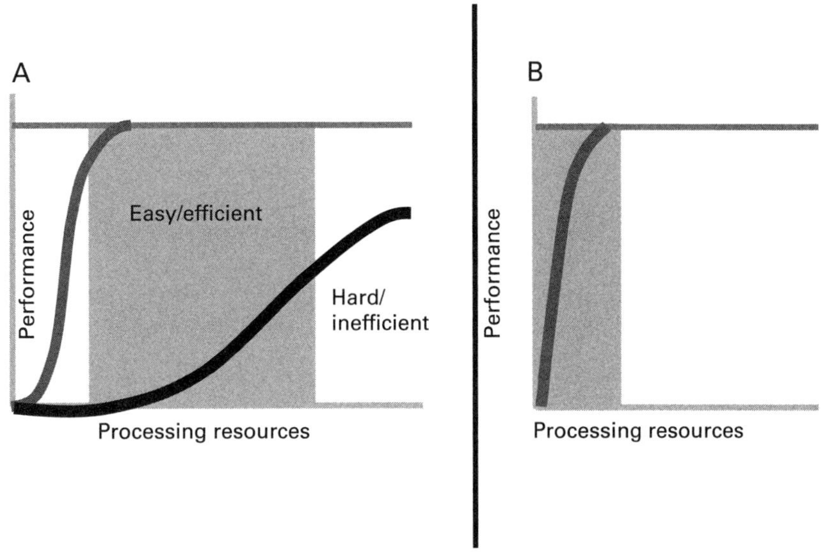

Figure 4.4
Idealized graphs of processing resources against performance (Norman and Bobrow 1975). (A) For an easy task, very little expenditure of attention brings subjects to ceiling performance. For a hard task, the slope of the function is much shallower and considerable effort may not be sufficient to bring performance to ceiling. One can imagine a continuum of scenarios between the two shown here. The gray rectangle represents the range of resources manipulated in typical laboratory tasks. (B) Where processing is so efficient that performance reaches asymptote before resources are taxed at levels typically employed (as in panel A), more extreme manipulations (represented by the gray rectangle) may be required to reveal capacity limitations. Adapted with permission from Nakayama and Joseph 1998.

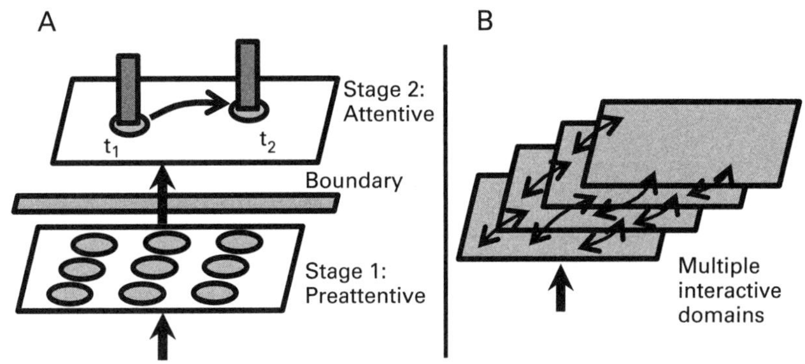

Figure 4.5
Visual processing models. (A) Traditional model in which preattentive and attentive stages are separated by a "boundary." Attention is required for focused processing, which proceeds in a serial manner from time 1 to time 2 (t_1 to t_2). (B) Dynamic model in which communication between "domains" is bidirectional and context dependent.

the early visual system (e.g., orientation), in the *dynamic* alternative framework, visual processing is neither preattentive nor attentive but takes place through the iterative exchange of information involving multiple "domains" that are bidirectionally interconnected (figure 4.5B; for a related idea, see Di Lollo et al. 2001).The proposed framework posits the existence of a *gradient* of *processing efficiency*; within this framework, the question of whether something can be processed without attention (i.e., at the lower end of the continuum) loses its "special" status.

The efficiency gradient is based in part on the properties of early visual areas but, critically, also on how they are dynamically configured based on task demands. Such configuring is thought to depend, in large measure, on frontal and parietal cortex (Desimone and Duncan 1995). A crude way to summarize my proposed framework is in terms of a series of "gates" (figure 4.6) that are controlled by goal- and stimulus-bound factors. Earlier gates, such as those in retinotopically organized visual cortex, may be more strongly permeable, such that they are *less* susceptible to capacity limitations, that is, to the effects of attentional manipulations. In this sense, it may be extremely difficult to demonstrate capacity limitations to basic visual properties (e.g., orientation) that are processed in early visual cortex. Later gates are more

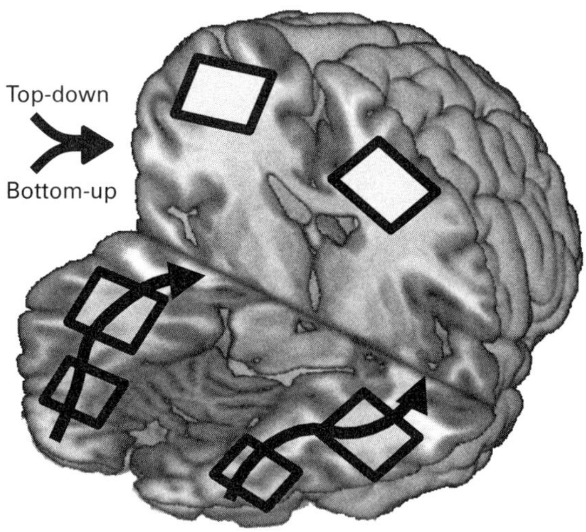

Figure 4.6
Processing efficiency and "gates." Multiple selection processes operate during vision and can be viewed as gates of different permeability. Processing efficiency of early stages is sufficiently high, such that they are less susceptible to capacity limitations. The gates are thus more permeable (illustrated via relative transparency). Later stages exhibit stronger capacity limitations and have gates that can be more thoroughly shut (less transparent). Some stages may function as "bottlenecks," such as illustrated for the parietal cortex (opaque gates).

susceptible to being more thoroughly shut. This organization is thus consistent with a gradient of attentional effects from early to later visual regions, with larger effects of attention being observed in later regions (Kastner and Ungerleider 2000). Beyond visual cortex, in parietal and frontal cortex, true "bottlenecks" appear to exist and, indeed, severe capacity limitations are observed in these territories (Marois and Ivanoff 2005; Tombu et al. 2011)—in this sense, "gates would be shut" under some conditions.

To illustrate some of the elements of the proposed framework, let us briefly reconsider one of the studies discussed previously (Fenker et al. 2010). Under low-demand conditions, the effects of task-irrelevant fearful faces were observed both behaviorally and by MEG. Under high-demand conditions, although an effect of valence was observed in the MEG signals, no concomitant behavioral changes were detected. Regardless of the issue of statistical power (only eight subjects were investigated), this study illustrates the possibility of dissociations between behavior and neural events. I suggest that capacity limitations may have operated more vigorously here at sites that were closely tied to determining behavior. At the same time, as measured by MEG, effects were expressed at sites not robustly modulated by attentional demand. More generally, according to my framework, capacity limitations operate at multiple sites in the brain and correspond to multiple *selection processes*. In this manner, task performance and perceptual encoding are subject to selection mechanisms that operate in parallel, or in close temporal succession (see figure 4.6). Affective processing will therefore be expressed neurally and impact behavior in ways that are highly dynamic and task dependent. And effects of affective items will be discerned or not, depending on processing efficiency.

Although the ideas described here need to be developed in greater detail, the proposed processing efficiency framework may help explain some of the findings not readily explained by simpler versions of attentional demand (see "Emotional Perception Does Not Require Attention, Again" above). In other words, the reported effects of valence may have involved processing stages quite resistant to modulation (i.e., unpliable stages).

Role of Goal Relevance

Attention is usually discussed as being influenced by both exogenous and endogenous factors. "Exogenous factors" are those related to physical features of the stimulus, such as contrast, color, and motion. For example, all else being equal, attention will be captured by a moving stimulus more robustly than by a static one. "Endogenous factors" are most often associated with current or temporary goals. For example, if one seeks to find a specific friend at a particular train station, that goal will facilitate one's finding the friend.

When exogenous factors are seen to function automatically (e.g., contrast in a pop-out display), whereas endogenous ones are seen as subject to control, an asymmetry arises, with exogenous factors appearing to be *stronger* than endogenous ones. Accumulating evidence, however, shows that goal-based attention, though viewed as an "endogenous factor," can be quite powerful. Moreover, the deployment of attention to events relevant to current goals has been shown to occur in an involuntary manner (Folk, Remington, and Johnston 1992; Moskowitz 2002; Vogt, De Houwer, and Crombez 2011; Vogt et al. 2010). Indeed, the effectiveness of goal-based attention has been demonstrated across diverse contexts. For example, involuntary attention to abrupt onsets is contingent on "control settings" (Folk, Remington, and Johnston 1992; see also Gibson and Kelsey 1998). In an elegant series of experiments, Julia Vogt and colleagues have characterized several properties of goal-based attention, which were manifested even when the activated goal was not central to task performance. In one study (Vogt et al. 2011), they tested whether aversive emotions also bias attention toward stimuli that support the alleviation of the emotion. They induced disgust by having subjects touch fake disgusting objects (controls touched nondisgusting objects). The results of a subsequent attentional task revealed, not surprisingly, that attention was oriented to disgusting pictures. Interestingly, subjects in the disgust condition also oriented toward pictures representing cleanliness. The deployment of attention during the aversive condition thus appeared to be guided not only by the emotional state, but by the goal to alleviate this emotional state.

But what happens when goals compete with emotional content? If goals strongly govern attentional priority, goal-relevant items should be able to effectively compete with emotional stimuli. To probe this question in a later study, Vogt and colleagues (2012) examined whether goal-relevant information still evokes an attentional bias when it competes with threatening stimuli. In three experiments, subjects performed a dot-probe task, which evaluates mechanisms of spatial attention, combined with a separate task that induced a temporary goal. Their attention was oriented to goal-relevant pictures even when the stimulus was simultaneously displayed with threatening pictures. This was observed even in a group of high-anxious subjects, and in the presence of a more powerful threat (a colored patch signaling the presentation of an aversive noise).

The studies above have several important implications. Goal-relevant items acquire properties similar to those observed for stimuli that involuntarily capture attention, such as abrupt onsets and emotional stimuli. In this regard, endogenous attentional processing may act like "automatic" processing. Thus the dichotomy of exogenous/stimulus-driven/automatic versus endogenous/

goal-driven/controlled breaks down—and terms like "bottom-up" and "top-down" can only be used descriptively. In particular, goal-relevant items are powerful enough to win the competition against simultaneously presented emotion-laden items, illustrating that emotional stimuli do not have a unique, "special" status. How do goal-relevant stimuli acquire their competitive advantage? One idea is that templates prime visual mechanisms to facilitate detection (Desimone and Duncan 1995; Grossberg 1980), as implemented in several computational models of visual search (Grossberg, Mingolla, and Ross 1994; Wolfe 1994).

Finally, although the literature on goal-relevance and attention has developed independently of the literature on motivation and attention, they are obviously closely related. For one thing, motivation is an effective way to induce goals, as when subjects are told they will be rewarded for correct performance in trials involving a specific stimulus (e.g., Kristjansson, Sigurjonsdottir, and Driver 2010; chapter 6 will review findings on the interactions of reward and attention in detail).

Related Proposals

My processing efficiency framework is related to several other proposals. For example, Jeremy Wolfe and colleagues (2011) have proposed that vision proceeds along both a capacity-limited pathway and a nonselective one not subject to the same limits, which can extract statistical features from the entire scene. The nonselective pathway enables semantic processing that supports "gist" perception, for example, but not precise object recognition.

A key distinction that dominated attention research for several decades was whether the "locus of selection" (where attentional limitations are observed) occurred before stimulus identification, as assumed in "*early* selection" theories, or after identification, as assumed in "*late* selection" theories (see Driver 2001). More recent work on attention has concluded that the locus of selection is more flexible (Yantis and Johnston 1990), operating at earlier or later stages depending on several factors, most notably, the extent to which available resources are taxed (Lavie 1995). Although some proposals (e.g., Yantis and Johnston 1990) have emphasized the existence of two loci (early and late), others have gone further. Thus, in his FeatureGate model, Kyle Cave (1999, 190) has proposed that there is no strict separation between preattentive and attentive mechanisms because the selection process is "distributed across the levels of the hierarchy." Sabine Kastner and Mark Pinsk (2004, 483) have proposed that "neural mechanisms of selective attention operate at multiple stages in the visual system and beyond and are determined by the visual processing capabilities of each stage. In this respect, attention can be considered in terms of a multilevel selection process." And Edward Vogel, Geoffrey Woodman, and Steven Luck (2005, 1907) have proposed that "attention selects relevant information at whatever stage or stages of processing are overloaded by a particular stimulus-task combination." Evidently, the processing efficiency framework described here is related to these ideas.

Multiple Waves and Processing Efficiency

What is the relationship between the multiple waves model described in chapter 3 as an alternative to the standard hypothesis with its purported subcortical pathway and the processing efficiency framework outlined here? At a basic level, the multiple waves model holds that affective visual processing should be understood from the perspective of the many cortical and subcortical routes that link areas in the visual system and connect it to the rest of the brain. If we think of attention as involving multiple selection processes, the multiple waves model and processing efficiency framework can be merged, provided each "wave" has its own selection characteristics, as developed next.

Some researchers have associated multiple levels of attentional selection with specific areas of visual cortex. For example, the lateral geniculate nucleus in the thalamus is an early stage at which visual information is modulated by attention, and may play an important role as an early "gatekeeper" of visual signals (McAlonan, Cavanaugh, and Wurtz 2008; O'Connor et al. 2002).[7] At intermediate processing levels, such as in area V4, mechanisms of attentional "filtering" highlight task-relevant information and filter out unwanted information (e.g., Reynolds, Chelazzi, and Desimone 1999). Regions of parietal cortex may function as processing "bottlenecks" (e.g., Marois, Chun, and Gore 2000). And so on.

Whereas it is natural to think of selection processes in terms of specific "stages" of the visual system, we can also think of them as constituting entire "waves" of processing. For example, consider two sets of visual signals traversing the visual stream: the first proceeding from the lateral geniculate nucleus to V2, to TEO in temporal cortex, to the amygdala; and the second, from the lateral geniculate nucleus, via V1, V2, V4, TEO, and TE, to the amygdala (see figure 3.1). Let us call these two sets "wave 1" and "wave 2," respectively. The signals arriving at the amygdala in wave 1 will be quite distinct from those arriving in wave 2, which will be considerably more refined and integrated and may have been subjected to attentional modulations not present in wave 1. An important aspect of this idea is that, even though processing in areas V2 and TEO was part of both waves, the timing of the waves may have subjected their signals to different attentional influences. In this regard, attention effects are known to grow over the time course of neuronal responses (Mehta, Ulbert, and Schroeder 2000). Moreover, the timing of attentional effects is earlier in area V4 and inferotemporal cortex, later in area V2, and later still in area V1—that is, their trend is *opposite* to that of response latency (Buffalo et al. 2010). Taken together, these considerations suggest that attentional selection can be viewed in terms of spatiotemporal profiles, and that the multiple waves model and the processing efficiency framework can be merged into a more general model, where processing across "stages" depends on attention to a greater or lesser extent based not only on selective computations but also on task demands and other contextual factors.

Conclusions

In reviewing the role of attention during the processing of emotion-laden visual information, this chapter has discussed evidence both for and against automaticity. In addition to presenting empirical data, it has emphasized that several of the terms most often used to describe affective processing—"preattentive," "automatic," and "unaware"—are saturated with both intended and unintended connotations. In the final section, I advanced a *processing efficiency* framework as a way out of the current impasse observed in the literature, a framework in which terms like "preattentive" and "automatic" do not exist because the binary dichotomies are replaced by a continuum of processing efficiencies (and associated "gates" or "bottlenecks"). (Indeed, I recommend that at least the term "preattentive" be eliminated from the affective lexicon.) My proposed processing efficiency framework and the more general multiple waves model that includes it reflect the view that both limited and *un*limited processing camps are partly right, depending on one's usage of the terms "preattentive" and "automatic." But in a manner less appealing to the *un*limited processing camp, they incorporate important capacity limitations that moderate the impact of emotional information in the brain and behavior. The value of the more continuous outlook espoused here may be its insistence on careful characterization of the mechanisms by which affective processing is made more or less efficient—a level of characterization that both camps can certainly benefit from.

Notes

1. Strictly speaking, all features of automaticity can be measured in many different ways, which impacts potential conclusions.

2. It remains to be determined how the hypothalamus and the periaqueductal gray receive their signals. If they receive signals from cortical sites such as ventral insula and orbitofrontal cortex, as proposed by Pichon, de Gelder, and Grèzes (2012), processing in these regions presumably would also need to be independent of attention.

3. For approximately half of the subjects in Lim, Pamala, and Pessoa 2009, a picture of a house was initially paired with shock; for the other half, a picture of a building other than a house.

4. Gamma-band activity refers to brain wave signals in the 30–60 Hz range (though the range varies), which are thought to be important during many effortful tasks.

5. The "magnocellular system," which is particularly sensitive to moving stimuli, responds at the lower end of this range, whereas the "parvocellular system," which is sensitive to form information, responds more slowly still.

6. In any case, "automatic yet requires attention" will sound like a contradiction in terms to many people.

7. Another "early" neural mechanism involves the spinal cord, which modulates responses to pain (Sprenger et al. 2012).

5 Cognitive-Emotional Interactions in Prefrontal Cortex

Prefrontal cortex (PFC), the large sector of frontal cortex forward of motor cortex, has a central role in cognition (Fuster 1997; Luria 1966; Stuss and Knight 2002), one critical for the maintenance and manipulation of information. Notably, it participates in cognitive control operations that regulate the flow of information during nonroutine, challenging situations. Since at least Walle Nauta's synthesis of frontal lobe function (1971), however, it has been suggested that prefrontal cortex plays a chief role in emotion as well. Indeed, Nauta suggested that it could be considered "the major—though not the only—neocortical representative of the limbic system," by which he meant that prefrontal cortex could be expected to have a marked influence on "affective and motivational responses to [a person's] environment" (Nauta 1971, 182).

It is now accepted that prefrontal cortex is not a homogeneous structure but can be subdivided into several regions based on functional specialization, cytoarchitecture, and connectivity. Major PFC territories are involved in emotion—these include medial prefrontal cortex (within which is anterior cingulate cortex), orbitofrontal cortex, and adjacent regions of the insula.[1] Indeed, these areas figure prominently in several influential proposals of the brain basis of emotion (Damasio 1999; Davidson and Irwin 1999; Rolls 2005). Nevertheless, many views of prefrontal cortex tend to focus on its cognitive aspects (e.g., Cohen 2005). By minimizing the insights of researchers such as Walle Nauta (1971) and Karl Pribram (1967a, 1967b), the proposals portray prefrontal cortex's core function as cognitive. For instance, a well-known framework of PFC operation mentions emotion only in passing (Miller and Cohen 2001). Although several regions of prefrontal cortex are currently acknowledged as important for emotion, existing proposals frequently compartmentalize it into cognitive and affective regions (see Bush, Luu, and Posner 2000). It is particularly noteworthy that lateral prefrontal cortex is still viewed as a quintessential cognitive region, especially the portion that is loosely referred to as "*dorsal-lateral* prefrontal cortex."

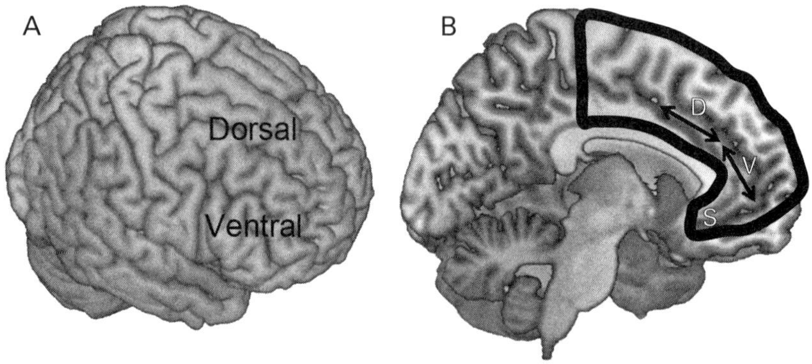

Figure 5.1
Frontal cortex anatomy. "Prefrontal cortex" refers to cortex "in front of motor areas," typically anterior to Brodmann area 6. (A) Lateral surface of cortex, showing dorsal and ventral sectors. (B) Medial surface of cortex, outlined in black, showing approximate locations of dorsal (D) and ventral (V) sectors. In the chapter, dorsal parts of medial prefrontal cortex also include parts posterior to the "D" arrow (such as presupplemantary and supplementary motor areas). "Ventral" is also called "rostral" (or toward the *rostrum*, "beak" in Latin). S, subgenual anterior cingulate cortex.

This chapter will review functional MRI studies that have investigated cognitive and emotional processing in human prefrontal cortex (see also Dolcos, Iordan, and Dolcos 2011) to explore how emotion and cognition, domains traditionally thought of as mutually antagonistic, interact there. As described below, cognitive and emotional signals are combined in complex ways in prefrontal cortex, such that both influence observed activity—and ensuing behavior. Indeed, a central theme of the present chapter is whether the mapping of cognition and emotion onto prefrontal cortex is better characterized by a *segregation* or an *integration* model (see Shackman et al. 2011). The chapter will focus on lateral and medial prefrontal cortex and will not discuss orbitofrontal cortex, whose contributions to emotion are well accepted (Zald and Rauch 2007). And because it will discuss *where* in prefrontal cortex certain processes are suggested to occur, readers less familiar with neuroanatomy should consult figure 5.1.[2] To anticipate some of the results discussed below, on both lateral and medial surfaces of the brain, dorsal areas will often be linked with cognition; and ventral areas, with emotion. As mentioned in the introduction to the book, however, the terms "emotion" and "cognition" will be used descriptively.

Emotional Conflict

In the Stroop task, subjects are shown words and asked to identify their ink color while ignoring their meaning. Because reading is a highly learned process, the word "red" in green ink, say, poses a conflict that needs to be resolved for correct performance. Although the monitoring of conflict and its resolution are proposed to rely on cognitive control (Botvinick et al. 2001), several versions of the standard Stroop task have been developed to explore the affective processing of conflict. For instance, in an emotional version of the Stroop task, subjects are asked to identify the ink color of words that are either emotionally neutral (e.g., "apple") or emotionally salient (e.g., "blood"; Mathews and MacLeod 1985; McKenna 1986). In the neuroimaging literature, one of the central goals is to determine whether brain areas involved in regular, nonemotional conflict processing play similar roles during emotional conflict. The broader aim is to understand the relationship between cognitive and emotional systems in the brain.

In one study, Paul Whalen and colleagues (1998) compared the counting Stroop task to an emotional version of the task. Subjects were asked to report the number of words on the screen. In the counting Stroop task, an incongruent stimulus (e.g., "two, two, two") was contrasted to a congruent one (e.g., "two, two"). In the emotional Stroop task, stimuli containing negative words (e.g., "murder, murder, murder") were contrasted to those containing neutral words (e.g., "chair, chair, chair"). Word meaning was thus task irrelevant in both cases. Behaviorally, both types of interference were associated with increased reaction time in incongruent and emotional conditions. During the counting Stroop task, incongruent trials evoked stronger responses than congruent trials in the *dorsal-posterior* portion of medial prefrontal cortex (Bush et al. 1998). During the emotional counting Stroop task, emotional trials evoked stronger responses than neutral trials in the *ventral-anterior* portion of medial prefrontal cortex. These results suggested that medial prefrontal cortex is composed of subregions that are *differentially* involved in cognitive and emotional tasks (as discussed in depth below; Bush, Luu, and Posner 2000; Devinsky, Morrell, and Vogt 1995).

Here let me note that I use the term "medial prefrontal cortex" to *include* anterior cingulate cortex. In the neuroimaging literature, the labeling of the anterior cingulate is fairly inconsistent; indeed, in a quite unfortunate practice, activation sites outside the cingulate gyrus proper (e.g., those in medial portions of Brodmann areas 8 and 9) are labeled as "anterior cingulate cortex."[3] Moreover, unless otherwise noted, I will refer to dorsal-*posterior*-medial prefrontal cortex simply as "dorsal-medial prefrontal cortex" and to ventral-*anterior*-medial prefrontal cortex simply as "ventral-medial prefrontal cortex" (see figure 5.1). Finally, whereas for some researchers posterior parts of medial *frontal* cortex, such as the cingulate motor areas, presupplementary motor area, and supplementary motor area are not considered *prefrontal* cortex per se. I make no such distinction here.

Further support for a *dissociation* between the neural substrates of interference processing comes from studies by Tobias Egner, Amit Etkin, and colleagues. In one study (Egner et al. 2008), during the emotional conflict task, subjects categorized faces according to emotional expression (e.g., happy vs. fearful), while ignoring emotionally congruent or incongruent words (e.g., "HAPPY," "FEAR") written across the faces. During the nonemotional conflict task, subjects performed a gender task, while ignoring gender words written across the faces. Based on an analysis of sequential effects (e.g., contrasting incongruent trials that followed either an incongruent or a congruent trial), the authors proposed that the *resolution* of emotional conflict was handled by ventral-medial prefrontal cortex, whereas the resolution of nonemotional conflict was handled by lateral prefrontal cortex. In particular, resolution of emotional conflict involved interactions with the amygdala: decreased activity in the amygdala was correlated with increased activity in ventral-medial prefrontal cortex across subjects (see also Etkin et al. 2006). The authors also indicated that conflict *monitoring* engaged dorsal aspects of medial prefrontal cortex in *both* emotional and nonemotional conditions. Their findings suggest that the neuroanatomical networks recruited to overcome distraction (the resolution part) vary with the nature of the conflict, but that a common mechanism for registering conflict (the monitoring part) is also at play.

Rebecca Compton and colleagues (2003) also compared emotional and nonemotional versions of interference tasks to determine the extent to which neural circuits are shared between these two domains. They employed the traditional nonemotional color-word Stroop task and the emotional Stroop task, in which subjects also had to identify the ink color of a word and the meaning of the distractor word was either neutral or emotional. Their results showed that activity increased in left dorsal-lateral prefrontal cortex during both incongruent color-word trials and negative trials. The authors proposed that this region participates in maintaining attentional set *whether or not* the challenging task-irrelevant information is emotional. Thus, although one body of research suggests a segregation of the circuits involved in cognitive and emotional tasks, another provides evidence for common neural substrates.

Emotional Distraction

Studies of distraction have contributed to our understanding of attentional control during the processing of emotion-laden stimuli. Again, a central question addressed by these studies is whether attentional control is qualitatively different when emotional versus nonemotional items are encountered. The studies thus investigate the extent of overlap of activations observed in these two cases. The distracting information employed is frequently visual, such as emotion-laden pictures.

Hiroshi Yamasaki, Kevin LaBar, and Gregory McCarthy (2002) investigated responses evoked by neutral and emotional distractors. Subjects viewed a stream of square shapes of varying sizes and colors. Among this stream, both target stimuli (circles) and distractors (neutral and emotional pictures) were presented. Based on the pattern of responses, the authors proposed that attentional mechanisms of target detection were located in *dorsal*-lateral prefrontal cortex, and that the processing of emotional distractor stimuli took place in *ventral*-lateral prefrontal cortex.

The findings of several other studies are consistent with the dorsal versus ventral segregation described above. In a series of studies, Florin Dolcos and colleagues investigated emotional distraction in the context of working memory tasks. Subjects were shown sample stimuli that had to be remembered during a subsequent delay period during which they saw distracting stimuli, including neutral and emotional pictures. The findings of one of their studies (Dolcos and McCarthy 2006) are illustrated in figure 5.2. During the delay period, responses in *dorsal*-lateral prefrontal cortex (figure 5.2B) were highest for the "scrambled" (digitally scrambled versions of pictures), intermediate for neutral, and *lowest* for emotional distractors—a pattern of responses also observed in parietal cortex. In a parallel fashion, behavioral performance was best during the scrambled, intermediate for neutral, and *worst* for emotional distractors. Both the dorsal-lateral prefrontal and parietal cortex sites were similar to sites that are robustly engaged during working memory tasks (Pessoa and Ungerleider 2004c)—response magnitude at similar sites was found to be correlated with behavioral performance (e.g., Pessoa et al. 2002). It therefore appears that viewing emotional distractors during the delay period interfered with neural activity normally observed in these sites—activity that supports working memory performance. Indeed, on incorrect trials with emotional distractors, subjects exhibited the weakest responses in dorsal-lateral prefrontal cortex during the delay period.

Responses in *ventral*-lateral prefrontal cortex (figure 5.2C) followed the opposite pattern, namely, the strongest responses were observed during the viewing of emotional distractors. What is the role of this region during the task? First, subjects who exhibited greater activity there in response to emotional distractors tended to rate them as both less distracting and less emotional. Second, the larger the evoked responses, the better the performance during the working memory task. Together, these findings suggest that ventral-lateral prefrontal cortex contributed to *inhibiting* the distracting effects of stimuli presented

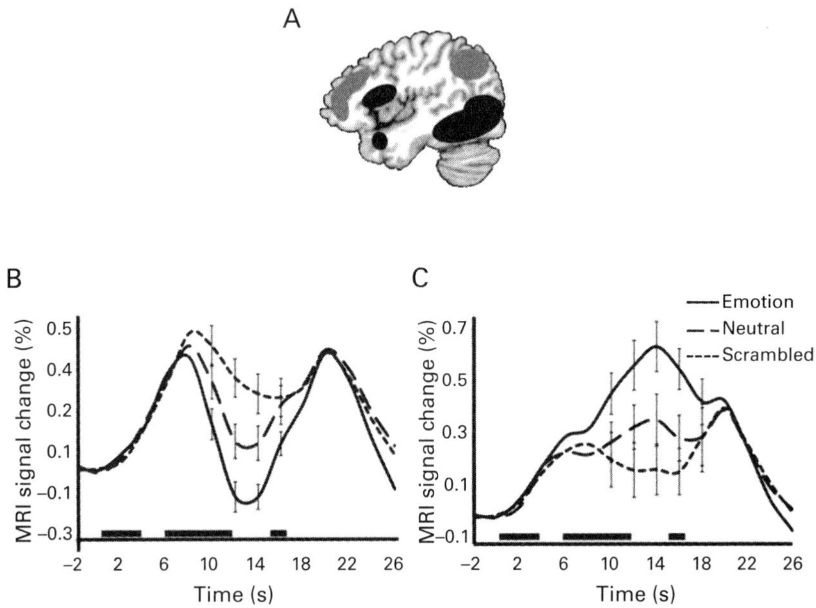

Figure 5.2
Emotional distraction during a working memory task. Subjects were shown scrambled, negative, or neutral distractor images during the delay period of the task. (A) Schematic representation of differential responses in brain. Regions where responses were stronger to scrambled than to emotional images are shown in light gray; regions where they were stronger to emotional than to scrambled images, in dark gray. (B) Time course data for dorsal-lateral prefrontal cortex. (C) Time course data for ventral-lateral prefrontal cortex. Horizontal bars in panels B and C correspond to onset and duration of sample stimuli, distractors, and probes, respectively. Time series plots kindly provided by Florin Dolcos, adapted from Dolcos and McCarthy 2006.

during the delay period, a function that was particularly needed when emotional pictures were viewed. Using a paradigm similar to that of Florin Dolcos and Gregory McCarthy (2006) to investigate the effects of distraction during working memory, Alan Anticevic, Grega Repovs, and Deanna Barch (2010) reported convergent findings. In their study, when subjects were faced with emotional distractors, responses in *ventral*-lateral prefrontal cortex were stronger for correct versus incorrect trials. No difference was observed in behavioral performance when neutral distractors were displayed, suggesting that this region is particularly important during emotional distraction.

Interestingly, when subjects saw negative distractors, Anticevic, Repovs, and Barch observed a negative relationship between the strength of amygdala responses and behavioral performance (across subjects)—that is, stronger amygdala responses were accompanied by lower performance. And the same pattern was observed when both neutral pictures and other distractor stimuli were shown. These results indicate that amygdala responses may have reflected the behavioral relevance of the distractor stimulus and, more generally, are consistent with the amygdala's playing a role during cognitive performance (see also Schaefer et al. 2006).

With regard to the much studied attentional control of nonemotional information, broadly speaking, two fronto-parietal systems have been proposed: a dorsal system involved in controlling goal-directed attention and a more ventral system involved in redirecting attention to behaviorally salient events (Corbetta and Shulman 2002). The studies reviewed above suggest that ventral-lateral regions of prefrontal cortex are strongly engaged by emotional distractors. It is thus possible that the frontal component of the ventral attentional system identified in nonemotional studies is attuned to information with emotional content. In this regard, it is noteworthy that ventral areas of lateral prefrontal cortex are more strongly interconnected with the amygdala than dorsal ones (although not as robustly as medial PFC areas; Ghashghaei, Hilgetag, and Barbas 2007); ventral-lateral prefrontal cortex is also interconnected with the anterior insula (Augustine 1996), another structure that influences valuation processes.

The ventral-lateral PFC site observed in distraction studies appears to overlap with areas of the ventral attentional system (which extends into the anterior insula). But here a possible contradiction arises. Whereas, in emotional distraction, the participation of ventral-lateral prefrontal cortex appears to counter the impact of distraction, its participation in the attentional reorienting network would be expected to have exactly the opposite effect, namely, to signal the need to reorient attention toward the distracting object (contributing to the "interrupt" function described in Corbetta and Shulman 2002). It is thus possible that ventral-lateral prefrontal cortex has spatially overlapping though separate neural populations that are engaged in these opposite functions, which are not readily apparent at the standard resolution of functional MRI (though multivariate pattern analysis may be a fruitful approach; see Haynes and Rees 2006; Norman et al. 2006).

A key question concerns the specificity of the engagement of ventral-lateral prefrontal cortex during emotional distraction. Put another way, what are the most critical dimensions of the stimulus? At the moment, the answer to this question remains unknown, and more research is needed. It is conceivable, for example, that the robust engagement of ventral-lateral prefrontal cortex is related to the *potency* of the distraction—which is certainly a property of the emotional pictures employed in several studies. Initial findings by Florin Dolcos, Alexandru Iordan, and Sandra Dolcos (2011) suggest that *arousal* is an important dimension of ventral-lateral PFC engagement insofar as they observed similar findings to those discussed above with high-arousal positive images (including but not limited to erotica).

Emotion Regulation

A body of research has documented the powerful ability people have to regulate their emotions (Gross 2007). Many regulation strategies exist, including "reappraisal" (e.g., reinterpreting the meaning of a stimulus), "cognitive distancing" (e.g., becoming a detached observer), and "behavioral suppression" (e.g., inhibiting external displays of emotion such as facial expression). Collectively, the research suggests that prefrontal cortex is broadly engaged during deliberative forms of emotion regulation, especially ventral-lateral prefrontal cortex (Berkman and Lieberman 2009; Ochsner and Gross 2005, 2008). In addition, about half of the emotion regulation studies find activity in the *posterior* portion of dorsal-medial prefrontal cortex (Berkman and Lieberman 2009), and a few studies also find bilateral activation in dorsal-lateral prefrontal cortex. Furthermore, reappraisal also *decreases* activity in regions such as the amygdala, insula, and medial orbitofrontal cortex that are needed for the evaluation of emotional significance. Responses in these regions also increase when subjects are instructed to enhance the emotional impact of a stimulus.

An important finding of the emotion regulation literature is that not only are responses in prefrontal cortex increased and those in the amygdala (and medial orbitofrontal cortex) decreased during regulation, but these responses are also *inversely* related (Kim and Hamann 2007; Ochsner et al. 2004; Phan et al. 2005). In one study (Banks et al. 2007), the strength of negative correlation of responses in the amygdala with those in dorsal-medial prefrontal cortex predicted the extent of *attenuation* of negative affect reported by subjects following reappraisal (figure 5.3). In other words, the greater the task-dependent negative correlation between these two regions, the more effective the reappraisal strategy, as noted by attenuated intensity of negative affect (see also Urry et al. 2006).

Although, as noted, several studies have reported a negative correlation between responses in prefrontal cortex and the amygdala during reappraisal, their relationship appears to be more context dependent. For example, Tor Wager and colleagues (2008) observed a *positive*, not negative, correlation between responses in ventral-lateral prefrontal cortex and the amygdala during reappraisal of negative images. And the stronger this correlation, the *less* successful the reappraisal. In their study, successful reappraisal involved dampening, not increasing, the positive correlation between ventral prefrontal cortex

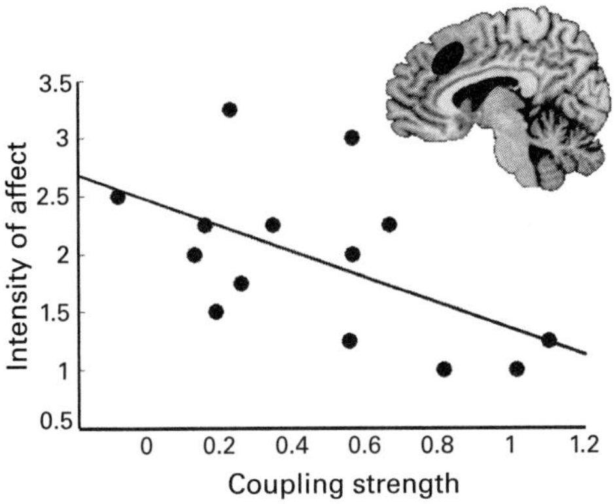

Figure 5.3
Emotional regulation. Upper right: Task-dependent "coupling" (correlation) with the amygdala was observed in dorsal-medial prefrontal cortex, schematically shown in black, as well as other regions not shown. The coupling was stronger in the "reappraise" condition than in the "maintain" condition. Lower left: Strength of the coupling was inversely related to the intensity of negative affect reported by subjects (each dot indicates a subject's correlation value). Data kindly provided by Luan Phan from Banks et al. 2007.

and amygdala. Thus the particular strategy adopted may play a key role in the region interactions engendered by emotion regulation. Whereas Wager and colleagues asked their subjects to produce a positive interpretation of the negative scene so as to reduce its emotional impact, Banks and colleagues (2007) asked their subjects to reinterpret the content of negative pictures so that they would no longer elicit a negative response. If, indeed, the precise instructions alter the associated circuits, the relationship between prefrontal cortex regions and the amygdala may be malleable and context dependent during regulation.

In all, deliberate emotion regulation is a process by which subjects attempt to alter their experience while viewing emotional stimuli. This process engages several prefrontal territories along lateral and medial aspects of prefrontal cortex, involving both dorsal and ventral sites. In parallel, in many cases, a set of regions whose function is linked to stimulus evaluation is engaged as well, including the amygdala, medial orbitofrontal cortex, and the insula. Because the observed relationship between prefrontal cortex and evaluative regions is typically reciprocal, these interactions have often been described in terms of the "cognitive suppression of emotion." The study by Wager and colleagues (2008) suggests, however, that the coupling between prefrontal cortex and other "regulated" regions may be context dependent, and that multiple, parallel circuits may operate to determine the fate of regulatory attempts. Finally, although the link between prefrontal cortex and the amygdala (or other regions) has been investigated across subjects, a goal for future research will be to establish this link on a trial-by-trial basis.[5] Thus, on a trial in which response magnitude in prefrontal cortex is stronger, one should observe decreased response in, for example, the amygdala, in conjunction with decreased negative affect. Such a trial-by-trial relationship would establish a stronger link between responses in prefrontal cortex and the amygdala, on the one hand, and subjective experience, on the other.

Emotion-Cognition Push-Pull

In an important paper, Wayne Drevets and Marcus Raichle (1998) noted that regional blood flow during attentionally demanding cognitive tasks *decreased* in regions such as the amygdala (see also Shulman et al. 1997), parts of orbitofrontal cortex, and ventral-medial prefrontal cortex, whereas blood flow *increased* in these regions during specific emotion-related tasks. Conversely, blood flow during experimentally induced and pathological emotional states *decreased* in regions such as dorsal-medial and dorsal-lateral prefrontal cortex, whereas blood flow *increased* in these regions during some cognitive tasks. These reciprocal patterns of activation suggested to Drevets and Raichle (1998, 370) that emotion and cognition may engage in *competitive* interactions: "Deactivation of emotion-related areas may reflect a relative reduction in the resources devoted to emotional evaluation or experience" during an effortful cognitive task. Moreover, they proposed that "during intense emotional responses to a threat, suppressing areas devoted to working memory and deep processing for visuospatial or semantic information may permit more rapid automatic responses to govern behavior. Nevertheless, such deactivations may have associated 'costs' of reducing the functions of these areas" (Drevets and Raichle 1998, 375).

To explore the relationship between cognitive and emotional processing, in a subsequent study, Simpson and colleagues (2000) compared their subjects' responses while viewing unpleasant images to those while performing effortful cognitive tasks. They found that performing effortful cognitive tasks reduced activity in portions of ventral-medial prefrontal cortex and other regions.[6] When subjects were asked to determine the number of humans in unpleasant pictures, however, the reduced activity typically found during cognitive tasks was *not* observed, suggesting that emotional processing *engages* these regions. The overall pattern of results indicated to the authors that some brain regions may be tonically active, and that their function "is necessary for the ongoing detection and evaluation of environmental and internal stimuli of relevance to the motivational state of the individual" (Simpson et al. 2000, 166), although they did not explicitly refer to emotion. Furthermore, an attenuation of this function takes place during the performance of tasks requiring focused attention.

The notion of a *push-pull*, antagonistic organization of neural circuits involved in cognition and emotion has been considered by other researchers, too. Thus, to test the reciprocal relationship between emotion and cognition circuits, Vinod Goel and Ray Dolan (2003) evaluated brain responses during "hot" and "cold" reasoning. They asked subjects to perform logical judgments about arguments that varied in emotional salience but had the same general logical structure. Emotionally salient arguments involved terms such as "Nazis" and "child molesters"; neutral arguments included terms such as "reptiles" and "mammals." Subjects were required to determine whether an argument's conclusion followed logically from its premises. "Cold" reasoning trials resulted in *enhanced* activity in dorsal-lateral prefrontal cortex and *decreased* activity in ventral-medial prefrontal cortex. By contrast, "hot" reasoning trials resulted in *enhanced* activation in ventral-medial PFC and *decreased* activation in dorsal-lateral PFC. In other words, Goel and Dolan observed a reciprocal relationship in dorsal and ventral prefrontal cortex regions that reflected the degree to which reasoning was "hot" or "cold."

As suggested by Drevets and Raichle (1998), a push-pull relationship has also been observed in amygdala responses. For example, Lotte Van Dillen, Dirk Heslenfeld, and Sander Koole (2009) presented their subjects easy or difficult arithmetic problems immediately following unpleasant images. Whereas the more challenging problems produced stronger responses in lateral prefrontal cortex than did the easier problems, the reverse pattern was observed in the amygdala, that is, stronger responses to emotional images during easy versus hard problems, for instance. Because of the sequential nature of their task, in which presentation of arithmetic problems followed presentation of emotional images, their study was particularly well suited to reveal the dynamic nature of the evoked responses. Thus performing a difficult task suppressed amygdala responses (figure 5.4).

My colleagues and I obtained related findings in two studies that, interestingly, did *not* involve emotional stimuli. In one study (Hsu and Pessoa 2007), we compared our subjects' amygdala response during a difficult search task (searching for a target letter among distractor letters) to that during an easy search task (searching for a target letter among an array of "O"s). We observed decreased amygdala response in difficult versus easy search conditions, showing that task difficulty by itself impacts responses in this structure—that is, independently of activation by emotional stimuli. In another study (Pessoa,

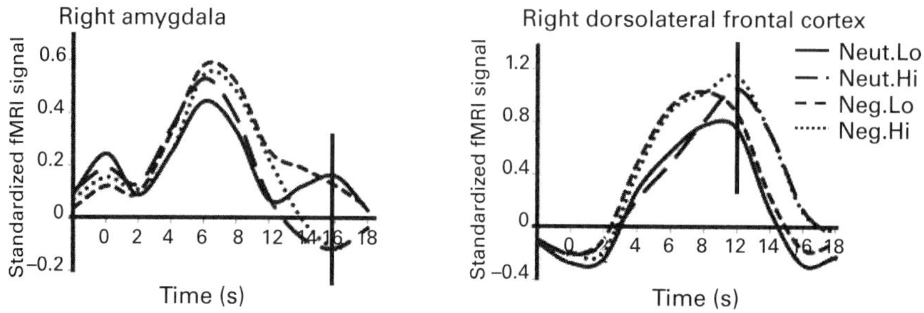

Figure 5.4
Cognitive-emotional push-pull. Neutral/negative pictures were presented at t = 0 s, and low-/high-load arithmetic problems were presented at t = 4 s. Responses in the right amygdala, though initially higher for negative versus neutral pictures (see values at 6–8 s), were considerably dampened during the high- versus low-load arithmetic task (see values at 12–16 s; vertical bar indicates the location of the troughs). In contrast, responses in right dorsal-lateral prefrontal cortex *increased* during the high- versus low-load arithmetic task (see values at 10–12 s; vertical bars indicate the location of the peaks). The responses of these two regions are consistent with a push-pull relationship in which the cognitive load in prefrontal cortex subsequently dampens responses in the amygdala. Figure kindly provided by Lotte Van Dillen from data of Van Dillen, Helsenfeld, and Koole 2009.

Padmala, and Morland 2005), we asked subjects to perform a difficult bar-orientation discrimination (no emotional stimuli were presented). Again, relative to fixation, responses in the amygdala decreased. Notably, these decreased responses in the bars-only condition appeared to combine *additively* with the increased responses we observed during the viewing of fearful faces. To clarify this finding, two additional experimental conditions need to be described. In the attend-*face* condition, subjects presented with the same peripheral bars and an emotional face at fixation were asked to indicate whether the face was male or female. As expected, we observed increased amygdala responses to fearful versus neutral faces. In the attend-*bar* condition, subjects presented with the same face-plus-bars stimulus were asked to determine whether the bars were of the same orientation. Amygdala responses in the attend-bars condition corresponded roughly to those in the attend-face condition *minus* the bars-only condition. These findings also reveal a type of push-pull relationship for amygdala responses: viewing fearful faces increased responses, whereas performing a difficult visual discrimination decreased them.

The existence of a push-pull relationship between emotion and cognition has also been suggested in clinical studies. For instance, Helen Mayberg and colleagues (1999) reported that normal sadness—experienced following a mood induction procedure—was associated with *increased* blood flow in subgenual cingulate cortex (Brodmann area 25) and *decreased* blood flow in right dorsal-lateral prefrontal cortex.[7] In contrast, remission of depression in patients was associated with *increased* blood flow in right dorsal-lateral prefrontal cortex and concomitant *decreased* blood flow in subgenual cingulate cortex. Mayberg and colleagues concluded that experiencing sadness results not only in activation of ventral-medial prefrontal cortex but also in the simultaneous deactivation of dorsal-lateral cortex and other cortical regions known to mediate attentional processing.

A theme that emerges from the studies reviewed in this chapter is that a dorsal versus ventral distinction can be found in sites sensitive to cognitive and emotional manipulations, respectively. This distinction echoes models of the organization of medial prefrontal cortex, a complex brain region involved in many functions (Vogt 2008). In an often cited paper, George Bush, Phan Luu, and Michael Posner (2000) proposed that the organization of medial prefrontal cortex can be understood in terms of ventral-anterior and dorsal-posterior sectors that are linked to emotional and cognitive processing, respectively (see figure 5.1). Their proposal was based on an informal meta-analysis of data from neuroimaging and lesion studies as well as on anatomical connectivity. Although it did not emphasize a reciprocal relationship between emotion and cognition, Bush, Luu, and Posner's proposal was highly influential in highlighting an anatomical distinction between cognitive and emotional circuits based, at least in part, on a dorsal-ventral axis of organization.

To investigate ventral and dorsal circuits, researchers have employed functional connectivity analysis (see chapter 8 for further discussion), in which the pattern of correlations between brain regions reflects the strength of their coupling (which may involve indirect anatomical connections). A common procedure is to choose a specific "seed" region and to characterize the region's interactions with the remainder of the brain during the "resting state" (when no task is executed; Greicius et al. 2003). During "rest," amygdala signals are *positively* correlated with those from ventral brain regions such as ventral-medial, orbitofrontal, and insular cortex (Anticevic, Repovs, and Barch 2010; Roy et al. 2009). In contrast, responses in the amygdala are *negatively* correlated with responses in dorsal brain regions, including sites in lateral prefrontal and parietal cortices. Although these results are in broad agreement with the dorsal versus ventral organization reviewed previously, it is noteworthy that the functional connectivity of the amygdala appears to differ depending on the amygdala subregion considered (Roy et al. 2009). That said, because some of the subdivisions probed are quite small, additional studies performed at higher spatial resolution are needed to confirm these initial findings (see note 1 to chapter 2).

Beyond the Dorsal versus Ventral-Medial Dichotomy in Prefrontal Cortex

The proposed organization of medial prefrontal cortex is an important component of the idea of a dorsal-ventral axis for cognition and emotion and has had a strong impact on research. This section will touch on specific studies that challenge the proposed dorsal versus ventral segregation and will discuss more general arguments against the dichotomy, including results from both informal and formal meta-analyses.

Specific Studies

Dean Mobbs and colleagues (2010) examined how brain responses vary as a function of perceived threat proximity. In an unusual experimental manipulation, they had their subjects, when inside the functional MRI scanner, place a foot into a custom-built box containing six compartments separated by sliding partitions. They then asked them to watch a video of a live and active tarantula placed randomly into five of these compartments at varying distances from the foot (unbeknownst to the subjects, the video was actually prerecorded). Parametric increases in evoked responses to perceived threat as a function of proximity were observed in several brain regions, such as bilateral insula, basal forebrain (possibly including the bed nucleus of the stria terminalis), ventral striatum, and midbrain (figure 5.5). Notably, extensive parametric activation was observed in dorsal-medial prefrontal cortex (see also Mobbs et al. 2007, 2009).

Thomas Straube and colleagues (2009) also found that activity increased parametrically in dorsal-medial prefrontal cortex as a function of how anxious subjects said they were in anticipation of mild electrical shocks. Of note, the relationship was only observed during high, but not low, levels of perceived threat. The authors suggest that this relationship may ensue during "states characterized by enhanced executive functions, including hyperscanning of the environment and increased drive for action" (Straube et al. 2009, 980; see also Straube et al. 2007).

In a third study, my colleagues and I (Pereira et al. 2010) proposed that dorsal-medial prefrontal cortex, specifically, a region called "midcingulate cortex," is involved in the interaction between negatively valenced and motor signals in the brain. The background for the study was an earlier behavioral investigation (Pereira et al. 2006) in which we reported that viewing unpleasant pictures in a blocked fashion (lasting approximately 3 minutes) interfered with the performance of a basic nonemotional visual detection task. In both the neuroimaging and the behavioral study, we asked subjects to perform a simple visual target detection task while viewing unpleasant or neutral pictures. Our functional MRI results (Pereira et al. 2010) revealed that an unpleasant emo-

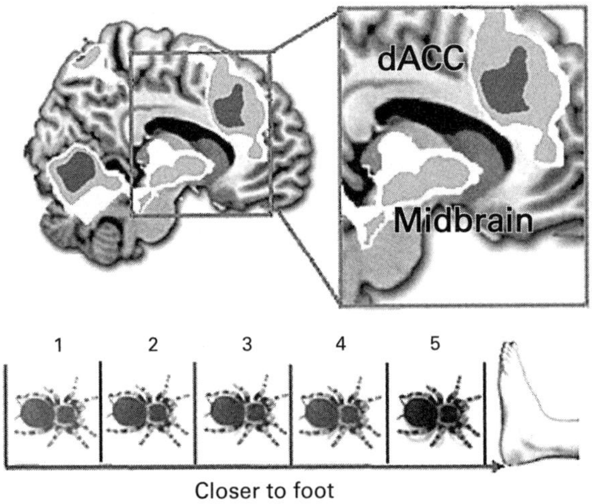

Figure 5.5
Threat and medial prefrontal cortex responses. Proximity to threat, schematically represented by the distance of the spider to the foot (lower panel), was associated with responses in several brain regions (upper panels), including medial prefrontal cortex, thalamus, and midbrain. dACC, dorsal anterior cingulate cortex. Reproduced with permission from Mobbs, D., Yu, R., Rowe, J. B., and Eich, H., FeldmanHall, O., and Dalgleish, T. (2010). Neural activity associated with monitoring the oscillating threat value of a tarantula. *Proceedings of the National Academy of Sciences of the United States of America*, *107*(47), 20582–20586.

tional context (associated with the blocked conditions) modulated responses in several brain regions involved in target detection. In particular, dorsal-medial prefrontal cortex was recruited when subjects performed visual detection tasks while viewing unpleasant pictures, and responses in this region closely mirrored the pattern of behavioral interference. The findings thus support the interpretation that dorsal-medial prefrontal cortex is a key site for the interaction between emotional and motor signals. Importantly, the findings, like those of the other studies reviewed here, show that the dorsal-medial prefrontal cortex is engaged by emotion in ways that run counter to the notion of segregated territories.

Informal and Formal Meta-Analyses

Like others before me, I have suggested that emotion and cognition share "mental resources," such that manipulations of emotion will interfere with cognitive performance and vice versa. As an initial approach to understanding resource consumption by cognition and emotion, I assessed the correspondence of brain sites that are sensitive to manipulations of attention *and* threat (Pessoa 2009). The attentional network has been extensively researched and involves fronto-parietal regions, including sites along dorsal and ventral portions of the lateral surface of prefrontal cortex, as well as dorsal-medial prefrontal cortex (Corbetta and Shulman 2002; Kastner and Ungerleider 2000; Pessoa and Ungerleider 2004b,c). In an informal meta-analysis to assess brain regions that are sensitive to high levels of threat, I reviewed activation sites reported in aversive conditioning studies (Pessoa 2009).[8] The activations were obtained by contrasting subjects' responses to aversive stimuli (i.e., stimuli paired with an unconditioned stimulus) with those to safe stimuli. Surprisingly, activation sites were repeatedly reported not only in the amygdala but also in frontal cortex, sites that overlapped with those indicated above—consistent with findings from formal meta-analyses (Etkin and Wager 2007; Mechias, Etkin, and Kalisch 2010). Which is to say, threat engages central nodes of the attentional network. Importantly, large extents of prefrontal cortex are engaged during the processing of threat, including *both* dorsal and ventral brain regions that, critically, encompass dorsal-medial prefrontal cortex (see below and chapter 7 for further discussion of emotion and cognition in the context of mental resources).

In a recent attempt to understand the organization of medial prefrontal cortex and its role in emotion, Amit Etkin, Tobias Egner, and Raffael Kalisch (2011) reviewed both human and nonhuman animal literatures. Their review does not support the dichotomy that dorsal regions are involved in cognition, whereas ventral regions participate in emotion. Rather, the authors suggest that sites in *both* dorsal and ventral medial prefrontal cortex make prominent contributions to emotional processing. For instance, they point out that dorsal-medial PFC activity correlates with fear-conditioned skin conductance responses and with increases in heart rate induced by social threat. Furthermore, given that direct electrical stimulation of dorsal-medial prefrontal cortex can elicit subjective states of fear, they propose that this region is involved in generating fear responses. Based on other data, they also suggest that fear appraisal relies on dorsal-medial prefrontal cortex (see their review for the specific references to the work mentioned in this paragraph).

An insightful formal meta-analysis of human neuroimaging studies (Shackman et al. 2011) further demonstrates the considerable overlap of sites in medial prefrontal cortex engaged during negative affect and cognitive control. As can be seen in figure 5.6, activation sites of negative affect studies encompass a broad expanse of medial prefrontal cortex, as do those of cognitive control studies, although the latter exhibit a discernible concentration of sites in more dorsal regions. Indeed, studies of cognitive control observed activation sites in dorsal-medial PFC more often than they did in ventral-medial PFC. Notably, however, in studies of negative affect, activation of dorsal and ventral

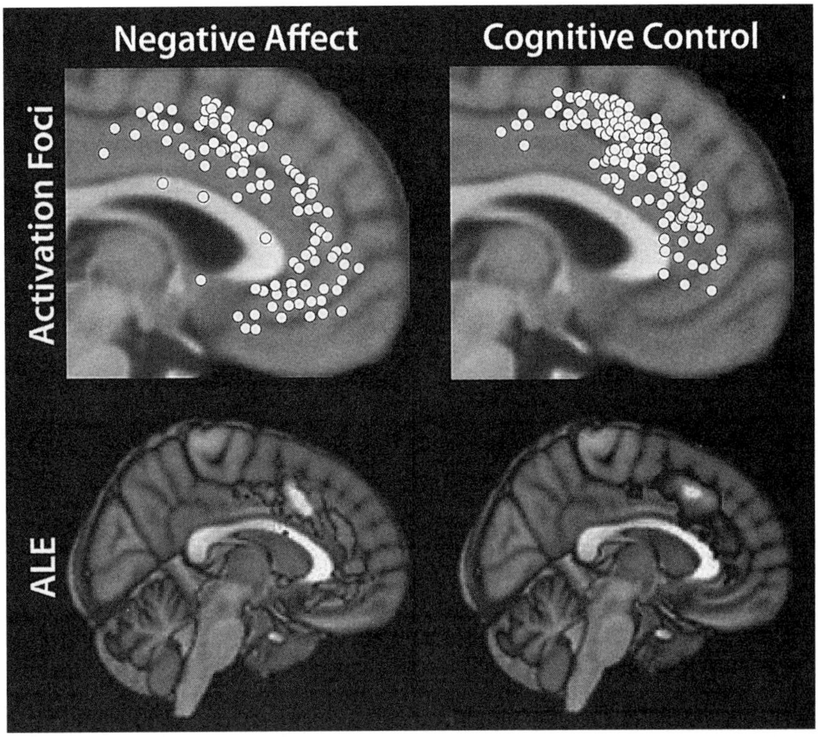

Figure 5.6
Cognition and emotion in medial frontal cortex. (Top) Foci of activation across studies of negative affect and cognitive control. Extensive overlap between emotion and cognition was observed in dorsal-medial prefrontal cortex. (Bottom) Formal meta-analytical activation maps. ALE, activation likelihood estimate. Figure kindly provided by Alex Shackman and adapted with permission from Shackman et al. 2011.

prefrontal cortex was equally likely. These findings, which are based on nearly 200 separate studies, strongly undermine the notion of a strict segregation between emotion and cognition in medial prefrontal cortex.

Summary

Although still influential, as reviewed here, the segregation model of medial prefrontal cortex organization is no longer tenable, as different research groups now argue (e.g., Etkin, Egner, and Kalisch 2011; Pessoa 2009; Shackman et al. 2011). Individual studies of threat and fear conditioning provide clear instances that run contrary to the segregation model. Importantly, when large numbers of studies are considered jointly, the weight of their findings strongly favors an organization of medial prefrontal cortex that is *not* segregated into affective and cognitive compartments, but, instead, is shared by cognitive and affective domains in a way that allows medial prefrontal cortex to support the *adaptive control* of complex behaviors (Pessoa 2008; Shackman et al. 2011). The next section argues further that, even though relationships between emotion and cognition in the brain are at times push-pull, a model that characterizes cognitive-emotional interactions more broadly is required when one considers all the available findings.

Beyond Push-Pull: When Emotion and Cognition Work Together

If emotion and cognition are not always antagonistically related, how do they interact in prefrontal cortex? This section will illustrate cognitive-emotional interactions by reviewing experimental designs involving diverse cognitive functions, including response inhibition, working memory, and response conflict. In several cases, the experimental manipulation requires subjects to perform an executive function when they are faced with either neutral or emotional items.

Empirical Findings

Response Inhibition An important dimension of executive function is inhibiting and controlling behavior. Response inhibition, which involves canceling an intended action, engages several prefrontal regions, such as dorsal-lateral, ventral-lateral, and medial prefrontal cortex (Aron, Robbins, and Poldrack 2004; Rubia et al. 2003). This mechanism is often investigated by using "go/no-go tasks" in which subjects are asked to execute a motor response when shown the "go" stimulus (e.g., "Press a key as fast as possible when you see a letter stimulus"), but to withhold the response when shown the "no-go" stimulus (e.g., "Do not respond when you see the letter 'Y'"). Typically, the go and no-go stimuli are presented as part of a rapid stream of stimuli, such as a sequence of letters. Martin Goldstein and colleagues (2007) investigated the interaction between the processing of emotional words and response inhibition. They asked their subjects to silently read every word in a rapid stream and to press a button for every word in regular font but to refrain from pressing it for words that were italicized. Response inhibition when encountering negative words (e.g., "worthless") engaged dorsal-lateral prefrontal cortex. Interestingly, this region was not recruited by negative valence or inhibitory task demands per se; instead, dorsal-lateral prefrontal cortex was sensitive to the explicit interaction between behavioral inhibition and the processing of negatively valenced words.

Working Memory Evidence for cognitive-emotional interactions also comes from working memory studies. A previous section reviewed findings of studies using emotional items as distractor stimuli. But what is the impact of the emotional content of items to be remembered on the maintenance of working memory? In one study (Perlstein, Elbert, and Stenger 2002), when subjects were asked to keep in mind neutral or emotional pictures, maintenance-related activity in dorsal-lateral prefrontal cortex was modulated by the valence of the picture, with pleasant pictures increasing activity and unpleasant pictures decreasing activity relative to neutral ones. Interestingly, emotional pictures did not affect responses in dorsal-lateral PFC in a second experimental condition during which subjects were *not* required to keep information in mind, indicating that the modulation of sustained activity by emotional valence was specific to the experimental context requiring active maintenance.

In another working memory study (Gray, Braver, and Raichle 2002), subjects watched short videos intended to induce emotional states, including clips from uplifting, sad, and neutral movies. After each video, subjects were scanned while performing a "3-back" working memory task that employed either word or face stimuli (they were asked to press a target button if the stimulus currently on the screen was the same as the one seen three trials back). Task-related activity in lateral prefrontal cortex in both hemispheres showed a statistical interaction of emotional state by stimulus type, with no main effects. For trials involving face stimuli, evoked responses were strongest in the pleasant condition, weakest in the unpleasant, and intermediate in the neutral condition; for trials with word stimuli, the reverse pattern was observed. Notably, the crossover interaction activity patterns in lateral prefrontal cortex were correlated with behavioral performance across subjects: those with a stronger effect of emotion behaviorally also exhibited stronger effects in terms of lateral PFC responses. In sum, lateral prefrontal cortex activity on both hemispheres reflected *equally* the emotional and working memory task components, stemming not from the working memory task or from mood alone, but from an interaction between emotion and cognition.

As a final working memory example, a study by Susanne Erk, Anna Kleczar, and Henrik Walter (2007) also investigated emotional distraction, much like the studies by Florin Dolcos and Gregory McCarthy (2006) and Anticevic, Repovs, and Barch (2010) discussed previously (for related studies, see Mather et al. 2006; Qin et al. 2009). Presented an array of letters, subjects were asked to remember either one letter (low-demand condition) or six (high-demand condition). During the delay period, distractor images were shown (neutral, positive, or negative). In contrast to the results of the studies previously reviewed, Erk, Kleczar, and Walter (2007) did *not* observe deactivations of dorsal-lateral prefrontal cortex. Instead, this region responded to both working memory demands *and* emotion. Specifically, responses in this region *increased* both when working memory demands were high versus low and when the task-irrelevant stimulus during the delay period was negative versus positive or neutral. Thus negative stimuli acted much like a task-demand effect, which reliably engages dorsal-lateral prefrontal cortex. In another critical difference between their results and those of studies previously reviewed, Erk, Kleczar, and Walter found that negative images actually *improved* behavioral performance, whereas they impaired performance in the other studies. It is possible, however, that the observed differences in the results of these studies are related to the type of material to be remembered since task-irrelevant emotional items disturb visuospatial but not verbal working memory (Lavric, Rippon, and Gray 2003; Shackman et al. 2006).

Response Conflict Cognitive-emotional interactions also have been investigated during response conflict tasks in which a single stimulus primes two separate motor responses—as in the classic Stroop task. Sarah Hart and colleagues (2010) investigated the impact on subjects of viewing emotional pictures *prior* to performing the counting Stroop task (see also Blair et al. 2007). Both aversive and neutral images were employed. Stroop stimuli involved congruent (e.g., "22"), incongruent (e.g., "222"), and control (e.g., two star-shaped geometrical shapes) conditions. Reaction times during incongruent trials were slower when subjects viewed aversive versus neutral stimuli (no effects of valence were observed during congruent or control trials). Hart and colleagues reasoned that viewing aversive pictures would be associated with decreased neuroimaging responses in lateral prefrontal cortex due to a push-pull emotion-cognition type of relationship. Although they observed some reduction of evoked responses during *congruent* trials that followed aversive pictures, however, they did not detect decreased responses during *incongruent* trials.

In a recent study, Jong Moon Choi, Srikanth Padmala, and I also investigated cognitive-emotional interactions in the context of response conflict (Choi, Padmala, and Pessoa 2012). We hypothesized that shock anticipation would consume processing resources required for executive control (see Pessoa 2009), and thereby increase response conflict since fewer resources would be available to handle response conflict. Trials started with the presentation of a cue that indicated trial type (figure 5.7). During threat trials, subjects received an unpleasant, though not painful, electrical shock in one-third of the trials; during safe trials no shock was administered. Because our goal was to probe

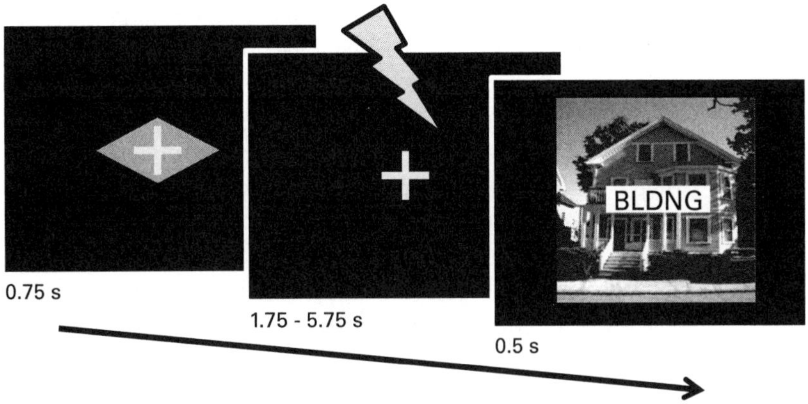

Figure 5.7
Task design. Subjects performed a response-conflict task in two conditions, safe and threat. In the threat condition (shown here), a cue stimulus signaled that a mild electric shock could occur during the delay period following cue offset and prior to the target display. Subjects were instructed about the meaning of the cue stimuli prior to task execution. Adapted with permission from Choi et al 2012.

the effect of threat monitoring, not shock delivery, on task performance, we discarded shock trials from the main analyses. During the subsequent target phase (which occurred 2–6 s after the cue), we asked our subjects to determine whether images contained a house or a building (other than a house) while ignoring task-irrelevant words superimposed on them. The strings "house," "bldng," and "xxxxx" were used to created congruent, incongruent, and control trials.

To assess how emotion affected response interference, we compared reaction times on incongruent and control trials (the former being slower than the latter). Behaviorally, response interference *increased* in the threat condition, that is, we detected an interaction between threat and cognitive condition. This suggests that subjects found it harder to handle incongruent trials following threat cues. Notably, the interference relationship was linearly related to state anxiety, with larger interference for subjects with higher anxiety scores. We saw a confluence of the effects of cognitive task condition, threat processing, and state anxiety in the anterior insula that paralleled our behavioral results. Given the push-pull relationship reported in other studies, we were also interested in probing potential activity *decreases* following threat versus safe cues. Intriguingly, although we did find decreases, they did not coincide with sites in lateral prefrontal cortex previously reported (Anticevic, Repovs, and Barch 2010; Dolcos and McCarthy 2006). Instead, the observed deactivation exhibited considerable overlap with regions that are also deactivated during effortful tasks (see previous discussion of Drevets and Raichle 1998 and below).

Prefrontal Cortex and the Relationship between Emotion and Cognition

As we have seen, one type of interaction between emotion and cognition, observed across experimental paradigms such as those involving emotional distraction and emotion regulation, among others, relies on a push-pull relationship between more dorsal brain regions important for cognition and more ventral regions needed for emotion. However, as illustrated in the previous section, cognitive-emotional interactions take diverse forms that go beyond a straightforward antagonistic relationship. I suggest that lateral prefrontal cortex, which has well-documented roles in cognitive processing, is a *convergence site* for cognitive and emotional signals (see also figure 7.7).

A prominent question regarding the function of lateral prefrontal cortex during cognitive-emotional interactions is whether, during a cognitive task, emotional information decreases or enhances the region's responses. The studies by Dolcos and McCarthy (2006) and by Anticevic, Repovs, and Barch (2010) showed instances during which emotional distractors do lead to *decreased* responses in parts of lateral prefrontal cortex that are important for cognitive tasks, including dorsal-lateral prefrontal cortex. However, decreased activation has not been universally observed during cognitive tasks. In many studies with emotional distractors, researchers have not observed decreases of activation in dorsal-lateral prefrontal cortex. For example, in conditions of emotional distraction, Erk, Kleczar, and Walter (2007) observed increased responses to emotional stimuli in dorsal-lateral PFC that were similar to those associated with an increase in working memory load. In another study, Seung-Lark Lim, Srikanth Padmala, and I observed increased dorsal-lateral PFC responses when subjects detected an affectively significant target in resource-limited conditions (Lim, Padmala, and Pessoa 2009), although the fact that the emotional stimulus was task relevant may have played a role (see chapter 6). Thus, during many tasks, emotional stimuli generate *increased* responses in dorsal-lateral prefrontal cortex.

In addition, decreased responses during threat processing do not always involve the same frontal and parietal areas that are recruited by effortful tasks. As described above in the context of our response conflict task (Choi, Padmala, and Pessoa 2012), when subjects viewed a cue stimulus that signaled potential shock, we observed extensive cortical *deactivation* (figure 5.8)—but in regions that are active *at rest*. Overall, the pattern observed in our study was distinct from that typically interpreted as an emotion-cognition push-pull.

The relationship between lateral PFC responses to emotional stimuli and task performance is complex as well. For instance, the working memory study

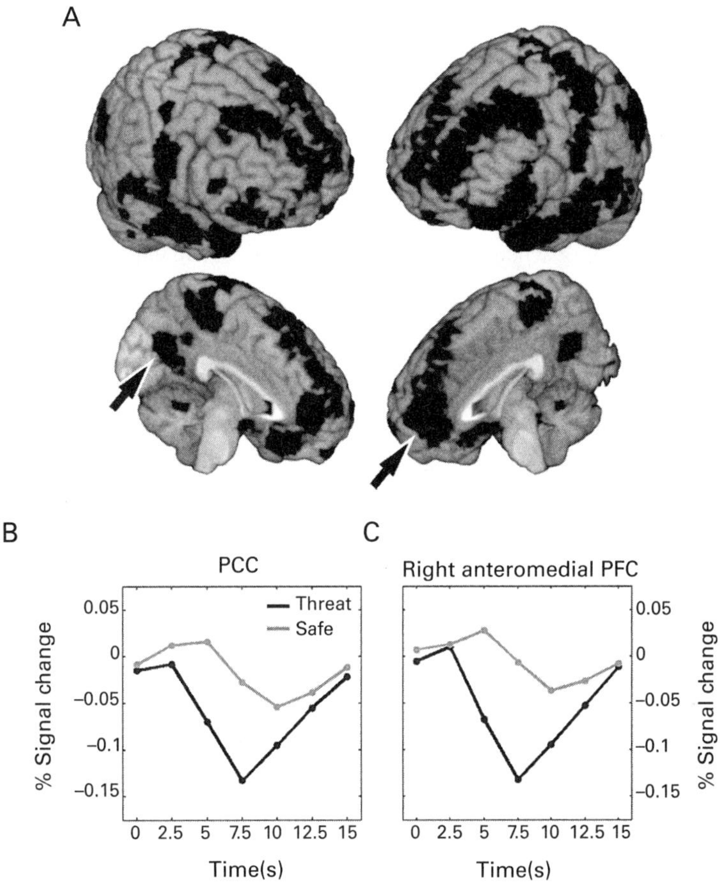

Figure 5.8
Responses to the cue stimulus. (A) Voxels showing stronger responses during safe than threat trials. (B, C) Mean estimated responses from the posterior cingulate cortex (PCC panel B; arrow at lower left in in panel A) and right anterior-medial prefrontal cortex (panel C; arrow at lower right in panel A) during safe and threat trials. Reproduced with permission from Choi, Padmala, and Pessoa 2012.

by Dolcos and McCarthy (2006) showed that decreased responses in dorsal-lateral prefrontal cortex were associated with impaired task performance. However, I argue that, because it is unclear whether increased responses reflect greater capacity to utilize the region, neural inefficiency, or increased effort, in most cases, researchers cannot uniquely interpret the functional significance of the direction of lateral PFC engagement (i.e., increases vs. decreases). Thus the interpretation of the pattern of responses in prefrontal cortex in the context of cognitive-emotional manipulations is far from straightforward. Increases or decreases cannot be understood simply based on the direction of the responses; they need to be *anchored* in terms of behavior. For instance, emotional distractors displayed during the delay period of working memory tasks were associated with both decreased activation in dorsal-lateral prefrontal cortex and impaired task performance (Anticevic, Repovs, and Barch 2010; Dolcos and McCarthy 2006). In general, the pattern of activation can be understood more clearly only when the impact of a behavioral manipulation is determined in both behavior and brain responses.

The studies reviewed in this section illustrate that the effects of emotion on cognition, and vice versa, are best viewed not as a simple push-pull mechanism, but as interactions that result in processes and signals that are neither purely cognitive nor emotional. Instead, their "cognitive" or "emotional" nature is blurred in a way that highlights the integration of these domains in the brain (Pessoa 2008). Chapter 7 will present the dual competition model, which describes the interactions between emotion and cognition in terms of competition at both perceptual and executive levels.

A final point to be made here is that lateral prefrontal cortex is also engaged by manipulations that are described as "purely emotional." This is relevant given the strong cognition-centered view of this brain region. One study (Somerville, Whalen, and Kelley 2010) investigated threat monitoring in a task where a subject's probability of shock varied up and down continuously over time. The authors observed activity in several structures hypothesized to be involved in threat monitoring, such as the basal forebrain (likely including the bed nucleus of the stria terminalis) and the right insula. The dorsal-lateral prefrontal cortex was also activated, and responses there increased linearly as a function of shock probability (for other studies of lateral PFC involvement in related manipulations, see Dalton et al. 2005; Simmons et al. 2004).

Anxiety, Executive Function, and Prefrontal Cortex Responses

Temperamentally anxious subjects exhibit enhanced perception of threat-related information, as observed during behavioral tasks (e.g., MacLeod et al. 2002; Mogg, Garner, and Bradley 2007; Holmes, Nielsen, and Green 2008). Anxiety, here understood in a broader, nonclinical sense, is also believed to alter cognitive processing. In fact, temperamentally anxious subjects show impairments during tasks that lack explicit threat-related material. For instance, trait anxiety impairs working memory and complex reasoning (Darke 1988; Derakshan and Eysenck 1998; Richards et al. 2000). This suggests that anxiety can modulate executive functions even when the stimuli themselves are not explicitly emotional.

Interest in the neural bases of anxiety-affected changes in cognition has quickened in recent years. In the context of the present chapter, a question of particular interest is whether cognitive control areas in prefrontal cortex, including dorsal-lateral prefrontal cortex, are *under-* or *overactivated* in anxious individuals. This question is relevant given the belief that anxiety is particularly associated with *reduced* processing efficiency. To maintain comparable levels of task performance, anxious individuals must exert greater cognitive effort (Eysenck et al. 2007), which is proposed to be linked to *increased* activation in brain regions involved with cognitive control, a pattern consistent with neuroimaging studies that interpret relatively greater activation in cognitive control areas (given equal performance) as a sign of reduced processing efficiency (e.g., Gray et al. 2005).

In contrast, some studies have reported that anxiety is associated with *under*activation in cognitive control circuits (Bishop 2007, 2009; Bishop et al. 2004), a finding that has also been interpreted in terms of deficient recruitment of cognitive control. In one study (Bishop 2009), when a nonemotional attentional task was not very demanding, the performance of high-anxious subjects was significantly worse with incongruent distractors than with congruent ones. Notably, high-anxious subjects tended to have *reduced* activation in left dorsal-lateral prefrontal cortex when faced with incongruent distractors, whereas low-anxious subjects tended to have *increased* activation in the same region. Thus the impaired performance of high-anxious subjects in the presence of incongruent distractors was possibly tied to a failure to use attentional control mechanisms when this was necessary for adequate task performance (see also Basten, Stelzel, and Fiebach 2011; for additional discussion, see Eysenck and Derakshan 2011).

Examining under- or overutilization of a brain area, however, does not lead to an unequivocal interpretation of cognitive processing in anxious subjects, as argued by Christina Fales and colleagues (2008).This is because either reduced or enhanced neural recruitment may reflect differences in a host of factors, including efficiency, motivation, effort, or the capacity to activate regions when needed. The difficulties surrounding the issue of under- versus overactivation are mirrored by those encountered in the developmental literature:

> Conflicting arguments have been made with regard to what pattern of functional activity represents functional maturity. Some studies report higher magnitude activity as immature, reflecting neural inefficiency or increased effort. Other studies interpret higher magnitude activity as more mature, reflecting greater capacity to utilize the

> functionality of the region. Others have asserted that maturation is reflected in the refinement of activation extent, shifting from larger, diffuse activations to more focal activations with increasing age. (Somerville and Casey 2010, 239)

In the context of anxiety, Fales and colleagues (2008) have suggested that anxious individuals utilize an *alternate* neural route to obtain the same level of behavioral performance. They employed a challenging 3-back working memory task to investigate brain responses as a function of anxiety; their experimental design allowed them to probe both transient and sustained brain responses. Compared to low-anxious subjects, high-anxious subjects exhibited significantly *reduced* sustained neural activity across several brain regions, while at the same time exhibiting *increased* trial-related transient activity in other regions (with areas of overlap, too). The latter results would argue for lesser efficiency in the anxious group, consistent with a processing efficiency framework (Eysenck and Calvo 1992). But when considered together, their results suggest that, at least in some situations, anxiety shifts the profile of neural activation across multiple brain areas—a shift that may involve both sustained and transient signals.

Conclusions

This chapter has reviewed key aspects of our current understanding of the neural substrates for cognitive-emotional interactions in prefrontal cortex based on human functional MRI studies. The conflict between emotion and cognition has a long history, dating back at least to ancient India and Greece; Plato, as we have seen, juxtaposed the charioteer, representing intellect and reason, with the horse, representing the soul's irrational passions. This "antagonism" theme fits well with some of the results reviewed here.

Nevertheless, cognitive-emotional interactions assume diverse forms and are not limited to mutual suppression. An emerging theme in functional MRI studies is that lateral prefrontal cortex is a focal point for cognitive-emotional interactions, which have been observed across a wide range of cognitive tasks. Importantly, whether responses in lateral prefrontal cortex (including dorsal-lateral PFC) increase or decrease in the face of emotional manipulations varies across contexts. As argued, the direction itself (increases vs. decreases) is not diagnostic with respect to the functional nature of the interaction, for example, whether increased activation signifies more or less efficient engagement of prefrontal cortex.

For the sake of brevity, I have not reviewed other models of emotional processing in prefrontal cortex. It is worth mentioning one of these: specifically, a model that goes beyond a simple cognition-emotion dichotomy to suggest that cognitive-emotional *gradients* exist (e.g., Lewis and Todd 2007). Along these lines, Philip Zelazo and William Cunningham (2007, 145) propose that "rather than positing discrete systems for hot and cool EF [executive function], this model views hot–cool as a continuum that corresponds to the motivational significance of the problem to be solved." In view of the findings reviewed here, more continuous models of the mapping between emotion~cognition and the brain are welcome developments. At the same time, these models need to be extended to allow for interactions between emotion and cognition at all levels of the neuroaxis, including dorsal-lateral and all of medial prefrontal cortex.

Several models of the organization of prefrontal cortex have been advanced. Although none has been widely accepted, a key idea among them is that anterior-posterior axes of organization exist, and important regional distinctions along these axes have been described (e.g., Badre and D'Esposito 2007; Fuster 1997; Passingham and Wise, 2012). Thus, having made the case against a segregation model of prefrontal cortex with respect to emotion and cognition, I do not wish to suggest that prefrontal cortex should be understood as a monolithic structure (see chapter 8).

To conclude, as I have argued elsewhere (Pessoa 2008), both conceptually and empirically, it is unproductive to carve the brain into cognitive and emotional realms. The present chapter has further highlighted this point by illustrating the varied ways emotion and cognition interact in prefrontal cortex.

Notes

1. Although the insula is a separate lobe of the brain, its anterior sectors have been grouped with the orbitofrontal cortex by some researchers (e.g., Carmichael and Price 1994).

2. I have chosen to hyphenate terms describing brain locations, using "ventral-medial" instead of "ventromedial" and "cortico-pulvino-cortical" instead of "corticopulvinocortical," for example, in the hope it will make them easier to read.

3. In a recent editorial board meeting of the *Journal of Comparative Neurology*, one member ventured (only half tongue in cheek) that the journal could perform a valuable community service if it "fixed" the labeling of anatomical regions in the functional MRI literature.

4. For some researchers, posterior parts of medial *frontal* cortex, such as the presupplementary, supplementary, and cingulate motor areas, are not considered medial *prefrontal* cortex per se. I make no such distinction here.

5. Although Eippert and colleagues (2007) attempted a trial-by-trial analysis, collinearity in their regressors precludes a clear interpretation of the results.

6. These regions have been thought to constitute the "resting state" network (also called the "default" or "task-negative" network); the "resting state" is observed when subjects simply lie in the MRI scanner without performing an explicit task (Fox, Russo, and Dutton 2005; Greicius et al. 2003).

7. "Subgenual cingulate cortex" is the portion of ventral-medial prefrontal cortex just under the "bend" of the corpus callosum (thus it is quite anterior and ventral). Excessive activation in this area has been linked to depression (Mayberg et al. 1997, 1999).

8. I thank my former doctoral student Seung-Lark Lim for providing me with the results of these studies.

6 Motivation

The previous chapters described the interactions of emotion with both perception and cognition. This chapter and the next will expand the discussion to include interactions of *motivation* with perception and cognition. Even though emotion and motivation are closely related—both depend on the organism's relationship with its environment—for the most part, they have been treated independently of each other.

Thus one speaks of "positive" and "negative *affective states*" of the organism in the case of emotion, whereas one speaks of "approach" and "withdrawal" of the organism in the case of motivation. And whereas, for emotion, the emphasis is most often on the evaluative aspect of the organism-environment relationship, for motivation, it is most often on how the organism acts in a given situation. Such distinctions reflect how researchers generally use the two concepts, but, taken to a logical extreme, they would relegate emotion to a passive, "felt" mental state. It is obviously much more than that; indeed, most emotion researchers link emotion to ways the organism acts upon the world.

In reviewing how motivation interacts with perception and cognition, this chapter will not make a sharp distinction between emotion and motivation because doing so would be less than helpful. Like the terms "emotion" and "cognition," the term "motivation" will be used descriptively, such as when it refers to specific paradigms—for instance, a study deals with motivation when rewards are employed.

In the past two decades, there has been considerable interest in the neural basis of reward and motivation. This line of research has highlighted the importance of the "dopamine system" and its cortical projection sites in behavioral control (Schultz et al.1992), which in turn has led researchers to formulate computational models of valuation (e.g., Montague, Hyman, and Cohen 2004) and has contributed to the development of the field of neuroeconomics (e.g., Platt and Glimcher 1999; Berns et al. 2001). Vigorous progress in understanding the mechanisms of reward and motivation has stimulated research into how motivation interacts with cognitive systems broadly conceived. The reasoning is that, if valuation processes shape behavior, they can be expected to influence perceptual and cognitive circuits central to the production of behavior.

As argued by John Tsotsos (1990, 1997), perception selects which inputs to process because of its limited capacity. But, given finite processing power, how does the brain distinguish stimuli that deserve further attention from stimuli that are better ignored? Traditionally, both stimulus- and goal-driven mechanisms have been posited as potential solutions to the "limited processing-resources dilemma" faced by the brain. By using goal-driven mechanisms, the brain can more efficiently allocate its resources based on current behavioral goals and prior knowledge. At the same time, by using stimulus-driven mechanisms, it can preferentially shift resources to salient features of the environment based on stimulus features (see chapter 7). Both types of mechanisms are intimately linked to reward and motivation, as described below.

In discussing interactions between motivation and cognition, this chapter will draw findings from a wealth of recent studies on attention, although it will also review studies involving other processes, such as response conflict, task switching, and working memory. Its focus will be on findings that illustrate emerging themes in ongoing research.

According to traditional psychological models, motivation relies on a global, rather blunt *energization* factor to influence the vigor and frequency of behavioral output, though typically without more *selective* effects. Contemporary research challenges this view. As shown by the findings discussed below, motivation impacts task performance in several ways that are quite specific.

Motivation and Attention

This section will review the interaction between motivation and attention in terms of behavioral studies as well as studies using EEG and functional MRI in humans and electrophysiology in monkeys.

Behavioral Effects

Visual attention is studied in terms of *exogenous* manipulations, where a conspicuous task-*irrelevant* item (e.g., a bright stimulus) is used to capture attention, and *endogenous* manipulations, where a symbolic task-*relevant* item (e.g., an arrow) is used to instruct subjects to direct their attention to a specific location. In a study from my lab, Jan Engelmann and I investigated the effects of motivation on task performance by probing the impact of parametric changes in incentive value on behavior during a difficult spatial localization task

(Engelmann and Pessoa 2007b). We instructed our subjects to indicate the location of a target stimulus (a red dot) as quickly and accurately as possible. We manipulated attention exogenously by using a peripheral cue that predicted target location with 70 percent validity and an incorrect one 30 percent of the time. In such cases, subjects' performance during validly cued trials exceeded that during incorrectly cued ones. We parametrically manipulated motivation in a blocked fashion by linking reward and punishment to behavioral performance, thus subjects earned \$1 or \$4 if their performance during a block of trials was both fast and accurate but lost 50¢ or \$2 if it was not.

In theory, motivation could lead to indiscriminate responding, which would increase the number of both correct detections and false alarms. In terms of data analysis, the proper way to handle this situation is to compute d-prime, which takes into account both correct detections and false alarms simultaneously to determine *sensitivity* (Green and Swets 1966). Detection performance improved as a function of absolute incentive value (figure 6.1A) independently of unspecific influences, such as general *activation* (e.g., purely faster response times) or response bias (e.g., more conservative responses).[1] We observed the same basic pattern of behavioral results in distinct versions of the task that varied in level of difficulty, type of target and distractor stimuli, and type of cue (exogenous vs. endogenous, investigated in Engelmann et al. 2009, as discussed below).

Based on the above results, we argued that motivation improved visual attention. Because improvements can result from diverse mechanisms, however,

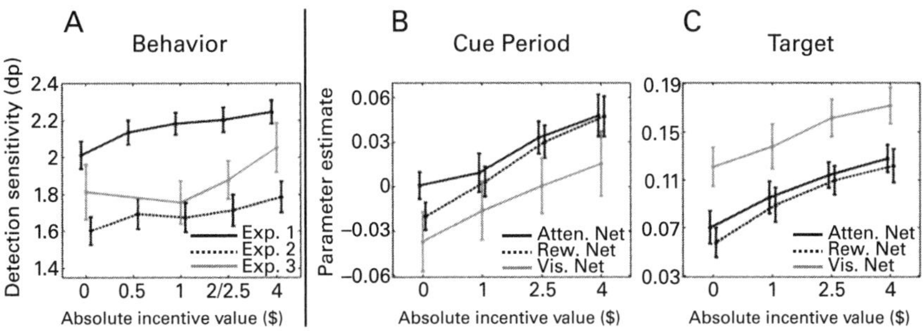

Figure 6.1
Behavioral and brain effects of incentive motivation. (A) In all experiments, the detection sensitivity measure, d-prime (dp), increased as a function of absolute incentive magnitude (the conditions included working to avoid potential losses). Experiments 1 and 2 by Engelmann and Pessoa (2007b); experiment 3 by Engelmann and colleagues (2009). (B, C) Responses observed by Engelmann and colleagues (2009) during the cue and target task phases in attentional (Atten.), visual (Vis.), and reward-related (Rew.) networks (Net; see figure 6.3 for some of the sites). Results were obtained by pooling responses from regions within the three networks. Reproduced with permission from Pessoa and Engelmann 2010.

the reasons for enhanced performance are difficult to ascertain conclusively. As pointed out by Ronald Hübner and Jan Schlösser (2010), incentives can speed up stimulus coding and motoric responding, and the saved time can then be spent to extend the response selection phase, which, in turn, can improve sensitivity. To further investigate this question, Hübner and Schlösser probed a version of the flanker task (see Eriksen and Eriksen 1974) under reward and control conditions. In the flanker task, which is a type of response interference task, subjects are asked to indicate the direction of a central arrow embedded in identical symbols or in arrows pointing to the opposite direction. Subjects who received a performance-contingent monetary reward had significantly higher accuracy than did those who earned a fixed reward. By varying response deadlines (i.e., the time by which a response had to be provided), Hübner and Schlösser were able to evaluate speed-accuracy trade-offs. Their results suggested that subjects had better information for decisions in the contingent reward condition—consistent with an enhancement in the *quality* of sensory coding (see also Baldassi and Simoncini 2011). However, they did not observe a reduction in the flanker effect (i.e., response interference) with contingent reward, indicating that reward did not narrow the spatial focus of attention (i.e., the distractors were not filtered out). Collectively, their findings showed a selective effect of motivation on sensory processing that likely did not alter the spatial breadth of the attentional focus (for further discussion, see Hübner and Schlösser 2010).

The two studies above illustrate the strategy of manipulating motivation in a manner that depends on performance: subjects are told in advance that they will receive extra monetary compensation if they perform at certain levels. The impact of motivation on behavioral performance has also been investigated when incentive levels are *dissociated* from task performance itself. One way to accomplish this is through *value learning*, which involves a learning phase when rewards are offered and a test phase when they may or may not be offered. Chiara Della Libera and Leonardo Chelazzi (2009) reported that, when tested after three training sessions on consecutive days, subjects found visual items often followed by high reward during training harder to reject as distractors and easier to select as targets. Conversely, they found visual items often followed by low reward during training easier to reject as distractors and harder to select as targets. The study reveals that visual processing is adjusted according to the reward history of the visual items encountered (see also Della Libera and Chelazzi 2006; Doallo, Patai, and Nobre 2013).

In a related value learning study, Brian Anderson, Patryk Laurent, and Steven Yantis (2011) reported that nonsalient, task-*irrelevant* stimuli previously associated with reward slowed visual search in a test phase during which no rewards were offered. Notably, the magnitude of slowing was spatially specific, such that when a target appeared in a location occupied by a high-value distractor on the previous trial, slowing was especially prolonged. Furthermore, such value-driven attentional capture operated on a basic stimulus feature, namely, color.

Value learning has also been employed in conjunction with the attentional blink (Raymond and O'Brien 2009; see chapter 2 for a description of the attentional blink task). During an initial learning phase, subjects viewed faces that were paired with low or high probability of monetary reward or punishment. Whereas their responses to faces associated with losses or zero gain/loss exhibited the standard attentional blink, their responses to those associated with reward did not. Therefore, association with prior reward counteracted the attentional blink, such that performance for rewarded stimuli was equivalent at short versus long lags between the first and second targets (i.e., performance at both showed reduced blinks).

A final example of the interaction between motivation and attention involved the "priming of pop-out" paradigm (Kristjansson, Sigurjonsdottir, and Driver 2010; "priming of pop-out" refers to the phenomenon that subjects perform better when a target color is repeated—even though the task is an extremely simple one). Subjects performed a visual search for a color-singleton target (e.g., a red item among green distractors), whose shape was then interrogated ("Is the small notch on the target at its top or bottom?"). They were told that they would be rewarded for fast and accurate performance, but without the relative values of specific target colors being explained. For one half of the subjects, a correct response for a red target earned 10 points on 75 percent of such trials and 1 point on the other 25 percent, and vice versa for a green target. For the other half, this was reversed. Incorrect responses earned 0 points. The findings revealed that the benefit of successive target repetitions (i.e., the priming of pop-out effect) was larger for the more highly rewarded target color. Thus priming of pop-out can be significantly enhanced for targets tied to reward. Moreover, in a second experiment, the extent of priming of pop-out dynamically tracked reversals in reward levels when rewards changed unpredictably; for instance, high-reward red on trial n became associated with low reward on trial $n + 1$ and low-reward green on trial n became associated with high reward on trial $n + 1$. Notably, behavioral changes were observed in as few as 5–10 trials after the reversal. The results of these two experiments are important (see also Shen and Chun 2011) because priming of pop-out is a basic process suggested to involve a low-level passive memory mechanism and is therefore believed to be largely immune to voluntary control or task priorities. The findings of the second experiment are noteworthy, too, since they indicate that implicit contributions were at play; upon debriefing, nearly all subjects reported that they thought the reward schedules were random.

To summarize, this section has briefly reviewed behavioral paradigms exploring how motivation affects attention. Early studies in signal detection (e.g., Green and Swets 1966) manipulated rewards and punishments, but focused on understanding changes in the response criterion. Researchers proposed that subjects adopt a stricter criterion in reward conditions so as to make fewer false alarms. Recent research has attempted to elucidate *how* motivation impacts performance in visual tasks. Does motivation enhance attention specifically or does it have only other, less specific effects, such as an overall speeding up of processing? Are the effects of motivation on attention related to changes in sensory mechanisms, or are they related to more decision-related operations? The answer to the first question is that motivation does indeed have *selective* effects. Examples include the increase in visual sensitivity (d-prime) in both endogenous and exogenous attention tasks, enhanced stimulus coding, and target/distractor effects. The impact of motivation on visual performance has been found both when reward was contingent on performance and when it was not. In the latter case, researchers employed value learning to show that the effects of reward are also manifested via associative learning mechanisms. Stimuli associated with reward in this way behave as if they had increased *salience* (see below). The second question may be better addressed by cognitive neuroscience studies than by behavioral studies, as reviewed below.[2] Nevertheless, Hübner and Schlösser (2010) have argued that the pattern of results of their speed-accuracy trade-offs indicates that "better information" is available for a decision in the reward condition, that is, the mobilization of attentional resources by reward improved the *quality* of sensory coding. Furthermore, findings that a basic feature, such as stimulus color, can affect visual performance when linked to reward even when task irrelevant, suggest that motivation alters sensory processing.

Brain Recording Results

Our understanding of motivation-attention interactions has been advanced by event-related potential (ERP) studies, which provide millisecond-level temporal resolution. These studies have capitalized on the characterization of several ERP response components (i.e., particular response peaks or troughs) that have been linked to psychological functions.

Monika Kiss, Jon Driver, and Martin Eimer (2009) used ERP to investigate the impact of reward on the same pop-out detection task described above. Subjects were asked to perform a visual search for a color-singleton target (e.g., a red item among gray distractors) and to evaluate its shape ("Is the small notch on the target at its top or bottom?"). They were told they could earn rewards for correct and fast responses to targets—a larger reward for a target in the high-reward color (e.g., red) and a smaller one for a target in the low-reward color (e.g., green). Measuring the "N2pc" response component, which has been linked to visual target selection and is observed approximately 200 ms after display onset, the authors found that it was earlier and stronger for high- than for low-reward targets. Notably, the N2pc amplitude correlated with the impact of reward level on performance efficiency (an index that takes into account both speed and accuracy). Later post-selection processing was also affected by reward level; specifically, a separate response component around 350–400 ms believed to be involved in further target processing, including its maintenance in visual short-term memory. Kiss, Driver, and Eimer's results demonstrate that visual selection of task-relevant items is rapidly modulated by reward-related priorities, even when the possibility of a reward during a trial is not announced in advance of the target by a cue stimulus (remember that in the task, the information about reward is part of the stimulus itself, namely, the target's color; for an example of cue stimulus that precedes a target, see figure 6.4).

In another ERP study of the interactions between reward and attention, Clayton Hickey, Leonardo Chelazzi, and Jan Theeuwes (2010) sought to more clearly dissociate the "strategic" and "incidental" effects of reward. In many experimental paradigms, as we have seen, subjects are explicitly told they may earn additional rewards depending on their performance in certain task conditions. Although reward effects have been documented in such circumstances, researchers have found it hard to separate the neural effects of reward from those of attention (see "Mechanisms of Motivational Effects: Conceptual Issues" below). Because subjects are more highly engaged and motivated to perform better when rewards are present, the concomitant effects may reflect their upregulation of attention. Therefore, to separate the effects of attention from those of reward more clearly, it is useful to devise tasks in which reward may have an effect due to "extra" engagement during reward trials.

To that end, Hickey, Chelazzi, and Theeuwes probed how reward in one trial affected visual processing in the next. Subjects were asked to detect a target among an array of distractors, and the target was defined in terms of a uniquely shaped object among same-shape distractors; importantly, target color was always task irrelevant. The impact of reward was investigated by determining performance and responses during trials that *followed* low or high reward; reward magnitude was actually randomized on a trial-by-trial basis. Following a high-reward trial, the P1 ERP response component, which occurs approximately

100 ms after stimulus onset, was found to be stronger contralateral to targets of the same (task-irrelevant) color rewarded on the *previous* trial. This response component is believed to reflect an amplification of early visual processing stages in extrastriate visual cortex (for review, see Hillyard, Vogel, and Luck 1998), thus also revealing facilitated responses based on previous-trial reward.[3] N2pc responses were found to be stronger as well, indicating that target processing was enhanced by the prior reward. Notably, however, P1 and N2pc effects were also observed on trials following high reward when a salient *distractor* was shown in the reward-paired color, indicating that additional visual resources were also allocated to this task-irrelevant object. Taken together, Hickey, Chelazzi, and Theeuwes's results reinforce the inference that reward has an impact on vision that is independent of its role in establishing goal-driven attention (e.g., when a subject deliberately increases attention in anticipation of reward). Furthermore, the observation of early effects in the P1 time range suggests that reward enhances visual *salience* (but see Sanger and Wascher 2011).

In a cell-recording study, Christopher Peck and colleagues (2009) also attempted to disentangle the contributions of attention and reward to neural signals. Monkeys performed a task in which a peripheral visual cue informed them about the possibility of reward, but, to be rewarded, they had to make an eye saccade to a separate target whose location was independent of the cue. In this manner, a cue predicting reward was spatially separate from, and could interfere with, a required action (figure 6.2, top). Cues predicting reward attracted attention to their location and evoked sustained excitation in posterior parietal cortex (figure 6.2, bottom; plate 6), whereas cues predicting no reward repulsed attention and evoked sustained inhibition. Thus, although the cues indicating the opportunity of reward had no operant significance (i.e., they did not specify the saccade location), they biased attention in a value-specific fashion. Critically, these biases were maladaptive since they interfered with the required saccade to the target. Michael Goldberg and colleagues (2002) have argued that posterior parietal cortex contains a *salience map*, such that the area of the visual field associated with the greatest activity corresponds to the locus of visual attention. Based on their results, Peck and colleagues (2009) proposed that posterior parietal cortex in the monkey contains a visuospatial map that takes into account reward expectations for the purpose of guiding attention. In this way, pairing a stimulus with reward enhances its salience.

Human functional MRI studies have also contributed to our understanding of attention-motivation interactions. In one such study (Weil et al. 2010), subjects performed a visual orientation discrimination task during low- and

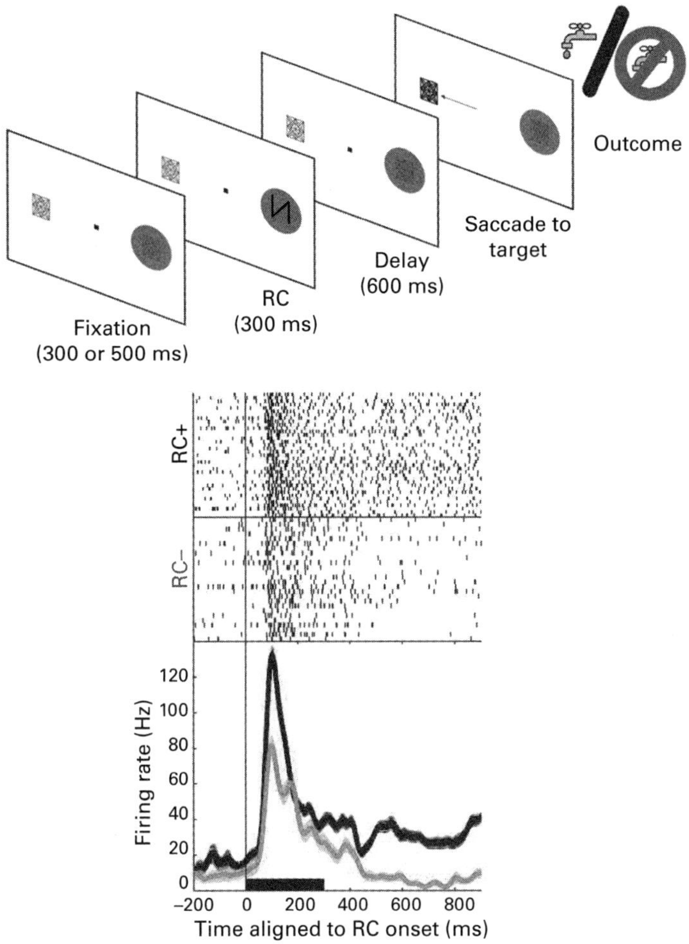

Figure 6.2 (plate 6)
Behavioral task and cell responses in the lateral intraparietal area. (Top) Two patterns, one to the right and one to the left of fixation, were shown throughout the trial. After a fixation period, a cue appeared (inverted "N"), followed by a delay period. At the end of the delay, one of the patterns was brightened (indicated by arrow), indicating the saccade target location. A reward was administered on correct reward trials. The gray circle indicates the area of the receptive field of the cell being recorded. (Bottom) Response of a representative neuron. Trial onset is aligned with the presentation of the cue stimulus. Graph shows average firing rates during reward (blue) and no-reward (red) conditions. The black horizontal bar denotes the cue duration. Reproduced with permission from Peck et al. 2009 and kindly provided by Christopher Peck.

high-reward trials, with a feedback phase to indicate the amount won on each trial. Behaviorally, task performance was better during high- versus low-reward trials but only for easier visual discriminations, not harder ones. Trial-by-trial analysis revealed that receiving reward on a given trial was associated with improved accuracy on the subsequent trial (for both difficulty levels). In terms of brain responses, early retinotopic areas V1, V2, and V3, and later nonretinotopic visual areas in temporal cortex showed distinct patterns of reward influence. Early visual cortex displayed stronger signals in response to the visual grating on the *next* trial, that is, after receipt of rewarding feedback on the previous one. In contrast, "later" visual cortex exhibited differential responses during the *current* trial, though only during reward feedback, a trial phase that actually contained an auditory stimulus. In other words, no effect of reward was detected during visual stimulation itself. Furthermore, brain responses to reward feedback were distinct from typical effects of spatial attention observed during visual tasks, and involved regions such as orbitofrontal cortex and the ventral striatum, which play prominent roles in stimulus valuation (see below).

In a second functional MRI study (Small et al. 2005; for a related study, see Mohanty et al. 2008), subjects were told that fast visual target detection could lead to monetary wins or avoidance of monetary losses. Endogenous attention was manipulated by using a central arrow cue that predicted target location with 80 percent validity. Behavior during *invalid* trials was used to assess mechanisms of disengagement, given that correct performance required subjects to shift their attention from the cued location to the opposite hemifield to process the target stimulus. Better performance during the *disengagement* of attention was associated with increased activity in the inferior parietal lobe in the vicinity of the temporo-parietal junction, a region implicated in the reorienting of attention (Corbetta, Patel, and Shulman 2008; Corbetta and Shulman 2002). Importantly, this effect was enhanced during trials on which subjects could win or avoid losing money, and was accompanied by activation in valuation-related regions, including orbitofrontal cortex. Of particular interest, responses in posterior cingulate cortex were correlated with the degree to which the cue improved performance, an effect that was enhanced by motivation. Given its anatomical connectivity with areas of the brain participating in attentional and motivational circuits, Diana Small and colleagues proposed that posterior cingulate cortex serves as a "neural interface" between motivation and the goal-directed control of attention (see also Engelmann et al. 2009; Pessoa and Engelmann 2010).

In a third functional MRI study, also investigating the interaction between endogenous attention and motivation (Engelmann et al. 2009), my colleagues and I asked subjects to detect a degraded face that could appear to the right or left of fixation. As in the second MRI study, we manipulated attention by using a central arrow cue that predicted target location (70% validity), and motivation was parametrically varied. The experimental design allowed us to estimate cue- and target-phase responses separately. In parallel with the behavioral findings, we observed parametric influences of incentive value on responses during both the cue (figure 6.1B) and target (figure 6.1C) periods (see also figure 6.7A, B). Interestingly, the results revealed effects of motivation across many brain regions, including occipital and temporal visual areas, parietal and frontal sites that are important for the control of attention, and several areas that participate in stimulus valuation, including the caudate.

In summary, over the past decade, our understanding of the neural bases of attention-motivation interactions has advanced greatly. ERP studies have demonstrated that the impact of reward on attention can be very fast, involving early response components (within 100–200 ms). Although the contributions of reward and attention to neural signals are often hard to disentangle, recent studies have made inroads into this question. For instance, in their monkey physiology study, Peck and colleagues (2009) were able to separate the spatially specific influence of reward from the behavioral requirements for successful task completion. Furthermore, findings from both monkey physiology and human ERP studies showed that, in several paradigms, reward acts in a manner similar to what would be expected for a physical change in stimulus salience. Broadly speaking, functional MRI studies have identified three types of brain region engaged during motivation-attention tasks (figure 6.3): (1) occipital and temporal visual cortical sites that are involved in sensory processing (including primary visual cortex); (2) parietal and frontal cortical sites that are involved in the control of attention (e.g., frontal eye field in frontal cortex); and (3) cortical and subcortical regions that are involved in stimulus valuation (e.g., orbitofrontal cortex).

Motivation and Sensory Processing

The impact of motivation on sensory processing described above is consistent with the findings of a growing number of human and animal studies. For example, Janaina Pantoja and colleagues (2007) investigated neuronal responses in rat primary somatosensory cortex during a tactile discrimination task. Stimulus-related information encoded by neuronal signals increased when the contingency between stimulus and response was crucial for reward, but not when reward was freely available. Related neuroimaging findings in humans were reported by Burkhard Pleger and colleagues (2008, 2009), who used a tactile discrimination task coupled with financial rewards. Changes

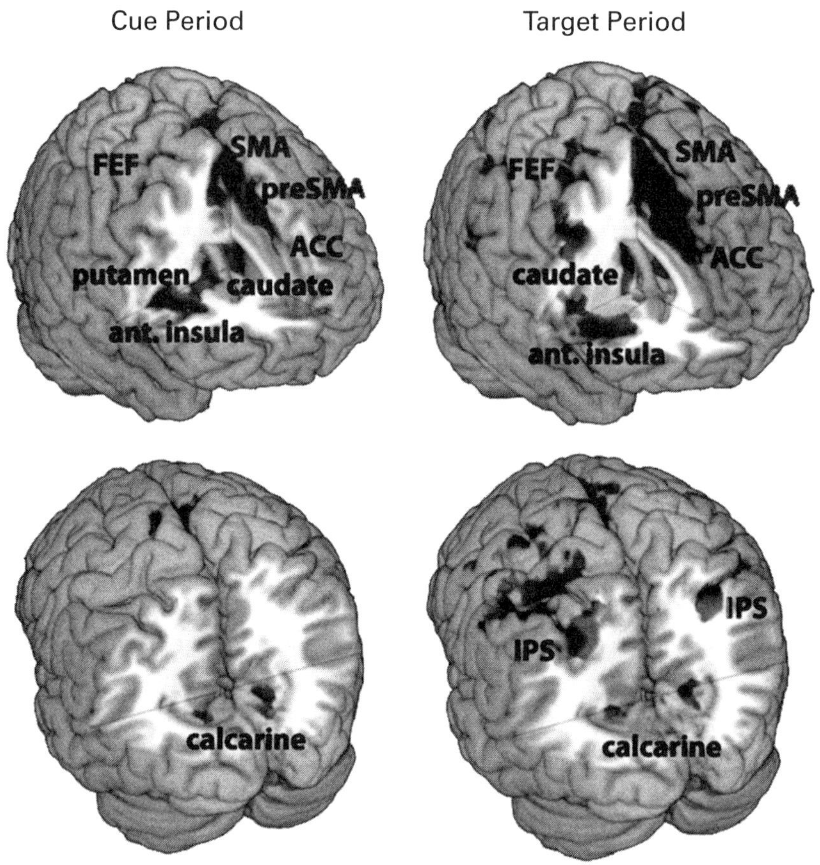

Figure 6.3
Brain regions exhibiting correlations with absolute incentive magnitude during cue and target task periods. Some of the attentional, visual, and valuation regions from Engelmann et al. 2009 are shown as viewed from the front (upper panels) and as viewed from the back (lower panels). The label "calcarine" refers to activations within and in the vicinity of the calcarine sulcus, where early visual areas V1 and V2 are found. ACC, anterior cingulate cortex; ant., anterior; FEF, frontal eye field; IPS, intraparietal sulcus; preSMA, presupplementary motor area. Reproduced with permission from Pessoa and Engelmann 2010.

in primary auditory cortex were also investigated (David, Fritz, and Shamma 2012).

As reviewed above, effects of motivation on early sensory cortex have been observed in the visual system. These effects are particularly interesting in the case of studies that functionally determined the location of *retinotopically* organized visual areas. Notably, even primary visual cortex was modulated by reward (Serences 2008; Weil et al. 2010; see also Shuler and Bear 2006 for evidence of reward signals in primary visual cortex in the rat).

Interactions between Motivation and Other Executive Functions

This section will discuss interactions between motivation and other executive functions, including response conflict, task switching, working memory, and long-term memory.

Response Conflict

Does motivation influence the selection of information? To answer this question, Srikanth Padmala and I investigated the effects of reward during a response-conflict task (Padmala and Pessoa 2011). As you may recall from chapter 5, in response-conflict tasks, the stimulus primes two competing motor responses (figure 6.4). Based on previous studies, we anticipated that motivation would enhance engagement of fronto-parietal attentional regions and, consequently, that these regions would be better positioned to exert goal-directed control affecting visual processing. This could be accomplished by

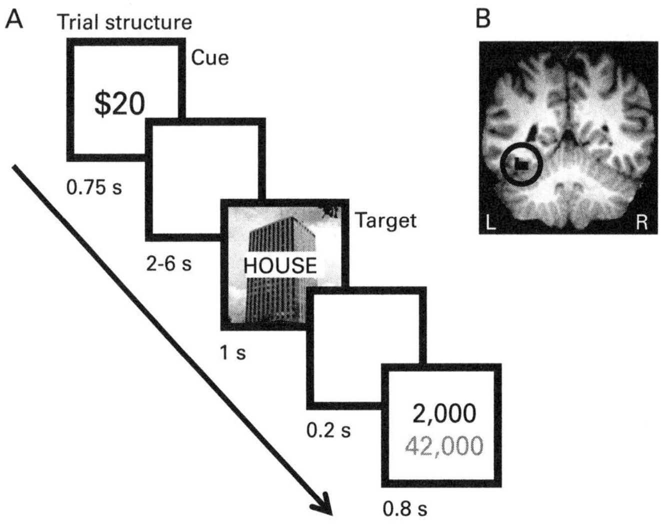

Figure 6.4
Response-conflict paradigm. (A) In the reward condition shown here, a cue stimulus ("$20") signaled that subjects would be rewarded for fast and correct performance; in the control condition (not shown here), a cue stimulus ("$00") signaled that there would be no reward. During the target phase, a stimulus picture of a house or building was presented together with a task-irrelevant word (an incongruent condition is illustrated here). After the target stimulus, subjects were informed about the reward and about the total number of points accrued. (B) Decreased activation in the left parahippocampal gyrus (circled area) in the reward condition was interpreted in terms of reduced word-related processing in visual cortex due to improved distractor filtering. Panel A reproduced with permission from Padmala and Pessoa 2011. Panel B was generated with data from Padmala and Pessoa (2011).

amplifying task-*relevant* information (Egner and Hirsch 2005; Polk et al. 2008) or by improving filtering of task-*irrelevant* information (Polk et al. 2008). As expected, we observed behavioral response interference: performance was slower on incongruent trials than on neutral ones. But response interference was reduced with reward. Given that reward also decreased response facilitation (i.e., the beneficial effect of a congruent task-irrelevant item), these results support the inference that motivation enhanced *attentional filtering*, thereby reducing the influence of the task-irrelevant word item.[4]

Our brain imaging results revealed that, during the cue phase (when subjects were told whether a reward was possible), responses in fronto-parietal regions were stronger with reward—consistent with increased attention. During the target phase (when subjects performed the actual task), we were interested in probing responses in dorsal-medial prefrontal cortex, a region suggested to be sensitive to response conflict (Botvinick et al. 2001). We assumed that, in the context of our task, responses in this region to the contrast of incongruent and neutral trials would provide an index of the amount of response-selection demand. As we had in our behavioral data, we observed a motivation by cognition interaction; the effect was found in dorsal-medial prefrontal cortex, such that interference-related responses *decreased* during reward trials. Notably, *larger cue*-related responses were associated with larger *decreases* in interference-related responses in dorsal-medial PFC during the subsequent *target* phase. The pattern of cue and target responses was thus compatible with the inference that the upregulation of control during the cue phase led to decreased interference during the target phase. How were cue and target responses related to the selection of visual information during the task? The relationship between cue and target responses was consistent with a *mediation* role for responses in visual cortex sensitive to word-related processing (in the left fusiform gyrus, a region responsive to word stimuli). As indicated in figure 6.4B, during the target phase, visual responses linked to distractor processing *decreased* in the reward condition. Collectively, our findings suggest that subjects were able to employ motivationally salient cues to upregulate attentional control mechanisms that influenced the selection of visual information in a way that *reduced* both behavioral conflict and related brain responses (see figure 6.5A).

At this juncture in discussing our study, it may be useful to discuss *functional connectivity* (a concept further elaborated in chapter 8). Whereas *anatomical* connectivity refers to whether two regions are physically linked, functional connectivity refers to the degree to which they are "temporally coherent" and can be estimated by their correlation, for example. Thus two regions whose responses are strongly correlated are said to display "*high*

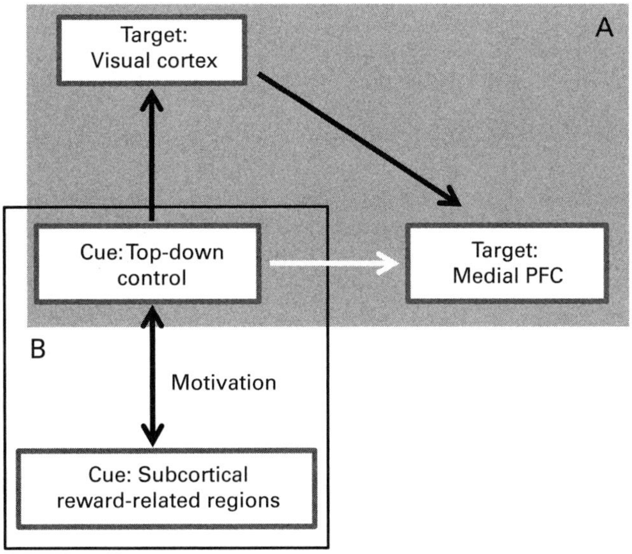

Figure 6.5
Hypothesized network interactions. (A) Predicted mediation by target/distractor processing in visual cortex of the relationship between attentional control implemented in fronto-parietal cortex during the cue phase and conflict-related activity in medial prefrontal cortex during the subsequent target phase (see white arrow). (B) Predicted effect of motivational context on functional interactions between fronto-parietal cortex and subcortical regions involved in reward processing. PFC, prefrontal cortex. Reproduced with permission from Padmala and Pessoa 2011.

functional connectivity"—sometimes described as "high coupling." Notably, when considering functional connectivity, whether two regions are physically connected—directly, indirectly, or not at all—is irrelevant. Functional connectivity is important because changes of this measure indicate changes of the *relationship* between regions. In particular, increases in functional connectivity are viewed as evidence for enhanced "communication."

We also observed responses to the cue in several subcortical sites that are engaged during reward-related processing, including the caudate and putamen in the dorsal striatum, the nucleus accumbens in the ventral striatum, and the midbrain. We reasoned that, if motivationally salient cues engage fronto-parietal regions more robustly during the cue phase, these regions should exhibit increased coupling with some of the above regions, which are sensitive to the motivational significance of the cues (figure 6.5B). And, indeed, we did observe increased functional connectivity as evaluated on a trial-by-trial basis between the intraparietal sulcus in parietal cortex and the putamen, caudate, and nucleus accumbens in the reward versus the no-reward condition (figure 6.6; for related findings during an antisaccade task, see Harsay et al. 2011).

Notably, the strength of the coupling between cortical and subcortical areas was linearly related to individual differences in reward sensitivity (as assessed through standardized questionnaires; see Carver and White 1994), indicating that the functional interaction between these regions was stronger for subjects who scored higher in this dimension (figure 6.6C).

In a pair of related behavioral and imaging studies, Ruth Krebs and colleagues investigated the impact of reward associations on conflict processing during the classic Stroop task, which asks subjects to determine the ink color

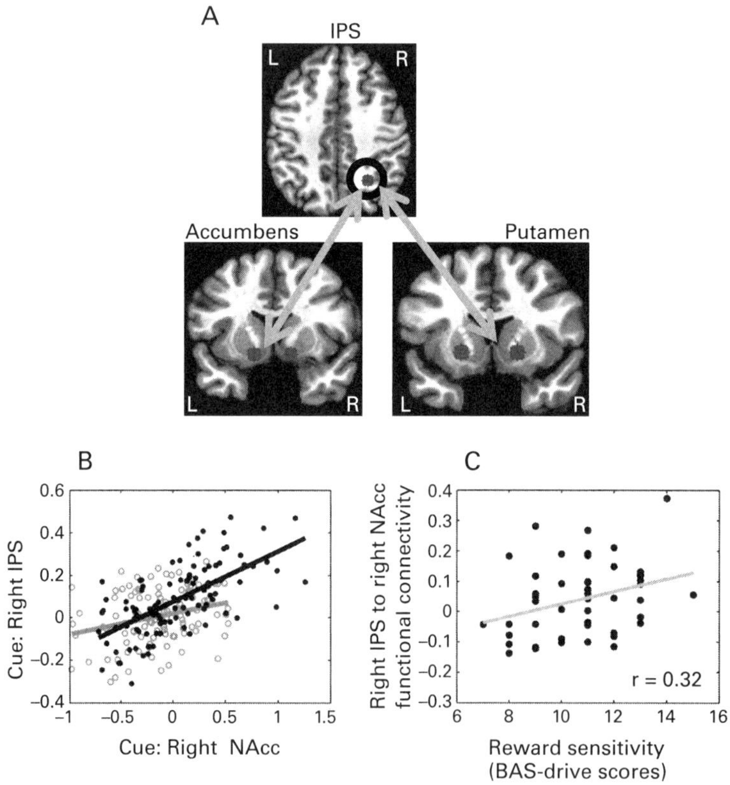

Figure 6.6
Functional connectivity during reward trials. (A) Regions exhibiting stronger functional connectivity with the right intraparietal sulcus (IPS) during the cue phase for reward trials. (B) Scatter plot showing the trial-by-trial relationship between right intraparietal sulcus and right nucleus accumbens signals during reward (black dots and line) and no-reward (gray dots and line) trials. Data are illustrated for a sample subject. (C) Scatter plot showing the relationship between "drive" behavioral activation system (BAS) scores and functional connectivity between right intraparietal sulcus and right nucleus accumbens. IPS,intraparietal sulcus; NAcc, nucleus accumbens. Subjects with higher drive scores exhibited increased functional coupling. Reproduced with permission from Padmala and Pessoa 2011.

of a word (Krebs, Boehler, and Woldorff 2010; Krebs et al. 2011). The words are displayed in different ink colors, and the two stimulus dimensions of word meaning and ink color at times conflict with each other (e.g., the word "red" in blue ink). In the study, one subset of ink colors was associated with the potential of reward (e.g., green and blue), while another subset was not (e.g., red and yellow); word meaning was not linked to reward. Krebs and colleagues observed that reward linked to the ink color led to overall faster responses and lower error rates. Importantly, the typical conflict-induced slowing of responses was *reduced* during trials containing reward-associated ink colors. In contrast, reward associations related to the *irrelevant* word-meaning dimension (e.g., the word "green" when the *color* green was rewarded) inflicted costs on performance if they were incongruent with ink color. Put another way, incongruent words semantically related to reward-associated colors (e.g., the word "green" in yellow ink) interfered more strongly with performance than incongruent words semantically *un*related to reward-associated colors (e.g., the word "red" in green ink). This occurred even when word meaning was always task irrelevant. In terms of brain responses, irrelevant words shown in reward-associated color increased activity in medial prefrontal cortex, possibly reflecting the greater response selection demands produced by these words.

Task Switching

What is the impact of motivation on the ability to change between tasks? Esther Aarts and colleagues (2010) investigated the influence of reward on task switching, during which subjects perform two or more trial types. If trial $n + 1$ is of the same type as trial n, it is called a "repeat trial"; when the task changes, it is called a "switch trial." Given that switching between tasks often leads to a cost (e.g., increased error rate), it is useful to compare performance during repeat versus switch trials. Although Aarts and colleagues did not find evidence for an effect of reward on switch cost in general, they observed an interaction with genotype. Dopamine circuits, which involve the midbrain and its subcortical and cortical target areas (e.g., nucleus accumbens and prefrontal cortex, respectively), have long been implicated in reward and motivation. Animal studies have highlighted a role for dopamine in the nucleus accumbens during preparatory response to signals that predict reward (Baldo and Kelley 2007; Berridge and Robinson 1998; Ikemoto and Panksepp 1999; Robbins and Everitt 1992; Schultz 2002). Aarts and colleagues were thus particularly interested in investigating the effects of genetic variation in dopamine transmission and potential impacts on behavior. They found that almost all carriers of a specific dopamine transporter gene that was associated with high levels of dopamine in the striatum (including the nucleus accumbens) showed a *smaller* behavioral switch cost when subjects were anticipating high versus low reward. And they observed an interaction with genotype in the responses of the nucleus accumbens as well.

Working Memory

Interactions between motivation and working memory have been the target of several neuroimaging studies (Beck et al. 2010; Gilbert and Fiez 2004; Jimura, Locke, and Braver 2010; Longe, Senior, and Rippon 2009; Pochon et al. 2002; Taylor et al. 2004). For example, during an "N-back" task (where subjects are asked to indicate whether the current item matches the N-th prior item in the stream), Jean-Baptiste Pochon and colleagues (2002; see also Longe, Senior, and Rippon 2009) observed sites in lateral prefrontal cortex that were sensitive to both working memory load and reward magnitude. In another study, Koji Jimura, Hannah Locke, and Todd Braver (2010) probed the contributions of reward during working memory by manipulating the reward context of individual trials. Motivation did not simply increase activation; it also shifted the timing of working memory responses. These effects were correlated with individual differences; subjects with higher reward sensitivity exhibited stronger early-transient responses. Jimura, Locke, and Braver suggested that, in the reward condition, subjects may have adopted a more *proactive* control strategy to aid performance—and thus increase their chance of reward. In other words, instead of a just-in-time *reactive* strategy, subjects may have, for instance, prepared a target response before target onset (for further discussion of proactive and reactive control, see Braver, Gray, and Burgess 2007; Braver 2012).

Long-Term Memory

Interactions have been also studied between reward and long-term memory. In a study by R. Alison Adcock and colleagues (2006), cues signaled high- or low-value reward for memorizing an upcoming scene. When tested 24 hours postscan, subjects were significantly more likely to remember scenes that followed high- versus low-reward cues. High-reward cues preceding remembered but not forgotten scenes activated both the hippocampus, a brain region centrally involved in memory processes, and the nucleus accumbens and midbrain (in a site possibly including the ventral tegmental area, which, like the accumbens, is an important dopamine-related site). Greater activation in these regions predicted superior memory performance across subjects. Furthermore, when subjects viewed cues indicating high reward, correlated activity between the hippocampus and midbrain increased, and the correlation strength predicted memory formation for the subsequent stimulus. This is another example of reward increasing the functional connectivity between brain regions—all the more interesting when connectivity is itself correlated with behavior.

Sustained Effects

Although the emphasis thus far has been on motivational effects that are relatively transient (albeit extending for a few seconds in working memory tasks), *sustained* signals reflecting motivational *state* have also been investigated. Thus, in the experiment by Jan Engelmann and colleagues (2009) described previously, monetary incentives were manipulated in a blocked fashion. Sustained (as well as transient) responses that varied parametrically based on incentive magnitude were observed in several brain regions, including sites in frontal cortex (e.g., frontal eye field), parietal cortex (e.g., intraparietal sulcus), and posterior cingulate cortex (figure 6.7). Related functional MRI results were reported by Hannah Locke and Todd Braver (2008), who described increased sustained activity across frontal and parietal regions on rewarded trials during which subjects performed a cognitive control task. And in the study reviewed above, Jimura, Locke, and Braver (2010) also observed sustained effects in lateral prefrontal cortex that correlated with individual differences in reward

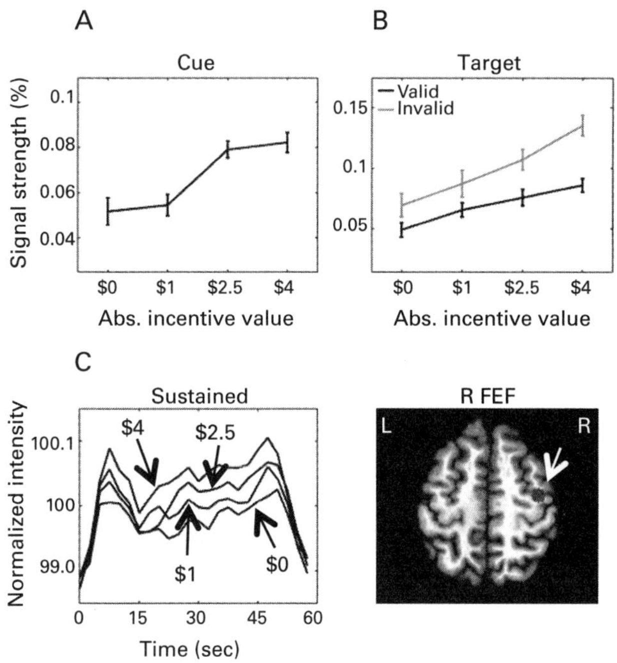

Figure 6.7
Parametric effects of absolute incentive magnitude on transient and sustained responses. (A) Cue responses. (B) Target responses. During valid trials (70%), the target appeared at the same location as the cue; during invalid trials (30%), the target appeared in an opposite location. The inset represents the location of the region of interest (ROI) from which the data plots are based. Abs., absolute; R FEF, right frontal eye field. Reproduced with permission from Engelmann et al. 2009.

sensitivity. These studies highlight the importance of studying both transient and sustained effects of motivation. Motivational states, which are reflected in increased sustained signals, may be associated with a more stable, heightened pattern of activity in regions associated with cognitive control that lasts for the duration of the task (Locke and Braver 2008).

The sustained effects described here are still on the order of seconds (or perhaps as long as a minute). Humans, in particular, are capable of "sustained motivations" of many hours (e.g., practicing a sport) or even years (e.g., preparing for the Olympics). How longer-term effects map to shorter-term ones, possibly via the action of personality traits, is a matter for future research.

Monkey Studies

Interactions between motivation and executive functions have been studied in many monkey cell-physiology studies. Not only do cells in lateral prefrontal cortex hold information of an object's shape and location, but they are also modulated by reward expectancy. Masataka Watanabe (1990, 1996) showed that activity of neurons in dorsal-lateral prefrontal cortex reflects both working memory and reward expectancy. For instance, some neurons exhibited stronger sustained activity when an eye movement to the left versus right had to be generated at the end of the trial. At the same time, their firing rate was modulated by reward, being highest for a raisin reward, intermediate for an apple reward, and lowest for a sweet potato reward (see also Leon and Shadlen 1999).

Further monkey studies confirm that motivational information influences cell activity in lateral prefrontal cortex where, critically, they reveal that cognition and motivation are *integrated*. For instance, during the delay period of a delayed–eye saccade task, some lateral PFC cells increased their firing if the monkey was initially cued to make a saccade to the preferred versus the opposite direction; these cells also exhibited increased firing during rewarded versus unrewarded trials (Kobayashi et al. 2002). Notably, during rewarded trials of saccades to the preferred direction, there was an increase of the amount of transmitted *information* with respect to target position, as quantified by information theory. In other words, the reward information increased the discriminability of target positions, leading to an enhancement of performance (see also Kobayashi et al. 2007).

Summary

This section described interactions between motivation and cognition across a range of tasks (for an example where reward impaired executive function, see Padmala and Pessoa 2010), in particular, those having *selective* effects on cognition. Traditional accounts describing motivation as a global activation independent of particular control demands (e.g., Duffy 1962; Hull 1943) have been echoed more recently by a functional MRI study in which Frédérique Kouneiher, Syvain Charron, and Etienne Koechlin (2009) argued that motivation and cognitive control can be regarded as two *separate* and *additive*—instead of interactive—factors. Although there is little question that motivation can have generalized, *activating* contributions to behavior (see Robbins and Everitt 2007; Salamone et al. 2009), the findings reviewed in the previous sections underscore the striking ability of motivation to shape behavior in specific ways, whether by reducing response conflict or task-switch costs, via selective effects on working memory, or by improving long-term memory. Another body of research demonstrating specific effects of motivation has investigated attentional effort, as described by Martin Sarter and colleagues (e.g., Sarter, Gehring, and Kozak 2006), and as will be discussed in chapter 7.

The question of the specificity of motivational effects is connected to the hypothesis that motivation enhances task processing so as to increase the likelihood of attaining a reward, a hypothesis consistent with the effects described in this chapter. Thus a change in the timing of the subjects' responses during a working memory task suggests a shift by them toward a proactive control strategy in reward contexts (Jimura, Locke, and Braver 2010). And the findings by Ruth Krebs, Carsten Boehler, and Marty Woldorff (2010) during response conflict indicate that the anticipation of reward enhances stimulus processing or reduces interference from conflicting information (see also Padmala and Pessoa 2011; Harsay et al. 2011). In these and other tasks above, it is highly unlikely that diffuse effects of motivation would have improved task performance and increased rewards in the same way.

Mechanisms of Motivational Effects: Conceptual Issues

What are the neural bases for motivation's effects on perception and cognition? It is instructive to first consider the relationship between motivation and cognition more abstractly. One possibility is that motivation has effects that take place independently of cognition (figure 6.8A). A second is that motivation modulates behavior by engaging the same set of functions that are used by cognition, in which case, the impact of motivation on behavior could be described as "mediated by cognition" (figure 6.8B). This mediation could be partial only, such that both direct (motivation-to-behavior) and indirect (motivation-via-cognition-to-behavior) effects take place. A third possibility is that cognition and motivation are more intertwined, such that they *jointly* guide behavior (figure 6.8C), in which case, although certain processes could

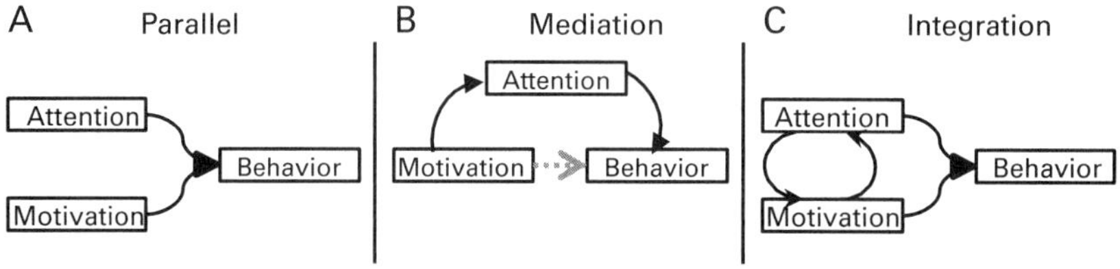

Figure 6.8
Three models of the relationships between attention and motivation. (A) In the parallel model, attention and motivation have independent effects on behavior. (B) In the mediation model, the influence of motivation on behavior is mediated via attentional systems. (C) In the integration model, attentional and motivational systems interact so strongly they cannot be decomposed. Adapted with permission from Pessoa and Engelmann 2010.

be described as "cognitive" and others as "motivational," the interactions between them are sufficiently high that their strict separation is more semantic than real. See Chelazzi et al. (2013) for related discussion.

Methodologically, disentangling the contributions of cognition and motivation to neural signals is far from easy (Maunsell 2004), especially when experiments involve goal-directed task manipulations. For example, in human studies, subjects may be instructed that a potential reward will result following a cue stimulus if their performance is both fast and accurate. In such cases, increased brain signals may reflect enhanced attention since subjects are more likely to engage attention when a reward is at stake. But whether the increased signals *actually* reflect greater attention is another matter, an issue forcefully described by John Maunsell (2004, 262–263) in the context of monkey physiology studies of attention:

> When the effects of spatial attention are examined, subjects are motivated to direct attention to one location or another only by expectations about which location is more likely to be associated with a reward. . . . Such reward manipulations reliably lead to shifts in attention. . . . However, these experiments typically provide no basis for assigning changes preferentially to attention or to expectations about reward. In most cases, attention-related modulation could equally well be described in terms of expectation about rewards because the two are inextricably confounded.

Referring to figure 6.8, the situation described by Maunsell could be portrayed in terms of the mediation model (B). In other words, *mechanistically*, effects of motivation are obtained via "attentional circuits." Whereas this relationship would presumably indicate that such motivational effects are less interesting, chapter 7 will argue that they are not. This is because *how* motivation recruits "cognitive" circuits is as important as which circuits it recruits. Thus one needs to characterize the *interactions* even in mediation-type situations (see below).

Conclusions

Earlier motivation and attention studies focused on investigating changes in the response criterion and whether subjects employed an optimal decision criterion (for further discussion, see Navalpakkam, Koch, and Perona 2009). The "decisional" effects they reported are qualitatively different from effects reported by more recent studies, however; for example, effects such as increased visual sensitivity (d-prime) in attention tasks and target/distractor effects. Indeed, recent findings fly in the face of traditional psychological models, which describe motivation as involving a global activation that varies independently of control demands and behavior direction (e.g., Duffy 1962; Hull 1943). As we have seen, the effects of motivation discussed in this chapter, far from being global, reflect selective mechanisms that are manifested both behaviorally and neurally.

Here it is worth revisiting the parallel, mediation, and interaction models diagrammed in figure 6.8. While acknowledging the challenge of correctly interpreting brain signals in studies that employ reward (Maunsell 2004), I suggest that the major issue is actually conceptual. By using separate boxes for "attention" and "motivation," the models of figure 6.8 describe motivation in counterproductive ways. As developed elsewhere in the case of emotion and cognition (Pessoa 2008), I propose that it is also counterproductive to carve the brain into "attention" (or "cognition") and "motivation." Chapter 7 (see also chapter 8) will outline how motivational signals are *embedded* into cognition (and perception) through multiple mechanisms of communication, and how, if one accepts this view, the "inextricably confounded" relationship reported by Maunsell (2004) ceases to be a problem and can be seen as a *property* of brain organization.

Notes

1. "Activation" is here employed in the sense of an "energizing" function, one that is linked to, for instance, the vigor and frequency of behavioral output. "Activation" is a better term than "arousal," which has several connotations that are associated with reticular systems involved in sleep-wake cycles (Robbins and Everitt 2007).

2. The sensory versus decisional distinction of the question is, at any rate, artificial since these processes are closely related (Churchland, Ramachandran, and Sejnowski 1994; see also Thielscher and Pessoa 2007).

3. "Extrastriate visual cortex" typically refers to early visual cortex beyond primary visual cortex.

4. In contrast, Hübner and Schlösser (2010) did not observe a reduction of response interference (during a flanker task) when reward was provided.

7 Dual Competition Model

Previous chapters presented evidence for the interactions between emotion and cognition, on the one hand, and between motivation and cognition, on the other. How do these interactions determine the flow of information processing in the brain? Several proposals have been advanced, focusing either on perceptual or on cognitive mechanisms (Duncan and Barrett 2007; Lewis 2005; Pessoa 2008; Phelps 2006). The present chapter refines and extends a proposed alternative framework in which both emotional and motivational signals are integrated with perception and cognition so as to effectively incorporate *value* into the unfolding of behavior (Pessoa 2009). To reflect the central idea that *both* emotion and motivation influence competition at *both* the perceptual and the executive level, I have named this framework the "*dual competition* model" (see also Pessoa and Engelmann 2010; figure 7.1). At the perceptual level, this chapter will discuss the case of vision, which illustrates many aspects of the model that would apply to other sensory modalities as well (albeit with different specifics, of course).

Perceptual and Executive Competition

Objects compete for limited perceptual processing capacity and control of behavior (Desimone and Duncan 1995; Pashler 1998). Because processing capacity for vision is limited (Tsotsos 1990), selective attention to one part of the visual field comes at the expense of neglecting other parts. Thus a popular notion is that there is *competition* for neural resources in visual cortex (Grossberg 1980). To understand the flow of information processing more generally, we need to go beyond perceptual competition and explicitly incorporate the role of executive control. Although, historically, it has been difficult to specify the composition of "the executive," operations involved in maintaining and updating information, monitoring conflict and errors, resisting distracting information, inhibiting prepotent responses, and shifting mental sets are all

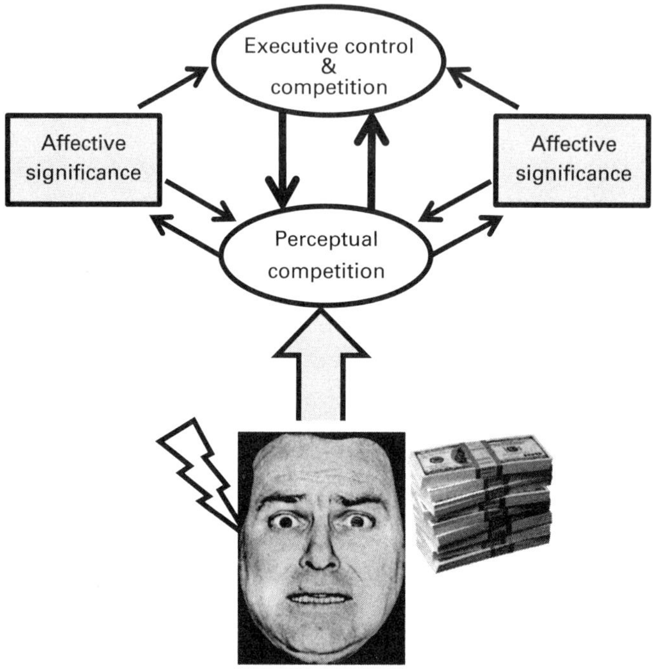

Figure 7.1
Dual competition model. Affective significance impacts the flow of information processing based on emotional and motivational factors, for both of which competition occurs at the perceptual and executive levels. Arrows denote functional pathways that do not necessarily map to specific anatomical connections. Individual differences in state or trait anxiety and sensitivity to reward are predicted to modulate the impact of affective significance on information processing.

thought to be important. Behavioral research indicates that executive control is not unitary and that different functions may have their own limited processing capacities, or *resources* (Kahneman 1973; Norman and Bobrow 1975). The standard strategy to study this question is by devising *dual-task interference* tasks. When performing function *X* does not interfere with function *Y* and vice versa—a "double dissociation"—it suggests that their resources do not overlap and that their resource pool is not unitary (Kahneman 1973).[1] Neuropsychological research also supports the dissociation of cognitive operations, consistent with the *fractionation* of the central executive (Norman and Shallice 1986; Stuss and Knight 2002). Here, again, findings of double dissociations are used to argue for the separation of the processes in question (for an extended discussion, see Shallice 1988; Shallice and Cooper 2011).

Yet ample evidence also suggests at least some *unity* of executive functions—certain mechanisms are shared across functions (Duncan et al. 1996; Miyake et al. 2000). For example, John Duncan and Adrian Owen (2000) have proposed that the processes underlying fluid intelligence are shared in key ways and overlap considerably in both lateral and dorsal-medial prefrontal cortex, as well as in parietal cortex. Capacity sharing has implications for human information processing because it implies *executive competition*: subcomponents of executive control are mutually interacting, such that multiple functions cannot be independently executed simultaneously. For example, handling conflict may make it especially challenging to deal with inhibition at the same time. This competition can be cast in terms of resources. Accordingly, even though some executive processes rely on partly independent mechanisms, they share a *common pool* of resources. Therefore, when a given function is needed, resources devoted to one operation will not be available for other operations, and behavioral interference will ensue. More generally, the question of the unity or fragmentation of executive control relates to the notion of cognitive architecture and modularity, as will be discussed in chapter 8.

In summary, to characterize the flow of information processing, it is useful to consider competition at two broad levels: perceptual and executive. This is not to imply, however, a strict separation between "peripheral" and "central" processes, which are strongly interlinked, and which for descriptive purposes are treated separately below.

Interactions between Emotion and Cognition

Perceptual Competition

Perceptual competition is affected by emotional content. As discussed in previous chapters, such influence has been extensively studied by using both behavioral and cognitive neuroscience techniques to illustrate how emotion-laden items are highlighted in perception (Pourtois, Schettino, and Vuilleumier 2012). What are the mechanisms by which affective significance influences perceptual processing and hence competition? Here the focus will be on vision, the perceptual system we know best of all; some of the ideas about affective visual processing introduced in chapter 2 will be elaborated here.

The first mechanism involves the amygdala (figure 7.2A; plate 7). Areas in anterior aspects of temporal cortex (called "TEO" and "TE") are the only visual cortical areas to project to the amygdala (their projections terminate mainly in its lateral nucleus; Freese and Amaral 2005; Amaral, Behniea, and Kelly 2003), whereas output connections from the amygdala, which are arranged in a strikingly different manner, reach nearly all of visual cortex. This highly asymmetric connectivity with visual cortex has led researchers to suggest that the amygdala exercises substantial modulatory control over

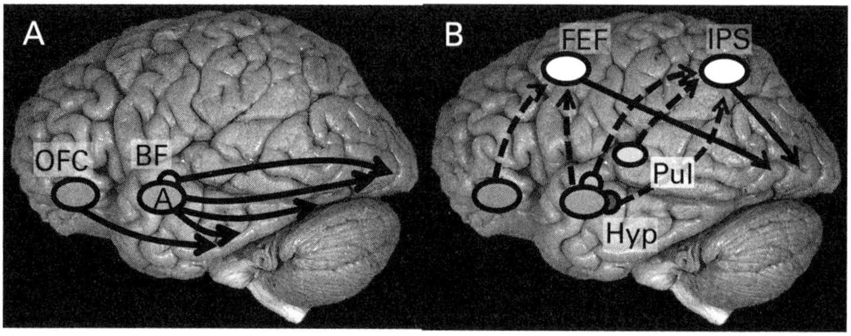

Figure 7.2 (plate 7)
Interactions between emotion and perception in the brain. (A) Visual processing is modulated by affective value via several mechanisms, including those involving projections from the amygdala, basal forebrain, orbitofrontal cortex, and possibly hypothalamus (not shown). (B) Other modulatory mechanisms rely on interactions between evaluative sites (colored ellipses and circles) and attentional control sites (white ellipses). Located in fronto-parietal cortex, the latter sites are known to exert top-down influences on visual processing. Dotted lines indicate possible indirect connections. All locations are approximate, and the positions of the amygdala, basal forebrain, hypothalamus, and pulvinar are shown on the lateral surface for schematic purposes. Labels from panel A were omitted to reduce clutter. A, amygdala; BF, basal forebrain; FEF, frontal eye field; Hyp, hypothalamus; IPS, intraparietal sulcus; OFC, orbitofrontal cortex; Pul, pulvinar. Reproduced with permission from Pessoa 2010a.

sensory processing along ventral visual cortex (e.g., Morris et al. 1998; Amaral, Behniea, and Kelly 2003; Vuilleumier 2005; Pessoa et al. 2002; Lang and Davis 2006).

Projections from the amygdala to visual cortex terminate preferentially in cortical layers I–II and V–VI (Freese and Amaral 2005), and avoid layer IV. This pattern is representative of "feedback" connections (e.g., from area V2 to area V1), in contrast to "feedforward" connections (e.g., from area V1 to area V2), which tend to mostly innervate layer IV. Because feedback connections are, in many cases, unable to drive neuronal activity (Girard and Bullier 1989), but have the ability to increase or decrease neural responses (Hupe et al. 2001), a common inference is that they exert a *modulatory* role. Thus the amygdala, through its connections to visual cortex, may well modulate information processing based on affective significance, although little is known about these connections. More generally, "modulatory" feedback connections play critical roles in visual processing. For example, at times, not only are they as fast as feedforward connections (Bullier 2001), but they also exert a major influence in determining neural response properties (Angelucci and Bressloff 2006).

A recent study investigating neuroanatomical pathways across the brain (Markov et al. 2011) has shed some light on the relative strengths of amygdalar connections to visual cortex. The authors found these strengths spanned several orders of magnitude; connections to visual areas V1, V2, and V4 were comparable in strength and, albeit relatively weak, were surprisingly strong for distant projections. But however often the amygdala's modulation of visual processing is emphasized in the emotion literature, five other modulatory mechanisms merit our consideration as well (see also Barrett and Bliss-Moreau 2009).

A second mechanism through which emotion may affect perception involves other valuation regions, most notably orbitofrontal cortex (Barrett and Bar 2009; figure 7.2A; plate 7) and possibly the insula (Markov et al. 2011). Orbitofrontal cortex is important for the evaluation of sensory stimuli (Zald and Rauch 2007) and is reciprocally interconnected with visual cortex, especially the more anterior portions of the ventral stream (Barbas 1995; Cavada et al. 2000; Rempel-Clower and Barbas 2000; Saleem, Kondo, and Price 2008). This region is thus capable of influencing responses in visual cortex based on affective value. Recently reported projections from the insula to visual area V4 (Markov et al. 2011) may also have a similar function, although these may be relatively weak.[2]

A third mechanism involves the basal forebrain (figure 7.2A; plate 7). As discussed in chapter 2, basal forebrain terminals influence visual processing through the release of acetylcholine. For example, cholinergic mechanisms affect the competition between attended and unattended stimuli (Furey, Pietrini, and Haxby 2000; Furey et al. 2008). In terms of emotional processing, it is noteworthy that the central nucleus of the amygdala has significant projections to several basal forebrain structures (Holland and Gallagher 2006; Holland, Han, and Gallagher 2000). Additional projections to these structures originate in orbitofrontal, insular, and cingulate cortex (Zaborszky et al. 1999). Thus several regions that participate in the evaluation of incoming inputs project to the basal forebrain, which is then able to modify information processing in visual cortex.

The basal forebrain is one of the largest neuromodulatory systems in the mammalian brain (Semba 2000). It influences not only sensory processing but also many other systems, evidence that the strict separation between perceptual and executive competition effects quickly breaks down, as is a fourth mechanism by which perceptual processing is biased by affective significance, one that involves the fronto-parietal "attentional network" (figure 7.2B; plate 7). This fourth mechanism is featured here, in the context of perceptual competition, because fronto-parietal regions provide top-down signals capable of biasing visual responses (Corbetta and Shulman 2002; Kastner and Ungerleider 2000), but it will also feature prominently in the discussion of executive competition below. Regions in lateral frontal cortex and parietal cortex are suggested to modulate visual processing according to an item's affective significance. In particular, both the frontal eye field and parietal cortex contain a *priority map*, namely, a representation of spatial locations that are salient (e.g., high-contrast stimuli) or relevant (e.g., stimuli connected to current goals; Fecteau and Munoz 2006; Serences and Yantis 2006). To embed affective significance into priority maps, fronto-parietal regions are suggested to work closely with evaluative sites, such as the hypothalamus, amygdala, orbitofrontal cortex, and anterior insula, to prioritize processing based on the emotional value of a sensory stimulus (figure 7.2B; plate 7; see "Priority Maps, Emotion, Motivation" below).

Evidence for interactions involving the fronto-parietal network was provided in a study by Aprajita Mohanty and colleagues (2009). In a variant of an endogenous cueing task, a central cue indicated both the expected location of an upcoming target face stimulus and its valence (negative or neutral). Responses in fronto-parietal regions were modulated by spatial *and* emotional cue information. When the cue signaled an emotional target, Mohanty and colleagues observed increased functional connectivity between these areas and the amygdala, which, they suggested, facilitated detection of the affectively significant item.

Still other evidence regarding fronto-parietal interactions with visual cortex comes from the attentional blink study reviewed in chapter 2 (Lim, Padmala, and Pessoa 2009). To elucidate how emotional content counters the blink, my colleagues and I evaluated both direct and indirect (via prefrontal cortex) potential contributions from amygdala to visual cortex by performing statistical path analysis and structural equation modeling. Given that areas in lateral prefrontal cortex have been suggested to be critical "bottleneck" regions during the attentional blink (Marois and Ivanoff 2005; see also Tombu et al. 2011), we were interested in evaluating whether the impact of the amygdala on visual cortex was mediated by prefrontal cortex. The results of our analysis were consistent with both amygdala and prefrontal cortex affecting visual cortical responses. Both influences supported the behavioral advantage of affectively significant items during the attentional blink task. Even though these results cannot be interpreted causally, they indicate that interactions between fronto-parietal and valuation regions provide a mechanism by which emotion participates in perceptual competition—in this case, a temporal competition mechanism as probed in the attentional blink paradigm.

The interactions between amygdala and components of the fronto-parietal network may rely on indirect anatomical connections since direct ones are sparse. But, as will be discussed in chapter 9, amygdala signals are capable of being rapidly broadcast across the entirety of prefrontal cortex, for instance. Thus a strong *functional* relationship does not have to depend on robust direct connections (chapter 8). Posterior cingulate cortex, which is both extensively interconnected with the amygdala and connected with lateral frontal areas, may also bridge the two systems (Mohanty et al. 2009).

A fifth mechanism involves the pulvinar complex of the thalamus (figure 7.2B; plate 7). As proposed in chapter 3, the importance of the pulvinar for affective processing is due not to its putative role as part of a subcortical pathway but, instead, to its connectivity with other cortical regions (see figure 3.1B). Thus the role of the pulvinar extends beyond standard attentional functions (see also Shipp 2004) to include mechanisms sensitive to affective significance. This proposal is consistent with findings by Robert Ward and colleagues showing the impact of pulvinar lesions on the processing of emotion-laden visual stimuli in humans (Ward et al. 2007; Ward, Danziger, and Bamford 2005), as well as with the findings of our attentional blink study (Padmala, Lim, and Pessoa 2010). Having observed a significant relationship between pulvinar responses and the ability to detect emotion-laden targets on a trial-by-trial basis (figure 7.3), we proposed that the pulvinar amplifies responses to stimuli of potential value to the organism during challenging sensory conditions (see figure 3.7B).

Although I have included the contribution of the pulvinar in affective processing under the umbrella of perceptual competition, I do not mean to suggest that this fifth mechanism involves direct connections between the pulvinar and visual cortex. Indeed, even though the inferior and lateral pulvinar connect

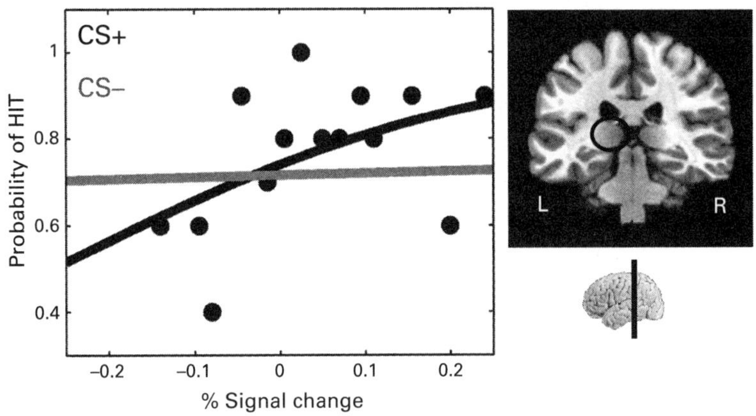

Figure 7.3
Pulvinar responses and behavioral performance during the attentional blink. A robust relationship was observed during the affectively significant condition (CS+) but not during the neutral condition (CS–). The inset shows the location of the left pulvinar. Reproduced with permission from Padmala, Lim, and Pessoa 2010.

extensively with visual cortex, the mechanism outlined here involves the *medial* pulvinar, which is densely interconnected with frontal, parietal, cingulate, and insular regions—as well as with the amygdala (see figure 3.6). Thus the pulvinar influence on visual processing is envisaged to depend on pulvinar-cortical circuits.

A sixth potential mechanism was recently described by Basilis Zikopoulos and Helen Barbas (2012), who report a pathway from the amygdala to the reticular nucleus of the thalamus and suggest that the connection is important for the capture of attention by emotion-laden stimuli. Notably, pathways from amygdala and orbitofrontal cortex overlap in the reticular nucleus. Their proposal was inspired by the idea that the reticular nucleus has key attentional functions (Crick 1984), an idea that has gained recent empirical support (McAlonan, Cavanaugh, and Wurtz 2008).

In summary, during the past decade, researchers have described a central modulatory role for a projection system emanating from the amygdala that reaches nearly all levels of the ventral visual system. Although this system is sometimes highlighted as the sole modulatory mechanism for visual processing, I propose that at least five other mechanisms need to be investigated as well. These mechanisms, which include both cortical and subcortical structures, involve complex network interactions that sculpt how visual signals evolve in response to the behavioral and affective significance of sensory stimuli.

Executive Control and Competition

"Perceptual competition" refers to processes that shape the ways visual items compete with one another. In essence, it speaks to how stimulus features or objects are selected so that they will affect behavior. Here we turn to *executive competition*, which deals less with sensory aspects of competition and more with the ways functions such as *inhibition*, *shifting*, and *updating* are influenced when emotional information is encountered.

Emotional content influences executive control, first, so that strengthened sensory representations will receive prioritized attention (figure 7.1; see chapter 4), and, second, so that affective information will be conveyed to structures involved in control (figure 7.1). But, because emotion can either enhance or impair cognitive performance, to see how emotional content impacts executive control, we must consider at least two key factors: the strength or *arousal* of the stimulus (or manipulation) and *task relevance* (see also Mather and Sutherland 2011).

When arousal is "low" and affective significance is task *irrelevant*, interference with the main task may be observed and the behavioral effect will be typically small. When, however, arousal is "high" and the stimulus/manipulation is task irrelevant, resources are more fully *diverted* toward the processing of the emotional item and, because the mobilization of resources is more pronounced, the effects on behavior are greater (Lang, Davis, and Öhman 2000; Panksepp 1998). For example, in our investigation of cognitive-emotional interactions, Jong Moon Choi, Srikanth Padmala, and I observed that response conflict increased on trials with the possibility of shock, suggesting that the impact of emotion on behavior comes in part from the more vigorous recruitment of attentional/effortful control required to prioritize the processing of high-arousal items (Choi, Padmala, and Pessoa 2012; see chapter 5). Attentional/effortful control involves resources that are shared across executive functions and, because situations associated with high levels of arousal are expected to recruit some of these resources (see also Bishop 2007; Eysenck et al. 2007; Mathews and Mackinstosh 1998), interference with other executive functions will ensue (figure 7.4A). The impact of emotion on performance thus occurs because of limited processing capacity and competition for "common-pool resources." For example, performance during a response inhibition task that required subjects to withhold responses to "stop" stimuli was compromised when they viewed high-arousal pictures (Verbruggen and De Houwer 2007).

What about the situation when the emotional stimulus is task *relevant*? Two outcomes are possible. If the affective intensity is "low," task performance might improve because control will be mobilized in the service of handling the task at hand, and the executive functions needed for task completion will more effectively compete for resources. In all, task performance will be enhanced. If, however, the affective intensity is sufficiently high, task performance might be compromised. Thus, in a study of response inhibition, my colleagues and I employed emotional stimuli as stop signals (Pessoa et al. 2012). Subjects had to perform a simple discrimination task but to withhold responding when they saw the stop signal. We found that, when we used both fearful and happy faces as low-arousal stop signals, response inhibition was enhanced relative to neutral faces, but when we employed high-arousal emotional stimuli (previously paired with mild shock) as stop signals, response inhibition was impaired relative to neutral stimuli. Thus inhibition performance was degraded even though emotional content was task *relevant*. We conjectured that processing the emotional stimulus consumed resources needed for inhibition.

Processing Resources Although the concept of resources so often invoked in accounts of the limits of human information processing has been criticized in the past (e.g., Navon 1984; Logan 1988; Neisser 1976) and has not been mechanistically specified, further insight into it can be gained by examining

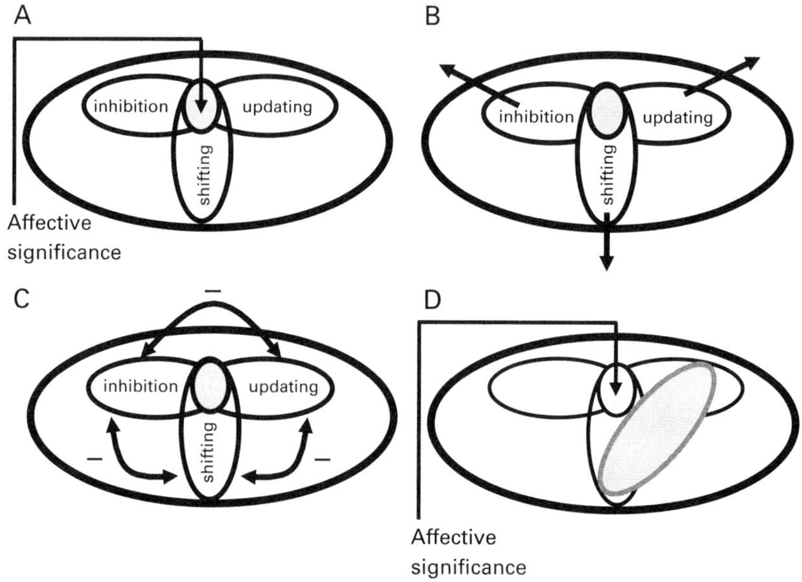

Figure 7.4
Executive control, competition, and processing resources. (A-C) Processes are proposed to share resources called "common-pool resources" (smaller ellipses in gray), such that the engagement of one will detract from the processing of the other. Common-pool resources are necessary for general functions of attentional/effortful control. (A) High-arousal emotional stimuli recruit common-pool resources that allow their processing to be prioritized, thus detracting from other mechanisms sharing those resources. (B) These stimuli also trigger executive functions, such as updating, shifting, and inhibition, to handle the challenges to the organism, as indicated by the arrows emanating from attentional/effortful control. (C) Competition for resources during cognitive and emotional manipulations can, at times, produce push-pull-like interactions. (D) Certain executive control mechanisms may be less affected by emotional manipulations. The three unfilled ellipses now represent unspecified executive mechanisms. Two of these (overlaid by the gray ellipse) share resources that are also consumed by threat processing; the remaining one does not. Panel A reproduced with permission from Pessoa 2009.

brain regions sensitive to changes in the distribution of resources. Researchers have identified a set of regions consistently involved in attentionally demanding tasks; referred to as the "attentional network," the set includes sites in lateral and dorsal-medial prefrontal cortex, anterior insula, and parietal cortex (Corbetta and Shulman 2002; Kastner and Ungerleider 2000).[3] These regions are suggested to generate and maintain endogenous signals based on current goals and task-relevant contingencies, and to send top-down signals that bias the processing of appropriate stimulus features and locations in sensory cortex, among other functions (Corbetta, Patel, and Shulman 2008).

To investigate the neural bases of capacity limits more directly, researchers have probed attentional *bottlenecks* observed during tasks such as the attentional blink (discussed in chapters 2 and 4) and the phenomenon known as the "psychological refractory period." When subjects perform two speeded sensorimotor tasks separated by a variable delay, the refractory period refers to the increase in reaction time to the second stimulus as the interval between the two tasks decreases. The effect is thought to result from a central attentional bottleneck responsible for operations such as response selection (Pashler 1994). Indeed, René Marois, Michael Tombu, and colleagues have proposed the existence of a "unified" attentional bottleneck that involves several regions of the fronto-parietal attentional network (Tombu et al. 2011).[4]

If robust emotional manipulations indeed consume processing resources, then they should engage sites implicated as "bottleneck areas." As described in chapter 5, an informal meta-analysis of activation peaks in aversive conditioning functional MRI studies revealed sites throughout lateral and medial prefrontal cortex (figure 7.5a,b; plate 8), in addition to the anterior insula (Pessoa 2009). Formal meta-analyses also have implicated these regions during aversive conditioning (Etkin and Wager 2007; Mechias, Etkin, and Kalisch 2010). Thus attentional bottleneck regions are consistently recruited during emotion processing. If this recruitment prevents them from being adequately engaged when neutral task-related processing is required, we should expect to see behavioral impairments.

To further explore these ideas, I reanalyzed data from a previous study (Pessoa 2009). In the original experiment (Lim, Padmala, and Pessoa 2008), subjects performed a visual search task under low and high attentional demands (figure 7.5c; plate 8), which were contrasted to determine brain sites sensitive to the availability of resources. Differential responses were observed in fronto-parietal regions overlapping with the attentional network. In the same study, subjects were also shown task-irrelevant threat (previously paired with shock) and safe faces (figure 7.5c; plate 8). Notably, increased responses to threat versus safe faces were observed in several of the same fronto-parietal regions. To test the hypothesis that additional processing resources were in fact recruited during the viewing of threatening stimuli, responses in regions modulated by attentional demand were correlated with behavioral accuracy during the task. The reasoning was that, if threat increased processing demands, task performance on the search task should *decline*. As illustrated for the case of the region in dorsal-medial prefrontal cortex (figure 7.5d; plate 8), the stronger the response during the threat condition, the worse the behavioral performance. A similar pattern of results was observed in regions in lateral prefrontal cortex and anterior insula. In addition, consistent with their increased processing, shock-paired stimuli had a stronger impact on behavior and associated brain

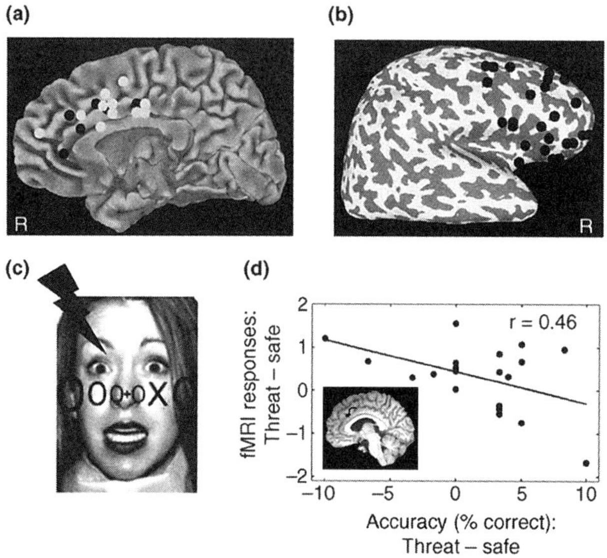

Figure 7.5 (plate 8)
Processing resources and threat processing. Summary of results from thirty-four PET and functional MRI studies of conditioning from 1995 to 2008, illustrating the coordinates provided for the contrast of threat (CS+) versus safe (CS–) conditions. (a) Peak activation sites observed in medial prefrontal cortex are shown for the left (green) and right (red) hemispheres (coordinates were projected onto a midline view for display purposes). (b) Peak activation sites observed for the right lateral surface are shown on an "inflated" surface, including the middle frontal gyrus, inferior frontal gyrus, and anterior insula (note that the surface inflation "pushed up" some of the activation sites relative to their standard anatomical positions). (c) Subjects viewed an array of letters superimposed on task-irrelevant faces and were asked to report whether the target letter "X" was present. In the low-demand condition shown here, the target appeared among a uniform array of distractors (pop-out condition). In the high-demand condition (not shown), a nonuniform array of letters was employed (search condition). In the threat condition, faces were previously paired with mild electrical shock. (d) Differential responses to task-irrelevant faces (threat–safe) were inversely related to behavioral performance, suggesting that the processing of threat captured processing resources needed for task execution; and the greater the differential response magnitude, the worse the performance. Results are shown for a region of interest in dorsal-medial prefrontal cortex that was defined in terms of a separate contrast of high versus low attentional demand (shown in the inset). Data reanalyzed from Lim, Padmala, and Pessoa 2008. Reproduced with permission from Pessoa 2009.

signals than unpaired stimuli during a subsequent implicit-memory task, where subjects were simply asked whether faces (half of which they had already seen, and half not) were neutral or fearful (see Lim, Padmala, and Pessoa 2008). Combined, these findings are consistent with the inference that the consumption of resources engaged by task-irrelevant threat faces (as indicated by responses in dorsal-medial prefrontal cortex, for instance) impaired performance on the main task. The results provide some support to the proposal that emotional processing shares resources with executive function. In particular, in the study described here, a simple type of sharing may have been involved, namely, threat faces captured attention and compromised task performance (see Bishop et al. 2004).

Triggering Additional Functions

Supporting the idea that shared resources are engaged by emotion are findings that emotion interferes with a wide range of cognitive operations. Because executive functions share common mechanisms, I suggest that emotion also acts on this common pool. A distinct type of impact is due to the influence on *specific* resources. As reviewed in the beginning of the chapter, neuropsychological research reveals dissociations between executive functions, which support the existence of resources that are particular to a given executive function (Shallice 1988; Shallice and Cooper 2011).

Dealing with an emotional stimulus requires the types of behavioral adjustments that characterize executive functions. For example, to refresh the contents of working memory, to switch the current task set, and to cancel previously planned actions might require *updating*, *shifting*, and *inhibition*, respectively. Such adjustments recruit specific resources in the service of emotional processing (figure 7.4B) and, if these resources are temporarily unavailable for the task at hand, behavioral performance will be compromised—the more so, the stronger the emotional manipulation (see below). An example may help to illustrate. Suppose a subject is performing a cognitive task and a change in background color signals that the subject will receive a shock sometime in the next 30 seconds. The subject might *update* the contents of working memory to include the "shock possible" information, *shift* between the execution of the cognitive task and "monitoring for shock" every few seconds, and, if another cue indicated that the shock would be delivered in the next second, *inhibit* a response to the cognitive task to prepare for the shock. In other words, dealing with the emotional situation necessitates the same types of executive functions that are considered to be the hallmark of cognition.

Cognitive-Emotional Interactions versus Push-Pull

The framework developed thus far suggests that brain regions important for executive control are actively engaged by emotion. In contrast, push-pull studies (e.g., Dolcos and McCarthy 2006; Anticevic, Repovs, and Barch 2010) have demonstrated *reduced* signals in some of these regions when emotional stimuli are shown. Hence the two frameworks appear to make opposite predictions. The findings of Alan Anticevic, Grega Repovs, and Deanna Barch (2010; see also figure 5.2) provide a potential clue as to when we might expect push-pull interactions. Whereas, relative to neutral, negative distractors decreased responses in dorsal-lateral prefrontal cortex during the delay period of the working memory task, task-related distractors (stimuli similar to items to be remembered) actually *increased* responses, in much the way increases in working memory demand would. It thus appears that the two types of distractor, negative and task-related, had distinct impacts on working memory maintenance. What explains this difference?

Regarding the effect of the negative distractors, Anticevic, Repovs, and Barch (2010, 168–169) put it this way: "Trials in which a subgoal was processed (resolving emotion interference) led to loss or neglect of the primary task goal (memory set maintenance)." In other words, dealing with the negative stimuli during the delay period produced a momentary "neglect" of the memory maintenance. In contrast, because neutral task-related distractors were so similar to the to-be-remembered items, participants may in effect have also held them in memory so as to avoid matching the final probe stimulus to a distractor. Consequently the distractors may actually have *increased* the demands of the working memory task. I therefore suggest that cognitive-emotional push-pull interactions are related to a type of competition that directs processing *away* from the concurrently executed main task, thereby producing *decreased* activation (in relative terms) in some of the key frontal regions underlying the task at hand (figure 7.4C).[5] Which is to say, deactivations are the result of competitive interactions between resources required for executive functions. As such, they should be understood not in terms of a mutually suppressive relationship between emotion and cognition, but in terms of *executive competition*.

In broader terms, to predict whether emotion will entail increased or decreased responses in brain regions important for executive control, we need to consider, first, behavior and, second, the pattern of responses when *neutral* valence items are processed. In terms of the working memory studies considered here, first, negative distractors impaired performance relative to neutral ones. Second, we know from the literature that activation strength in dorsal-lateral prefrontal cortex is closely associated with maintenance performance; stronger signals correlate with successful working memory (Funahashi, Bruce, and Goldman-Rakic 1989; Fuster 1973; Pessoa, Gutierrez, et al. 2002). Thus we can interpret decreased signals during working memory maintenance as probably associated with the behavioral interference seen with negative distractors. Now let us consider the response-conflict task discussed in chapter 5 and above (Choi, Padmala, and Pessoa 2012; see figure 5.7). First, response conflict *increased* during threat versus safe trials. Second, in the *safe* condition, incongruent trials evoked *larger* responses than control trials (on which there was neither interference nor facilitation in the main task). Therefore, in the threat condition, the incongruent trials evoked *even larger* responses relative to control trials (see figure 5.7).

How Strong Are Laboratory Manipulations of Emotion?

Arousal was described above as admitting "low" and "high" values. Naturally, response strength can be viewed as a continuous variable. But what precisely constitutes "high" arousal? In the laboratory, for obvious reasons, researchers can employ only a limited range of manipulation intensities. Thus stronger manipulations rely on presenting images of mutilation, sounding loud bursts of noise, or applying mild electrical stimulation, for instance. But even though the use of stronger stimuli may result in task interference as discussed in chapter 5, a careful look at the literature also reveals that effects of such stimuli on task performance are often modest and at times may even go undetected (e.g., Qin et al. 2009; Robinson et al. 2011; Blair et al. 2007).[6] One must be careful, of course, not to overinterpret negative findings, but they clearly seem to indicate that cognitive performance is not so easily affected by emotional stimuli. A possible interpretation of negative findings is that cognition is more impervious to emotional interference than previously thought. Alternatively, the stimulus intensities that are feasible in a laboratory setting may simply not be strong enough to generate robust interference. An added complication is that subjects who volunteer for studies involving stronger emotional manipulations may be below average in terms of fearfulness and trait anxiety. In this regard, a fruitful strategy would be to use sample sizes sufficiently large to allow researchers to determine how individual differences in fear-related responses influence the behavioral data.

Additional Predictions of the Dual Competition Model

In "Executive Control and Competition," I illustrated the impact of emotional stimuli on cognitive performance by drawing on a recent response inhibition task that employed "low" and "high" arousal stimuli (Pessoa et al. 2012). The low-arousal stimulus improved inhibitory performance, whereas the high-arousal stimulus impaired it. But will low-arousal always improve performance and high-arousal always impair it? The answer is no. Every emotional stimulus has increased behavioral significance and, accordingly, must be handled in some fashion—its potential impact on the organism must be assessed. When stimulus intensity is high, behavioral adjustments will be needed, and "additional functions" such as updating, shifting, and inhibition will be engaged. Even when intensity is low, *some* kind of evaluation process will occur as long as processing capacity is not exhausted. But because it requires processing resources to evaluate and handle the incoming stimulus, when the *main task* is demanding and very little spare capacity is available, low-intensity stimuli may impair performance.[7] This could help explain why, in one study, negative items impaired both response inhibition and working memory performance even when the negative items were of low intensity (drawings of snakes and spiders; Lindstrom and Bohlin 2012).

In practice, this discussion shows how simple dichotomies such as low versus high arousal and low versus high task demand will always be insufficient to describe the nuanced way in which emotion and cognition interact. In effect, what is needed is a more quantitative or computational description of the underlying interactions, especially when multiple factors interact in complex ways: for example, when the outcome of behavior is a function of at once stimulus intensity, relevance, and task demands, among other factors. Unfortunately, we are nowhere near developing such a description.

The scheme illustrated in figure 7.4A suggests that threat has similar effects on distinct executive functions, although some functions may be affected more than others (see Eysenck et al. 2007). It is particularly noteworthy that studies have found that threat, including the risk of shock, has little or no effect on verbal working memory tasks (Lavric, Rippon, and Gray 2003; Shackman et al. 2006). These results are important because they indicate that at least some executive mechanisms may be less susceptible to high-intensity emotional manipulations owing to the limited extent of shared processes (figure 7.4D). Although some researchers (e.g., Baddeley 2003) have extensively studied how verbal working memory intersects with other executive functions, how it is affected by emotional manipulations deserves further investigation.

Thus far, the emphasis has been largely on how emotion affects cognition. The sharing of resources predicts, however, that cognition should also mitigate the influence of affective processing. This type of interaction was demonstrated by Noga Cohen and colleagues, who showed that interference due to emotional pictures was reduced during response-conflict processing (Cohen, Henik, and Mor 2011; Cohen, Henik, and Moyal 2012; see also Cohen and Henik 2012). Once more, to unravel the ultimate effects of emotion on behavior, we need to consider processes competing at the perceptual and executive levels and the multidimensional factors that influence them.

Neural Interactions

Cognitive-emotional interactions rely on the communication between "task networks" (e.g., the attentional network during attention tasks) and "valuation networks," which involve both subcortical regions, such as hypothalamus and amygdala, and cortical ones, such as orbitofrontal cortex, anterior insula, and

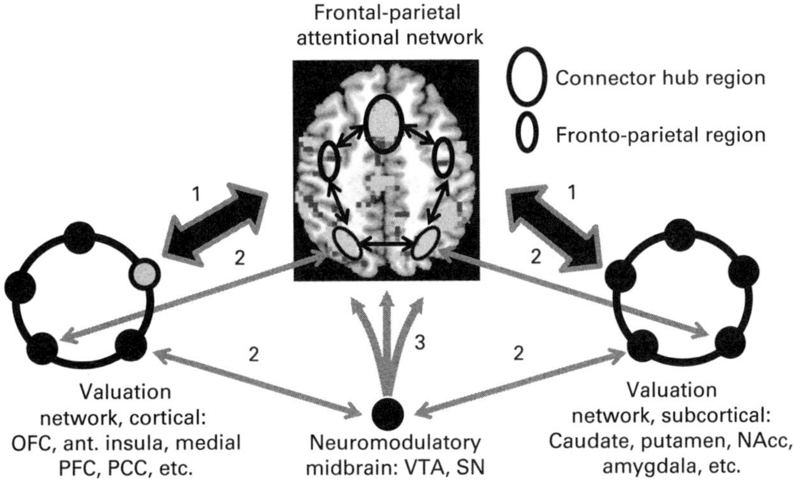

Figure 7.6
Modes of interaction between cognitive and motivation networks illustrated for attentional-motivational interactions. (1) Interactions rely on *hub* regions, such as those in dorsal-medial prefrontal cortex, which are part of both attentional and motivational networks (hub region in the slice and gray node in the cortical valuation network). (2) In addition, specific regions may link the two networks, either directly or via the thalamus. (3) Finally, motivational signals are further embedded within cognitive mechanisms through the action of diffuse neuromodulatory systems. ant., anterior; NAcc, nucleus accumbens; OFC, orbitofrontal cortex; PCC, posterior cingulate cortex; PFC, prefrontal cortex; SN, substantia nigra; VTA, ventral tegmental area. Reproduced with permission from Pessoa and Engelmann 2010.

medial prefrontal cortex. These interactions are suggested to take place via multiple *modes of communication*, as illustrated in figure 7.6.

The first mode involves *direct* pathways between task and valuation networks. One example is the pathway between orbitofrontal and lateral prefrontal cortex (Barbas and Pandya 1989). Orbitofrontal cortex is important for assessing the value of a stimulus, and lateral prefrontal cortex, as described earlier, is heavily engaged during cognitive tasks. Other examples are the pathways between the extensively interconnected lateral surface of the prefrontal frontal cortex (including dorsal-lateral PFC) and all cingulate regions (Morecraft and Tanji 2009). Thus direct pathways provide a substrate for cognitive-emotional interactions.

A second mode of communication relies on "*hub* regions" at the intersection of task and valuation networks. As will be further discussed in chapter 8, hubs are highly connected and central regions that play a key role in information communication between different parts of a network.

Dorsal-medial prefrontal cortex plays a prominent role as a hub region, or *common node*, of executive and emotional networks because of its participation in integrating inputs from diverse sources, notably cognitive and affective ones (e.g., Devinsky, Morrell, and Vogt 1995; figure 7.6).[8] This region may be involved in multiple executive functions, such as conflict detection, error likelihood processing, and error monitoring (Alexander and Brown 2011). It is also important for attentional processing more generally, including spatial attention (Pessoa and Ungerleider 2004c). As reviewed in chapter 5, dorsal-medial prefrontal cortex is also reliably engaged during conditions involving negative affect (see figure 5.6), as are all sectors of anterior-medial prefrontal cortex.

A second hub region, the anterior insula, is important for interoception, which involves monitoring the sensations needed for integrity of the state not just of the viscera (Craig 2002; Craig 2009) but of the entire body (Paulus and Stein 2006). Moreover, threat, uncertainty, and risk are all factors that engage the anterior insula (Singer, Critchley, and Preuschoff 2009), which is also reliably recruited by long-term memory, working memory, task switching, attention, and many other cognitive processes (Van Snellenberg and Wager 2010). Indeed, A. D. (Bud) Craig (2009, 65) observed that "no other region of the brain is activated in all of these tasks." And, in a recent analysis of the *functional diversity* of brain regions, the anterior insula emerged as one of the most diverse (Anderson, Kinnison, and Pessoa 2013; see also Uddin et al. 2013).

The anterior insula is thus robustly engaged by both cognitive and emotional tasks, and it provides an interface between the two, as exemplified by the findings of the response-conflict study reviewed in chapter 5 and above (Choi, Padmala, and Pessoa 2012), where, as we have seen, subjects were asked to indicate whether a picture was of a house or of some other building while ignoring task-irrelevant words (see figure 5.7). My colleagues and I observed interference-related (i.e., larger) responses in the anterior insula when subjects encountered an incongruent stimulus. And we also observed an interaction of the cognitive and emotional dimensions of the task in the anterior insula, namely, larger conflict responses when subjects encountered threat. This pattern of responses actually increased based on individual differences in trait anxiety, that is, high-anxious individuals exhibited increased interactions. This three-way statistical interaction (interference by emotion by trait anxiety) nicely illustrates the confluence of different signals in the anterior insula (see also Gu et al. 2012).

In summary, dorsal-medial prefrontal cortex and anterior insula are located at the intersection between emotion and executive function.[9] The importance of these regions in cognition-emotion interactions is captured by their proposed status as *hub* regions that play a prominent role in the *integration* of information by mixing signals that have distinct compositions (figure 7.6.) Consider, for instance, the representation of the entire body state that researchers have observed in the anterior insula. This representation involves signals from many parts of the body, including the viscera, signals reporting temperature, pain, itches, muscular sensations, touch, and other bodily sensations (Craig 2002). At the same time, the anterior insula is consistently engaged by executive functions. Thus it receives both interoceptive and executive signals. Of note, as will be further discussed in chapter 9, the anterior cingulate gyrus (including the midcingulate cortex as defined by Brent Vogt) probably has a more extensive descending projection system than any other cortical region (Vogt and Vogt 2009), with robust projections to autonomic regulatory structures. Now consider the dorsal-medial prefrontal cortex. Because of its roles in cognitive functions, this suggests that affective signals that mobilize the body are mixed with executive signals in the medial prefrontal cortex as well. Taken together, the anterior insula and dorsal-medial prefrontal cortex provide the substrate for ample cognitive-emotional integration that, in broad terms, includes both bodily "input" and "output" signals (see chapter 9 for further discussion). Moreover, these regions do not work in isolation. During cognitive-emotional interactions, they interact with lateral prefrontal cortex and parietal cortex (figure 7.6).

A third mode of communication depends on the diffuse action of neuromodulatory systems, including the action of dopamine and norepinephrine. Widespread modulatory connections originating from these systems reach large portions of the cortical surface and several subcortical areas, from which they are able to rapidly influence brain responses. Although neuromodulation can be viewed as another aspect of the network interactions described above, it is worth separating it from other network interactions because neuromodulatory interactions are often characterized as "regulatory" or "modulatory" functions. For example, neuromodulatory signals in isocortex (i.e., cortex containing six layers) target superficial (I–III) and deeper (V–VI) cortical layers, but tend to avoid layer IV (e.g., Raghanti et al. 2008), thus appearing to provide less of a "driving input" and more of a "*modulatory* input" (see note 2 to chapter 3) to ongoing information processing. Although neuromodulation will not be reviewed in detail here, its effects have been documented in stress studies (Arnsten 2009; Panksepp 1998). For example, exposing animals to acute stress leads to release of norepinephrine across a widely distributed brain network, including the prefrontal cortex (Arnsten and Li 2005; Joels et al. 2006). Acute stress also results in activation of the dopamine system, which has considerable effects on prefrontal cortex function and associated cognitive tasks (see below), particularly working memory (Arnsten 2009; Deutch and Roth 1990).

Interactions between Motivation and Cognition

The effects of motivation on information processing, like those of emotion, can be understood in terms of perceptual and executive competition (figure 7.1).

Perceptual Competition

Motivation influences perceptual competition in multiple ways. As reviewed in chapter 6, it increases visual sensitivity (d-prime) during both endogenous and exogenous attention tasks, enhances stimulus coding, and produces attentional capture effects during visual search, where it can either enhance or impair performance depending on whether the stimulus is task relevant or irrelevant. In many instances, the effects of motivation are comparable to those of emotion. For example, both reward-associated (Raymond and O'Brien 2009) and aversively conditioned (Lim, Padmala, and Pessoa 2009) stimuli counter the attentional blink. Increases in visual detection sensitivity have been reported for both reward manipulations (Baldassi and Simoncini 2011; Engelmann and Pessoa 2007) and aversive conditioning (Padmala and Pessoa 2008). And early effects on evoked responses, including the P1 ERP component, have also been observed for both motivation (Hickey, Chelazzi, and Theeuwes 2010) and emotion (Schupp et al. 2003b). Emotion and motivation thus have effects on visual perception that are qualitatively similar. Notably, both can be selective (e.g., they modulate a specific ERP response component, not the entire response), and both can be rapid.

Whereas some of the circuits of *affective attention* described above might be specific to emotion, several are likely to be common to both emotion and motivation, notably, the circuits between valuation networks and fronto-parietal regions important for attentional control. I propose that, even though the specific valuation regions recruited by emotion and motivation at any given time may differ, the interactions are similar. Which is to say, items of affective/motivational significance will redirect the flow of signals such that their processing is favored. I further propose that mechanisms involving the basal forebrain and the pulvinar also operate for both emotion and motivation (see figure 7.2; plate 7). More generally, despite the considerable differences between basal forebrain, pulvinar, and fronto-parietal mechanisms, each shapes visual perception by altering competition in visual cortex. Thus the idea is that their respective pathways may be engaged both during emotional and motivational conditions. It is intriguing to consider that, once they are engaged, the downstream effects on visual processing (and elsewhere) may be the same for both types of manipulation. In other words, once one of the pathways is engaged, the effect on perceptual competition is the same for emotion and motivation, and, from the perspective of visual cortex, it may be immaterial whether the effect resulted from "emotion" or "motivation."

Mechanisms Involving Associative Learning As reviewed in chapter 6, several studies have probed the effects of associative *value learning* during attentional tasks, on which reinforcement learning mechanisms may play a crucial role—possibly in ways related to those proposed by Yuka Sasaki, José Náñez, and Takeo Watanabe (2010) to take place during perceptual learning. The authors observed long-lasting improvements in the performance of perceptual tasks as a function of practice (e.g., practicing to detect a small misalignment of two line segments); this perceptual learning took place over a large number of trials, often requiring multiple experimental sessions. Although the previously reviewed value learning studies that included a training phase also had long experimental sessions, the number of their trials was relatively low compared to that in most perceptual learning studies. For instance, Brian Anderson, Patryk Laurent, and Steven Yantis (2011) employed 1,008 learning trials (or 240 in their shortest version) in their attentional capture study, and Jane Raymond and Jennifer O'Brien (2009) employed 600 trials in their attentional blink experiment, whereas trials for perceptual learning tasks typically number in the thousands (but for "fast" perceptual learning, see Poggio, Fahle, and Edelman 1992; Hawkey, Amitay, and Moore 2004).

That said, some perceptual learning reward studies have *not* included extensive learning phases, and at least one (Kristjansson, Sigurjonsdottir, and Driver 2010) reported initial effects of reward association in as few as six trials. In the study, effects were stable within fifteen trials of reversing color-reward contingencies. These effects would certainly necessitate mechanisms that operate at faster time scales than slower reinforcement learning processes. Interestingly, at least for complex objects, learning may be capable of increasing the salience of stimuli in anterior visual cortex relatively quickly (within 30–40 trials; Jagadeesh et al. 2001).

Although it is not known how reinforcement learning modifies sensory cortex in the types of studies discussed here, cholinergic action dependent on the basal forebrain is likely to play a prominent role. Such action influences signal-to-noise in sensory neurons and supports cortical plasticity during behavioral training and reorganization following sensory cortical lesions (Dykes 1997; Kilgard and Merzenich 1998; Weinberger 1995, 2003).

Mechanisms of Next-Trial Effects Another type of study discussed in chapter 6 investigated next-trial effects, where reward does not necessarily influence performance during the "current" trial, but instead improves performance during trials that follow correct performance coupled with reward *delivery*. In one such study involving somatosensory stimuli, Burkhard Pleger and colleagues (2009) revealed a causal role for dopamine in next-trial effects through pharmacological manipulations. Both the behavioral and neural impact of reward delivery was enhanced by a dopamine agonist and attenuated by a dopamine antagonist. More broadly, next-trial effects require processes that operate over several seconds, which would allow reward delivery to affect subsequent behavior. Notably, some dopaminergic mechanisms have a time course that can in fact extend for several seconds (Izhikevich 2007; Otmakhova and Lisman 1996). At least in the case of the primate visual system, however, because dopaminergic innervation of visual cortex is quite sparse, it is unlikely that monosynaptic effects are involved (Berger et al. 1988; Oades and Halliday 1987; see also Berger and Gaspar 1994). Next-trial effects on visual responses may thus depend on *source* regions in frontal and parietal cortex that exert top-down control on sensory processing. In other words, dopaminergic innervation of, say, frontal cortex would lead in turn to effects on visual cortex, much like the mechanisms involving the frontal eye field and visual cortex reported by Behrad Noudoost and Tirin Moore (2011a) and further discussed below.

Priority Maps, Emotion, and Motivation

Discussed in chapter 6, the concept of a salience map can be generalized to that of a *priority map* (Awh, Belopolsky, and Theeuwes 2012; Baluch and Itti 2011; Fecteau and Munoz 2006; Serences and Yantis 2006; Wolfe 1994). The version advanced here contains representations of spatial locations that are behaviorally important due to at least four types of contribution (figure 7.7): (1) physical salience, such as conferred by high contrast or motion; (2) endogenous relevance, such as that related to current goals; (3) affective significance, such as obtained by pairing an item with aversive stimuli; and (4) motivational significance, such as pairing an item with reward. Priority maps have been proposed to exist in the frontal eye field (Thompson and Bichot 2005), parietal cortex (Bisley and Goldberg 2010), and superior colliculus (Fecteau and Munoz 2006), among other regions (Shipp 2004). Mechanisms that embed affective and motivational significance into priority maps were discussed in the preceding sections. Thus, for example, emotion is integrated into priority maps through interactions between attentional sites in fronto-parietal regions and evaluative sites in regions such as the hypothalamus, amygdala, orbitofrontal cortex, and anterior insula important for determining affective significance (figure 7.2B; plate 7).

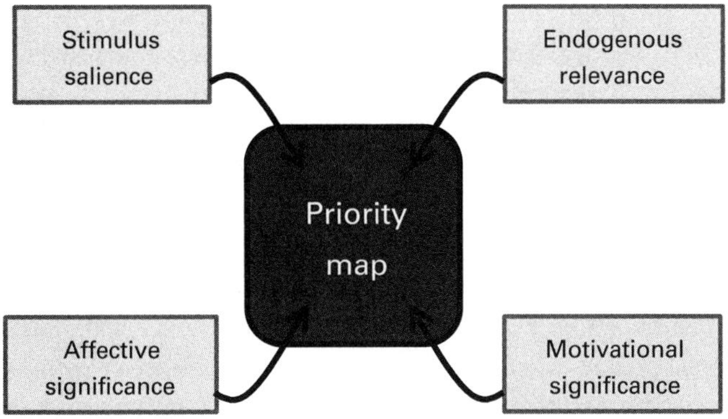

Figure 7.7
Priority map. At least four types of signals influence the determination of the locus of attention. In addition to the traditional components of stimulus-based and goal-based factors (e.g., stimulus salience and endogenous relevance), both affective and motivational factors (e.g., affective and motivational significance) need to be considered.

Executive Competition

Motivation influences executive competition, too. Chapter 6 described several examples studied during response-conflict, task switching, working memory, and long-term memory tasks. Two effects of motivation on executive functions are proposed here. In the first, motivation *sharpens* executive functions by enhancing them or by making them more efficient. A good illustration of this effect can be found in the working memory study by Okihide Hikosaka, Shunsuke Kobayashi, and colleagues described in chapter 6 (Kobayashi et al. 2002; see also Boehler et al. 2012). Sustained responses during the delay period had been shown previously to increase in trials involving reward (Watanabe 1996). In the study by Hikosaka and colleagues (Kobayashi et al. 2002), such enhanced activity during the reward condition actually reflected an increase in the *amount* of transmitted information regarding the item being maintained in memory. Thus reward amplified the "discriminability" of the working memory signal, leading to enhanced performance.

In the second effect, motivation *reallocates* resources available to executive functions, increasing the likelihood of reward attainment by improving performance. For example, in the study by Koji Jimura, Hannah Locke, and Todd Braver (2010), also described in chapter 6, brain responses appeared to reflect a shift toward a proactive control strategy that was beneficial to performance.

Absent a more precise description of resources, the two effects—sharpening and reallocation—could be mechanistically the same. One way to differentiate the two, however, is to investigate interference effects. Because of capacity sharing, the reallocation of resources should impact not only reward-relevant operations but also other operations that share the same processing resources (figure 7.8, dashed arrow). A resulting prediction is that motivation could influence executive functions in a way that actually impaired behavior. And, in our response-inhibition study (Padmala and Pessoa 2010), Srikanth Padmala and I found this to be so. We asked our subjects to make a simple visual discrimination ("go" trials) unless an auditory cue was emitted ("stop" trials). Subjects rewarded for accurate and fast performance on "go" trials exhibited impaired inhibitory performance, that is, they found it more difficult to withhold responding when encountering a "stop" signal. We reasoned that, to maximize reward, they may have allocated more resources to *go*-related processing, leaving fewer for *stop*-related processing. Motivation can thus be viewed as reallocating resources to prioritize implementation of the rewarded task component at the expense of unrewarded components (figure 7.8).

Neural Interactions

The same general architecture for cognitive-emotional interactions is proposed to mediate cognitive-motivational interactions as well (figure 7.6). In particular, the interactions between valuation networks and fronto-parietal regions important for attentional and executive control are suggested to be common to both emotion and motivation. Even though specific valuation regions involved at any given time may differ, the interactions are similar. Subcortical reward/valuation regions include the caudate, nucleus accumbens, midbrain, and the amygdala; and cortical regions include orbitofrontal cortex, anterior insula, medial prefrontal cortex, and posterior cingulate cortex.

The three modes of communication discussed in the context of cognitive-emotional interactions—direct and indirect connections, via hub regions, and

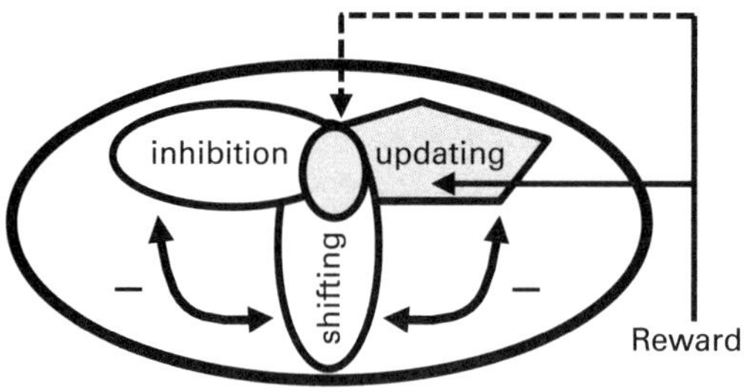

Figure 7.8
Executive control and reward. Motivation is proposed to have two key effects on executive function: first, it fine-tunes executive functions that are important for the task at hand (represented by the change of shape of the updating function; see solid arrow); and, second, it redistributes the allocation of common-pool resources (gray ellipse; see dashed arrow), and thus modulates how executive processes compete with each other. Reproduced with permission from Pessoa 2009.

neuromodulation—play analogous roles in the case of motivation. Connections between valuation and task networks are exemplified by pathways involving the caudate, which is connected with several regions of frontal cortex (including lateral sectors) and parietal cortex, in part via the thalamus (Alexander, DeLong, and Strick 1986).

Hub regions also play a central function during interactions between cognition and motivation. For example, M.-Marsel Mesulam, Aprajita Mohanty, and colleagues suggested that posterior cingulate cortex is important for the integration of motivational and spatial attention information (Mohanty et al. 2008; Small et al. 2005; see also Platt and Huettel 2008). Another key hub region is medial prefrontal cortex (including dorsal prefrontal cortex), already discussed in the context of emotion.

The literature describing electrophysiology data in monkeys suggests that dorsal-medial PFC neurons encode multiple aspects of reward. For example: reward-related responses in dorsal-medial prefrontal cortex depend on whether the outcome triggers a change in strategy, such as a switch from one action to another (Shima and Tanji 1998); dorsal-medial PFC neurons signal the proximity of a reward within an extended behavioral sequence (Shidara and Richmond 2002); and dorsal-medial PFC neurons may encode an estimate of the value associated with an ongoing task (Amiez, Joseph, and Procyk 2006; Sallet et al. 2007). Furthermore, monkeys with cingulate lesions exhibited relatively normal motor control, but showed deficits on a decision-making task that required the reward history to be integrated across several trials (Kennerley et al. 2006; see also below). In humans, a meta-analysis of neuroimaging studies revealed that reward and punishment consistently activate dorsal-medial prefrontal cortex (Liu et al. 2011). More generally, current views emphasize an *evaluative* role for medial prefrontal cortex, and argue that a primary function of this region is to monitor and adjust levels of control (Botvinick 2007; Holroyd and Coles 2002; Ridderinkhof et al. 2004; Rushworth and Behrens 2008; Shackman et al. 2011). This evaluative role is proposed to be especially relevant for assessing the costs associated with an action (Walton et al. 2007). In summary, medial prefrontal cortex is considered to be a critical component of the motivational system (Vogt 2008), leading Christopher Summerfield and Etienne Koechlin (2009, 1023–1024) to suggest that "the primary role of the MFC [medial frontal cortex] is *motivational control*, that is, a calculation of the requisite investment for optimal control over action selection."

In the literature, the anterior insula has been repeatedly implicated during the processing of negative events, such as during the anticipation and experience of aversive stimuli (Paulus and Stein 2006; Simmons et al. 2006), the handling of cues signaling monetary loss (Knutson and Greer 2008), and risk aversion (Kuhnen and Knutson 2005). But a growing number of studies have reported activation of anterior insula during appetitive conditions, for instance, during cues signaling potential monetary gains (Liu et al. 2011; Padmala and Pessoa 2011; Samanez-Larkin et al. 2007); the anterior insula also contributes to addiction (Naqvi and Bechara 2009). Furthermore, anterior insula neurons show increased responses when monkeys know they will, or might, receive a reward based on their performance (Mizuhiki, Richmond, and Shidara 2012). Here I propose that the anterior insula is a chief hub region for cognition-motivation interactions.

As in the case of emotion, a third mode of communication involves the widespread action of neuromodulatory signals, including those of dopamine and acetylcholine. Widespread neuromodulatory connections reach cortical and subcortical regions across the brain. Patricia Goldman-Rakic and colleagues (1989) suggested that a major function of dopamine is to control cortical excitability, thereby increasing the fidelity of signals computed within local networks. Indeed, the effects of dopamine appear to enhance the neuronal signal-to-noise ratio (Sawaguchi and Matsumura 1985), consistent with computational modeling results of the contribution of dopamine in working memory function (Gruber et al. 2006; see also Murphy and Sillito 1991 and Sato et al. 1987 for a related role of acetylcholine in enhancing signal-to-noise). It is thus possible that dopaminergic and cholinergic neuromodulation provides a key mechanism by which motivation sharpens executive control (and hence behavioral performance), for instance, by improving the signal-to-noise ratio of relevant neurons (for further discussion of cholinergic mechanisms, see below). In this fashion, motivation enhances processing efficiency in target cortical and subcortical regions.

Resources: Linking Human and Animal Literatures

Up to now, the concept of "resources" used here has been the one employed in the literature on executive function and capacity limits of human information processing. In motivation research in nonhuman animals, the concept is also linked to that of *effort* (Salamone et al. 2009), which relates both to the vigor and persistence displayed in pursuit of stimuli that predict reinforcement and to goal-directed changes in behavior. Such a goal-based aspect of motivation is well captured by Kent Berridge's *incentive salience* framework, where attributing incentive salience to an item converts its neural representation from that of a neutral, "cold" item into that of an attractive object animals will work to acquire (see Berridge and Robinson 1998). Such *wanting* (appetitive, goal-directed) aspects of motivation are suggested to be directly reliant on dopamine signaling via projections from the midbrain to targets in the dorsal and ventral striatum.

However independently the human and nonhuman literatures on resources may have developed, they are clearly connected in some ways. For example, Peter Redgrave and colleagues (Redgrave and Gurney 2006; Redgrave, Prescott, and Gurney 1999) have proposed that dopamine-related circuits in the striatum facilitate the *reallocation* of limited processing capacity toward unexpected events of behavioral significance, including rewarding ones. Thus, instead of simply providing a "reward signal," striatal activation drives the redistribution of available resources to salient events whose processing is then prioritized (see also Horvitz 2000; Zink et al. 2004).

A close link between resources as linked to capacity and resources as linked to effort is also found in research on *attentional effort* by Martin Sarter, William Gehring, and Rouba Kozak (2006), who propose that increased activity of those frontal sites receiving *cholinergic* signals represents an essential component of the neuronal mechanisms mediating increases in attentional effort. In their conceptualization, attentional effort is a function of the subject's motivation to perform, which is naturally sensitive to both positive and negative incentives. They further propose that increased prefrontal cholinergic activity contributes to the recruitment of goal-driven mechanisms (for a review of the role of the cholinergic system in both goal-driven and stimulus-related attentional processes, see Sarter et al. 2005). Such goal-driven mechanisms, which depend on fronto-parietal regions, act to enhance sensory processing and possibly also to attenuate interference effects. In this regard, cholinergic activity in humans was shown to selectively increase neural responses in visual cortex to task-relevant stimuli with reduced or no change in neural responses to task-irrelevant stimuli (Furey, Pietrini, and Haxby 2000; see also Furey et al. 2008). Hence cholinergic mechanisms alter signal-to-noise ratios and affect the competition between attended and unattended stimuli.

The major cholinergic site in the central nervous system is the basal forebrain, from which cholinergic projections to cortex are involved in diverse attentional functions, including sustained, selective, and divided attention (Sarter and Bruno 1999, 2000; Sarter, Bruno, and Turchi 1999; see chapters 2 and 9). The basal forebrain cholinergic system is also the recipient of substantial subcortical signals from the nucleus accumbens, which are in fact capable of controlling cholinergic signaling to cortex (Sarter, Bruno, and Turchi 1999). Indeed, interactions between the nucleus accumbens and cholinergic projections from the basal forebrain to cortex play important roles in the motivated recruitment of attention (St. Peters et al. 2011). As noted above, the ventral striatum is a significant recipient of *dopaminergic* signals from the midbrain. The contributions of dopamine and acetylcholine signaling thus appear to be closely intertwined during attention-motivation interactions (see figure 3 of

Sarter, Gehring, and Kozak 2006), highlighting the complex interactions between multiple circuits during motivated behaviors.

Endogenous Mechanisms

For a concrete example of an endogenous mechanism, let us focus on the cognitive function of attention. Both monkey electrophysiology and human neuroimaging research suggest that the control of selective/endogenous attention depends on a distributed set of fronto-parietal regions, including frontal eye field in frontal cortex and intraparietal sulcus in parietal cortex (Corbetta and Shulman 2002; Kastner and Ungerleider 2000). These regions, which in many cases appear to work together, are often thought to be *source* regions that exert control over sensory areas to help direct processing toward the most relevant object at a given time. One way to interpret the data from some of the attention studies described in chapter 6 is that attention-motivation interactions rely on the communication between the attentional and valuation networks (figure 7.6). More specifically, when subjects see a reward cue, their valuation regions are engaged and interact with cortical regions, including prefrontal cortex, that contribute to executive control. These interactions enhance cortical/subcortical processing, such that performance is improved.

Srikanth Padmala and I observed an example of such interactions in our study of reward and endogenous attentional control (Padmala and Pessoa 2011). The effect of reward during the cue phase on attentional control manifested itself a few seconds later during target processing (see figure 6.5 and associated text). Functional connectivity between motivation-related regions (including the caudate) and parietal cortex, which is important for spatial attention, increased based on reward and varied as a function of individual differences in reward sensitivity. As discussed earlier (see also chapter 8), functional connectivity is considered an indication of the "integration of information." Related findings were obtained by Helga Harsay and colleagues (2011), who studied the role of reward during the antisaccade task, where subjects must exercise deliberate/endogenous control in order to execute a saccade to the opposite side of a target stimulus (a briefly flashed asterisk in this case). Subjects performed faster and more accurate antisaccades when potential reward was involved. Harsay and colleagues also observed increased functional connectivity during the reward condition. The caudate exhibited increased correlation with several cortical regions linked to the execution of eye movements, including the frontal eye field (frontal cortex) and the intraparietal sulcus (parietal cortex). Notably, increased functional connectivity was predictive of higher benefits of reward on reaction time.

Several other examples of increased functional connectivity during reward conditions have been reported in the literature. For example, enhanced coupling between both ventral and dorsal striatum and fronto-parietal cortex was detected by Liane Schmidt and colleagues (2012) in a rewarded response-interference task. In their long-term memory study, R. Alison Adcock and colleagues (2006) observed increased correlated activity between the hippocampus and midbrain during high reward, and the correlation strength predicted memory formation. Finally, in a subsequent analysis of our response-interference task (Padmala and Pessoa 2011), my colleagues and I showed that caudate and accumbens increased their connectivity during reward conditions with more than a dozen cortical regions, many of which are known to participate in the task (Kinnison et al. 2012).

It is likely that the interactions described in this section involve dopaminergic sites in the midbrain, including the ventral tegmental area, as well as their targets in the ventral and dorsal striatum (Aarts, van Holstein, and Cools 2011). Related mechanisms between the midbrain and the hippocampus are important for memory (Shohamy and Adcock 2010; see also Luo et al. 2011). Noudoost and Moore (2011a) found that dopamine-mediated activity in the frontal eye field during a visual task, a frontal region important for attention, not only caused the monkey to select specific visual targets more frequently, but also led to enhanced and more selective responses of neurons in visual area V4. Notably, the effects in visual cortex were comparable in magnitude to the effects of goal-directed attention. Together, their findings suggest that dopamine contributes to the frontal eye field's control of visual signals via a mechanism analogous to attention, namely, by modulating long-range connections between frontal and visual cortex. Because Noudoost and Moore altered dopamine-related activity through local pharmacological manipulation, however, their findings do not elucidate the operations that normally occur during unaltered motivated behavior. Nevertheless, given the extensive dopaminergic innervation of the frontal cortex, related processes are likely at play during normal situations.

Reward and Threat

In many real-life contexts, rewards and threats may have to be *simultaneously* considered during behavior. For example, a chimpanzee may be motivated to consume a juicy piece of fruit, but know that doing so may trigger being smacked by a higher-ranking male nearby. Little is known about the neural mechanisms of these interactions.

Indeed, although it is unclear whether particular regions are engaged by appetitive-only, aversive-only, or both types of motivation, accumulating evidence indicates that many are engaged by *both* types. For example, a large corpus of studies supports the idea that midbrain dopaminergic regions and their projection sites in striatum and orbitofrontal cortex are implicated in appetitive processing (e.g., Delgado 2007; Haber and Knutson 2010; O'Doherty 2004; Schultz 2000). Yet these regions also participate in aversive processing, indicating that they are involved in both appetitive and aversive motivation (Bromberg-Martin, Matsumoto, and Hikosaka 2010; Morrison and Salzman 2011; Salamone 1994). In an analogous fashion but from the opposite valence, processing in the amygdala, bed nucleus of the stria terminalis, and anterior insula has been frequently linked with aversive events or stimuli (Craig 2002, 2009; Davis et al. 2010; LeDoux 2000). Yet these regions are engaged during appetitive processing, too (Everitt et al. 2003; Liu et al. 2011; Mizuhiki, Richmond, and Shidara 2012; Salzman et al. 2007), as indeed are several brain regions, including medial prefrontal cortex, during goal-directed behaviors.

In a recent study (Choi, Padmala, and Pessoa 2013), my colleagues and I used a task with cues signaling the chance of monetary reward or mild aversive shock to investigate the interactions between appetitive and aversive processing (figure 7.9A). Our task design allowed us to measure responses during the preparatory/anticipatory delay phase with minimal contamination from other task phases (as in the task depicted in figure 5.7, a delay separated the cue from the target stimulus; but in the current task, it lasted 12 seconds). Skin conductance response data revealed interactions between reward and threat during the delay phase. When reward and threat were jointly present, reward opposed threat and threat opposed reward. These push-pull interactions were detected across many brain regions—not only in regions that participate in valuation, such as the midbrain and striatum, but also in the anterior insula, dorsal-medial prefrontal cortex, and even in the frontal eye field (figure 7.9).

What accounts for the trade-off pattern of these interactions? One possible explanation is that reward was task relevant, whereas threat might have functioned as a distractor. Thus reward and threat might have acted against each other in a way that can be recast in terms of competition for limited processing resources (figure 7.4; Pessoa 2009). This explanation is appealing when sites such as the frontal eye field are concerned, given their association with attention, though perhaps less appealing when regions such as the midbrain are concerned. Although some researchers (e.g., Horvitz 2000; Salamone et al. 2009) have proposed that these regions can be viewed as associated with "effort," it is possible that the trade-offs observed in these regions reflected the organization of positive and negative systems into *opponent* motivational systems (Konorski 1967; Solomon and Corbit 1974)—which would operate in a push-pull fashion.

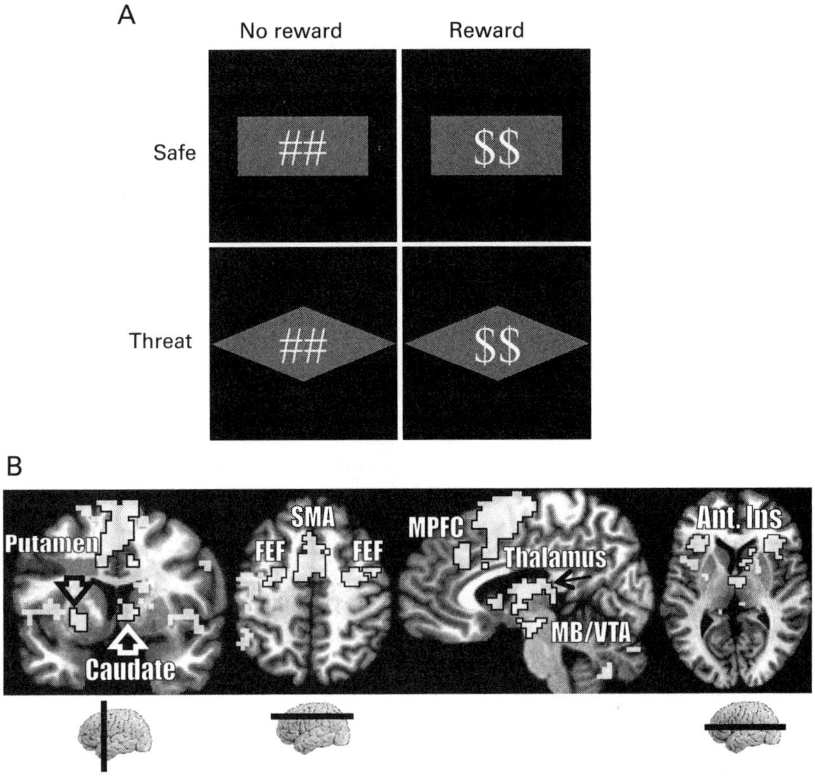

Figure 7.9
Investigating interactions between reward and punishment. (A) The cue stimulus indicated whether the trial involved the threat of shock or not (diamond vs. rectangle), as well as whether reward was possible ($$ vs. ##). The trial structure was similar to that employed in the task of figure 4.7, although the task involved a simple shape discrimination. (B) The activation maps display voxels with significant reward by threat interactions, namely, sites where reward opposed threat and threat opposed reward. Ant. Ins., anterior insula; FEF, frontal eye field; MB , midbrain; MPFC, dorsal-medial prefrontal cortex; SMA, supplementary motor area; VTA, ventral tegmental area.

Finally, a push-pull interaction was not observed in the amygdala, though it was detected in the bed nucleus of the stria terminalis, an intriguing discovery given the engagement of the bed nucleus during "anxiety-related" conditions (such as extended threat; Davis et al. 2010). The findings thus show that the bed nucleus is also affected by reward processing (see also McGinty et al. 2011).

Conclusions

Information with affective or motivational content can influence perception and cognition in powerful ways. The dual competition model described in this chapter outlines a framework that includes interactions at the perceptual and executive levels. In broad terms, it sees perception and executive functions as capacity-limited processes engaged in evaluating the behavioral relevance of stimuli and tasks in question. At the perceptual level, items with affective/motivational content act as if they had increased *salience*, which improves performance if they are task relevant but impairs it if they are task irrelevant. At the executive level, the behavioral interference of emotion often observed when high-arousal items are processed can be interpreted in terms of competition for finite available resources. Some effects of motivation on performance can also be thought of in terms of resources, especially when performance trade-offs are observed, such as in our recent response-inhibition study (Padmala and Pessoa 2010), although motivation's "sharpening" effect is not well captured by the idea of resources.

More generally, task performance is a function of at least three variables: (1) the pool of mental resources available at a given time; (2) "goal importance" (Kruglanski et al. 2012), where the subject may place greater importance on performing rewarded task components; and (3) the subject's skill level. Performance can be expressed in terms of these three variables—and, indeed, several others (for further elaboration, see Kruglanski et al. 2012; Sperling and Dosher 1986; Kahneman 1973). Accounts of the effects of motivation on behavior thus need to consider factors that go beyond processing resources.

Dorsal-medial prefrontal cortex featured prominently here as a neural basis for both cognitive-emotional and cognitive-motivational interactions. Alexander Shackman and colleagues (2011) have proposed that dorsal-medial prefrontal cortex implements domain-general processes of *adaptive control*, based on the region's extensive contributions to cognitive control, negative affect, and nociception. I suggest this proposal should be extended to incorporate motivation as well, which is to say, that dorsal-medial prefrontal context implements *motivated* adaptive control—where "motivated" is understood to include emotional processing.

The architecture portrayed in figure 7.6 implies that signals with emotional or motivational significance are not confined to specific brain regions but are broadcast extensively across the brain. A striking demonstration of this phenomenon is the finding by Timothy Vickery, Marvin Chun, and Daeyeol Lee (2011) that reward signals are distributed quite broadly in the brain even when reward was not paired with a specific visual stimulus or motor response. "This suggests an imperative," Vickery, Chun, and Lee state (2011, 175), "to study the effects of reinforcement and punishment in domains where they are not usually considered as important factors—from low-level sensory systems to high-level social reasoning. Such distributed representations would have adaptive value for optimizing many types of cognitive processes and behavior in the natural world."

A related example in the domain of emotion comes from Srikanth Padmala's and my analysis of the information content of functional MRI responses during a challenging fear detection task (Pessoa and Padmala 2007). Our results indicated that the representation of behavioral choice (associated with reporting the perception of a fearful face) was spread across several brain regions (see also Thielscher and Pessoa 2007). Such distributed encoding may help prepare the organism to appropriately handle emotional stimuli and regulate the associated emotional response.

As this chapter has shown, an important project for future research is to determine the relationship between how motivation, on the one hand, and emotion, on the other, affect sensory and cognitive processes and, in doing so, to elucidate both the similarities and differences of their mechanisms of interaction. Thus, in the case of positive stimuli, motivation often affects perception and cognition through the explicit strategic establishment of endogenous attention, whereas, in the case of negative emotional stimuli, its effects do not appear to be goal based in the same manner.[10] The present chapter has highlighted both dorsal-medial prefrontal cortex and anterior insula, hub regions where motivational, emotional, and cognitive signals come together, as crucial to both *motivation*-cognition and *emotion*-cognition interactions.

Notes

1. For limitations of the ubiquitous double dissociation research strategy, see, for example, Van Orden, Pennington, and Stone 2001.

2. Markov and colleagues (2011) combined all insular parts into a single region.

3. Lateral regions of frontal cortex involved in the attentional network include the middle and inferior frontal gyri and the frontal eye field (at the intersection of the precentral sulcus and middle frontal gyrus). Medial regions include the anterior cingulate cortex and presupplementary motor area.

4. See Dosenbach et al. 2008 for additional discussion of networks proposed to be involved in cognitive control, networks that include regions discussed here in the context of attentional and executive control, as well as other regions, such as the thalamus.

5. I am grateful to Alan Anticevic for discussions of this issue.

6. Intriguingly, in some cases, effects on brain responses are observed even when no concomitant behavioral impact is reliably detected. For a cogent discussion of this issue, see Wilkinson and Halligan 2004.

7. As described in chapter 4, if the main task nearly exhausts processing capacity, the impact of task-irrelevant items is likely to be greatly reduced and even eliminated.

8. Dorsal-medial prefrontal cortex includes anterior midcingulate cortex ("aMCC") as defined in O'Neill, Sobel, and Vogt 2009, as well as surrounding cortex not in the cingulate gyrus proper (such as the presupplementary motor area).

9. Because of its poor cortical lamination and its extensive connections with less differentiated cortex, the ventral portion of the anterior insula is sometimes suggested to be more closely aligned with emotional processing than the dorsal portion is (Mesulam and Mufson 1982a, 1982b; see also Dupont et al. 2003). Because, however, both ventral and dorsal anterior insula are *functionally* connected to the dorsal-medial prefrontal cortex during the resting state (Deen, Pitskel, and Pelphrey 2011), even ventral portions of the anterior insula might engage in interactions with regions of the medial prefrontal cortex important for cognitive and emotional tasks (see also Uddin et al. 2013).

10. Some authors have proposed, however, that negative emotion "is goal-directed and flexibly coordinates anticipatory responses that decrease the likelihood of future punishment" (Shackman et al. 2011, 160; see also Seymour, Singer, and Dolan 2007).

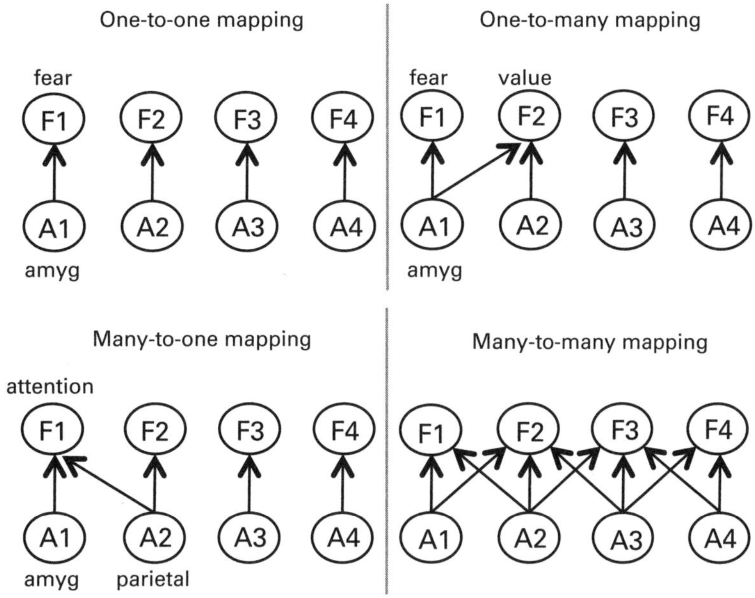

Figure 8.1
Structure-function mapping in the brain. Because the mapping from structure to function is *many* to *many*, understanding the instantiation of functions by the brain necessitates frameworks whose basic elements are networks, not regions. One-to-many mapping from the perspective of A1; many-to-one mapping from the perspective of F1. A, area; F, function.

functions and that many functions are carried out by many regions (figure 8.1). For instance, dorsal-medial prefrontal cortex is important for a diverse range of cognitive operations, as well as for emotional processing, and thus provides an example of a region involved in many functions, namely, an instance of a one-to-*many* mapping. Conversely, both frontal and parietal regions of cortex participate in attentional and executive processes, illustrating the situation of multiple regions carrying out a related function, an instance of a many-to-*one* mapping.

More generally, the mapping between structure and function is both *pluripotent* (one to many) and *degenerate* (many to one). "Pluripotentiality" refers to the ability of structurally the same element to perform many different functions, whereas "degeneracy" refers to the ability of structurally different elements to perform the same function, yield the same output (Edelman and Gally 2001), or complete a task. Notably, degeneracy should be distinguished from *redundancy*, which refers to structurally identical elements performing the same function (as in "backup" engineering systems). To the extent that pluripotentiality and degeneracy are accepted with respect to the mind-brain, the

8 Network Perspective on Brain Function

Whereas previous chapters covered specific aspects of how emotion and motivation interact with perception and cognition, this chapter will describe a general framework for understanding brain function, one that addresses the relationship between brain and behavior. To do so, it must first characterize the mapping between structure and function. Much has been written about the localizability of mental processes, a question that lies at the core of neuroscience as a scientific discipline. Even a cursory look at the history of the field suggests a continual swing of the pendulum between holistic and modular explanations of these processes (for a particularly lucid account, see Shallice 1998). The present chapter will discuss broad issues surrounding the link between structure and function in the brain that motivate a network perspective on brain function.

From Regions to Networks

The simplest way to conceptualize the relationship between a brain region and behavior is to assume a one-to-one mapping between a region and its function. For example, primary visual cortex is linked to visual perception (or a set of specific visual functions, such as "edge detection"). We can imagine extending such a strategy to other regions of the brain, although, of course, it becomes far less straightforward for regions farther from the sensory periphery. The end product would be a list of region-function pairs, with regions labeled as "perceptual," "cognitive," "emotional," "motor," and so on, based on their purported functions and how they are envisioned to ultimately shape behavior. For instance, we could describe the amygdala as "emotional," given its contributions to fear conditioning, and dorsal-medial prefrontal cortex as "cognitive," given its role in the processing of response conflict.

Disregarding for now the thorny issue of what precisely is meant by "region" and "function," it is readily apparent both that brain regions participate in many combination of the two indicates that there are no "necessary and sufficient" brain systems.[1] In particular, the existence of two or more degenerate systems that do *not* overlap precludes the existence of a single necessary region for a given function (Price and Friston 2002).

In the above discussion, I avoided the difficult question of what constitutes a brain region and the even more difficult question of what constitutes a function. Clearly, structure-function relationships can be defined at multiple levels, from the precise (e.g., "Primary visual cortex is concerned with edge detection") to the abstract (e.g., "Primary visual cortex is concerned with visual perception"), and depend on the specific level that is targeted. Some authors have suggested that, at some levels, a brain region has no more than *one* function. For instance, the left posterior fusiform gyrus in temporal cortex, which has been implicated in the processing of word forms, animal structures, and the like, can be described by a single, more abstract label of "sensorimotor integration" (see Price and Friston 2005). Cathy Price and Karl Friston suggest that whether a region can have more than one function depends on the level of the structure-function relationship, such that, at a sufficiently abstract level, a region will have a single function (note that, for this notion to be useful, the abstractness has to be relatively limited, and not simply a vague description such as "cognitive function"). Although the search for better conceptualizations of a region's functions is valuable, I propose below that the region level is *inadequate* to describe how brain structure is linked to mental function.

One way to restate the discussion thus far is to consider it in terms of two domains: psychological events (e.g., "functions," "behaviors") and physiological events (e.g., brain regions), which can be denoted Ψ and φ, respectively (Sarter, Berntson, and Cacioppo 1996). To understand how these two domains are related to each other, we need to know both $p(\Psi|\varphi)$, the probability of a psychological event given the involvement of some neural structure, and $p(\varphi|\Psi)$, the probability of a physiological event given a psychological one. In other words, we need to know the mapping between the psychological and physiological domains. In the ideal situation, both $p(\Psi|\varphi) = 1$ and $p(\varphi|\Psi) = 1$, which is to say, knowledge of the psychological perfectly predicts the physiological and knowledge of the physiological perfectly predicts the psychological. In real life, these two probabilities can differ dramatically: they are directly related to each other according to Bayes's rule (1763).

Casting the structure-function problem in the above terms is pertinent, given the fast accumulation of neuroimaging studies in the past two decades, which are now available to investigators through database services. Thus Russell Poldrack (2006; see also Poldrack 2011) evaluated the common practice in neuroimaging research of drawing *reverse inferences*, namely, reasoning backward, from the presence of brain activation to the engagement of a particular function (e.g., "If the amygdala was active, an emotion was involved"). The specific example considered by Poldrack assessed the ability to use activation in Broca's area (in left ventral-lateral prefrontal cortex) to predict the engagement of "language function." By applying Bayes's rule, which can be understood as a procedure to update one's prior beliefs based on new evidence, Poldrack found that P(Language|Activation in Broca's area) was .69 (based on activations available in the BrainMap database at the time), thus a relatively weak indication that the given activation in Broca's area involved language (but more likely than not). Note that Poldrack chose Broca's area and language because, if anything, this choice would be more favorable to the possibility of reverse inference, given the purported strong association of this area with language based on a long tradition of lesion work (Broca 1861; Shallice 1988). Even in such cases, however, $P(\Psi|\varphi)$ is nowhere near 1 (the one-to-one case; see "Understanding a Region's Function via Functional Fingerprints" below for related discussion.)

Degree of Isolability and Decomposable Systems

To understand the relationship between structure and function, it is instructive to consider architectural features that constrain the mapping between the two. At one extreme, there is the *module*, instantiated by a single brain region solely or mostly responsible for carrying out a certain function, for example, the fusiform gyrus in ventral visual cortex of the right hemisphere, proposed to be a "face module" (Kanwisher, McDermott, and Chun 1997). But there are at least four other architectures that inform the structure-function relationship, which we will now consider, drawing on the discussion by Tim Shallice (1988, chap. 11).

Processing Space Continuum Systems

Straightforward examples of the first of these four architectures are the retinotopic organization (i.e., orderly map of visual space) observed in visual cortex and the tonotopic organization (i.e., orderly map of sound frequency) found in auditory cortex. A more interesting example is the ventral visual system hypothesized by Rosemary Cowell and colleagues, not in terms of perceptual and memory "modules" in posterior and anterior regions of visual cortex, respectively, but in terms of a *hierarchical continuum* between the two types of visual processing (Cowell, Bussey, and Saksida 2010; McTighe et al. 2010). Cowell and colleagues propose that, given such a system, impairment of perceptual and memory functions following brain damage reflects the demand that each task places on stimulus representations rather than on dissociable cognitive modules.

Overlapping Processing Systems

In the second architecture, process P_A requires regions R_1 and R_C, whereas process P_B requires regions R_2 and R_C. Thus regions R_1 and R_2 are parts not of two isolable subsystems but of overlapping subsystems (the overlap involving common region, R_C).

Coupled Systems

In the third architecture, the *degree of isolability* of the coupled subsystems will depend on how strongly they interact with each other. If they only weakly interact, the operations of one subsystem can be established without considering those of the other, at least in some contexts. If, however, they interact more strongly or, in the extreme, quite strongly, the subsystems are only *partially* isolable, or *not at all* isolable.

Semimodular Systems

And, finally, in the fourth architecture, the degree of isolability of the system's semimodules will depend on the extent to which the system's function relies on internal versus external variables. The semimodules will be more isolable when internal variables prevail, and less so when external variables do.

More broadly, we can assess the degree of isolability of subsystems by linking it to the notion of *decomposable*, *nearly decomposable*, and *nondecomposable* systems (Bechtel and Richardson 1993; Simon 1969). At one extreme, a "*decomposable* system" is one in which each subsystem is highly modular and operates according to its own intrinsic principles, independently of the others. At the other, a "*non*decomposable system" is one in which the connectivity and interrelatedness of the subsystems are such that they are no longer clearly isolable. In between lies a continuum of possible organizations.

Networks versus Regions

The previous section considered the architectural features that inform structure-function mapping largely from an abstract, conceptual point of view. This section will briefly review proposals on the important ways networks of brain regions are involved in the implementation of behaviors; the section to follow will build on these proposals and describe a structure-function scheme based on networks. It should be stressed, however, that the objective of the present section is not to give a comprehensive historical account of the notion of networks, but simply to provide a few illustrations. Important treatments *not* covered here include those by Antonio Damasio (1989), Steve Grossberg (1980), Gerald Edelman (1978), and Malcolm Young and colleagues (1994), among many others. For a discussion of the modular viewpoint, see the recent treatment by Nancy Kanwisher (2010, 11168), who suggested that "the possibility is within reach of obtaining a cognitively precise parts list for the human brain." I take a different path, one that emphasizes coalitions of regions that, jointly, contribute to behavior.

Brain anatomy reveals that brain regions are interconnected in a rich and dense pattern, both locally and in terms of long-range connections. Traditionally, however, much of anatomy research has been devoted to describing the substrate for the elaboration of neural signals as they progress from sensory through associational to motor centers. This *hierarchical* scheme of cortical organization emphasized the convergence of information leading to integration in sites such as posterior parietal cortex, the superior temporal polysensory area, and prefrontal cortex. As Patricia Goldman-Rakic summarized in her influential paper (1988, 146): "The conclusion traditionally reached in virtually all comprehensive studies of cortical connections is that they are organized in a step-wise hierarchical sequence proceeding from relatively raw sensory input at the primary sensory cortices through successive stages of intramodality elaboration allowing progressively more complex discriminations of the features of a particular stimulus." In a departure from this scheme, Goldman-Rakic (1988) emphasized, instead, the existence of distributed processes carried out through several parallel systems; in this scheme, integrative functions emerge from the dynamics of the entire network rather than from computations performed at each nodal point.

In an influential series of papers spanning several decades, M. Marsel Mesulam has advanced a network approach to the localization of complex functions as "an alternative to more extreme approaches, some of which stress an exclusive concentration of function within individual centers in the brain and others which advocate a more uniform (equipotential or holistic) distribution" (Mesulam 1981, 309). In his 1981 proposal, Mesulam suggested that such a network approach would help reconcile some of the inherent problems with the "extreme approaches"; that a network involving posterior parietal, frontal, and cingulate cortex contributes sensory, motor, and motivational representations to attentional processes, respectively; and that reticular structures in the thalamus and brainstem are involved in arousal aspects of attention. In a subsequent paper, Mesulam (1998) advanced the notion that the human brain contains at least five "core" functional networks: (1) the spatial awareness/attention network anchored in posterior parietal cortex and the frontal eye field of frontal cortex; (2) the language network anchored in Wernicke's and Broca's areas; (3) the explicit memory/emotion network anchored in the hippocampal-entorhinal complex and the amygdala; (4) a face-object recognition network anchored in anterior temporal and midtemporal cortex; and (5) a working memory/executive function network anchored in lateral prefrontal cortex (and possibly inferior parietal cortex).

Helen Barbas (1995) has also emphasized the importance of networks in the domain of emotion and cognition, describing several anatomical features of prefrontal cortex that potentially underlie cognitive-emotional interactions. In particular, she proposed that pathways between the amygdala and both orbital and medial prefrontal cortex provide a means for sensory signals reaching prefrontal cortex to be integrated with emotional information.

A Conceptual Scheme

Building on these approaches, this section describes how structure-function relationships can be conceptualized as *networks* of brain regions (figure 8.2A). Such networks collectively support behaviors, with the network *itself* serving as the unit, *not* the brain region. Processes that support behavior are not implemented by an individual region, but rather by the interaction of multiple regions, which are dynamically recruited into multiregion assemblies. Thus Nico Dosenbach and colleagues (2008) have proposed that goal-directed control can be understood in terms of two networks, a cingulate-operculum network responsible for "set maintenance" (such as maintaining task focus over relatively extended periods of time) and a frontal-parietal network responsible for rapid adaptive control.[2] Thus, too, Steven Bressler and Vinod Menon (2010) have proposed that a salience network including the anterior insula and anterior cingulate cortex is involved in attention to external and internal worlds. Of course, these proposed networks and others like them are somewhat arbitrary, with a growing number of functions being ascribed to particular networks. In an interesting recent development, researchers have used meta-analysis to extract networks from a large body of studies. For example, based on the data of 825 studies, Roberto Toro, Peter Fox, and Tomás Paus (2008) investigated the pattern of *co*activation and identified at least three important functional networks: a fronto-parietal attention network, a resting-state network, and a motor network.

In most instances, networks are described in terms of unique, *non*overlapping sets of brain regions. Yet, more generally, networks contain overlapping regions, such that specific regions will belong to several intersecting networks (Mesulam 1990). In this manner, the processes carried out by a region will depend on its network affiliation at any given time, which highlights the importance of the *context* within which a brain region is operating (McIntosh

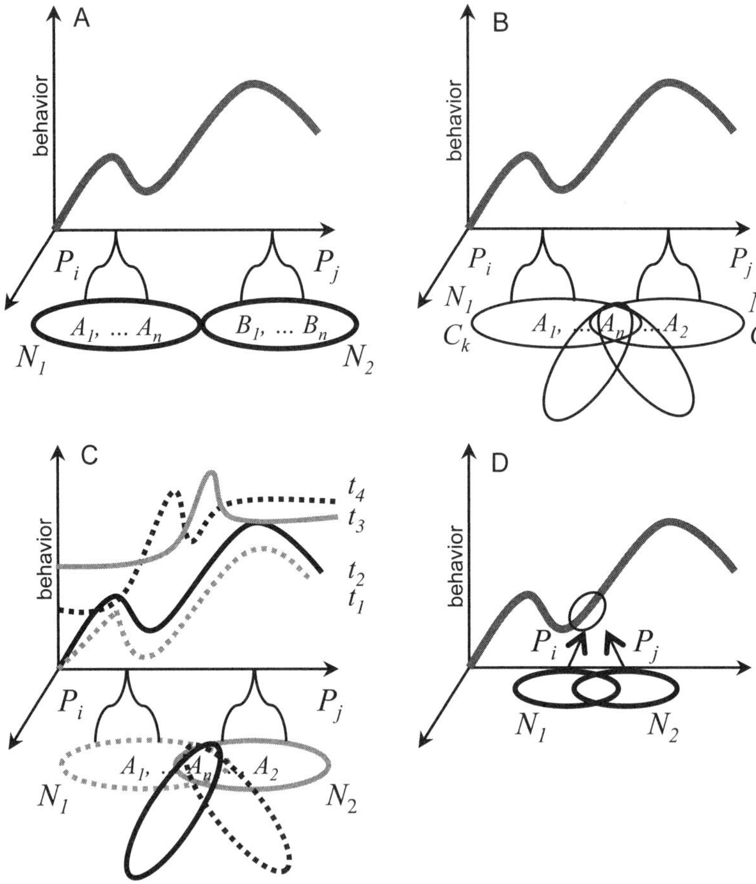

Figure 8.2
Structure-function mapping and networks. (A) The "landscape of behavior" depicts the multidimensional space of behaviors. A_1, A_2, A_N, B_1, and B_N = brain regions; N_1 and N_2 = networks; P_I and P_J = processes. (B) Intersecting networks. The networks C_K and C_L (and the additional ones) intersect at node A_N. (C) Dynamic aspects. Because region A_N will have network affiliations that vary as a function of time, the processes carried out by the emerging networks will evolve across time and lead to dynamic "landscapes of behavior." The four time points represented are such t_1 is close to t_2 but far from t_3 and t_4, which are close to each other. (D) Structure-function mappings in the case of networks. Two networks may instantiate similar processes, a case of many-to-one mapping. The reverse relationship is also suggested to apply to networks, namely, one-to-many mappings (see text).

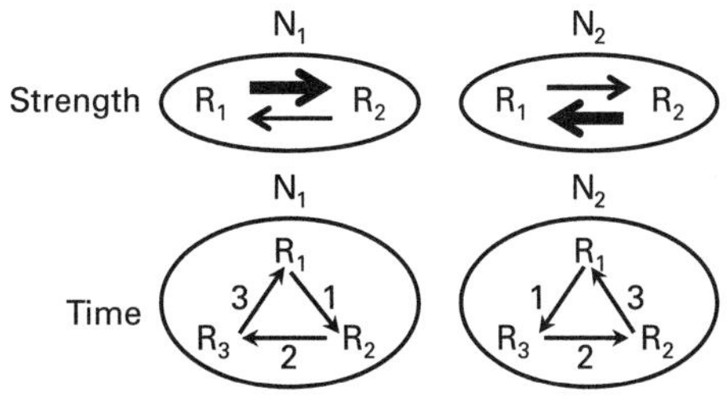

Figure 8.3
Network interactions. The same regions R may comprise distinct networks depending on how the regions interact both in terms of strength across time (top row) and time (bottom row). For instance, when the $R_1 \rightarrow R_2$ link is strong, network N_1 behaves differently from when the $R_2 \rightarrow R_1$ link is strong (in which case, the network is labeled N_2); this could occur due to plasticity. The bottom row illustrates that interregion interactions may follow a different temporal order, thereby leading to a different function. In both cases, the network label N_1 could have been used for the two scenarios, with the understanding that N_1 varies as a function of time.

2000). For example, in figure 8.2B, region A_N will be part of network N_1 during a certain context C_K, but will be part of network N_2 during another context C_L. The existence of context-dependent, overlapping networks also signifies that, from the perspective of structure-function mappings summarized in figure 8.2B, a given region will participate in multiple processes (I will return to the issue of overlap in a later section).

The importance of context emphasizes the need to consider *dynamic* aspects of structure-function relationships, to understand a network in terms of the interactions between multiple brain regions as they unfold temporally. In the extreme, two networks may involve the exact same regions that interact with each other in distinct ways across time (figure 8.3). Put differently, what matters in structure-function relationships is the profile of spatiotemporal activity: the structure-function mapping is a dynamic, not a static, property (Varela et al. 2001). More broadly, network affiliations evolve across time, such that structure-function mappings take place dynamically over several temporal scales—from hundreds of milliseconds to minutes to days (Bassett et al. 2011; Buzsáki 2006; Honey et al. 2007; Varela et al. 2001). So, too, regions are affiliated with networks and impact the behavioral landscape dynamically (figure 8.2C). Note that this strongly contrasts with the tendency to describe networks as static in much of the recent neuroimaging literature (but see Smith et al. 2012).

In a related vein, I have used the term "process" in describing the networks depicted in figure 8.2A, instead of "function." As an *emergent property* (a "process" is said to emerge from the interactions between regions; see Bressler and Menon 2010), a *process* serves as a useful external description of the operation of the network, as opposed to a fixed internal computation implemented by the network (Thompson 2007; Thompson and Varela 2001; Varela, Thompson, and Rosch 1992).

Another key issue is whether describing structure-function relationships in terms of networks allows for a one-to-one mapping. Thus, in proposing that a salience network involving the anterior insula and anterior cingulate cortex "mediates attention to the external and internal worlds," Bressler and Menon (2010, 285) note that "to determine whether this network indeed specifically performs this function will require testing and validation of a sequence of putative network mechanisms" (see also Moussa et al. 2011). Thus, too, in discussing the prospect of a *less* context-dependent architecture to describe network hubs, or regions of high connectivity, Randy Buckner and colleagues (2009, 1867–1868) observed: "An alternative possibility is that the hubs reflect a stable property of cortical architecture that arises because of monosynaptic and polysynaptic connectivity. Within this alternative possibility, the same hubs would be expected to be present all of the time, independent of task state, even when an active task is being performed."

I suggest that mapping structure to function one to one in terms of networks will be fraught with much the same difficulties as mapping it one to one in terms of brain regions (figure 8.1). For one thing, two distinct networks may lead to similar behavioral profiles (figure 8.2D). For another, a network's operation will depend on several more global variables, namely, an extended context that includes the state of several "neurotransmitter systems," arousal, slow wave potentials, and the like. In other words, a network that is solely defined as a "collection of regions" is insufficient to eliminate the one-to-many problem. What if we extend the concept of a network to accommodate these global variables? Then we encounter proposals such as the one by John Cacioppo and Louis Tassinary (1990), who suggest that psychological events can be mapped to physiological ones in a more regular manner by considering a spatiotemporal pattern of physiological events—denoted φ'' to suggest changes in space and time. The notion of a network is thus extended to incorporate other physiological events, for instance, the state of a given neurotransmitter. How extensive does this state need to be? Clearly, the usefulness of this strategy in reducing the difficulties entailed by many-to-many mappings will depend on how broad the context needs to be (Thompson 2007).

Networks versus Regions, Now with Network Theory

Because the brain is clearly not an equipotential mesh in which all regions play the same roles (Lashley 1930), we will always be wanting to unravel the impact of a given brain region on behavior. To do that, however, we need first to consider its connectivity pattern (Passingham, Stephan, and Kotter 2002). Intuitively, the amount of anatomical connectivity (also called "structural connectivity") will be a key element in determining the influence any given region has on brain processing. Accordingly, a region that connects to just a few others will have much less of an impact than one that is more richly connected (other things held constant). The topology of the connectivity will be another key element. A region with local connectivity will contribute to local computations, whereas a region with more widespread connectivity will have a broader effect.

The above considerations can be formalized using graph-theoretic concepts (Newman 2010). If brain regions are equated with *nodes* and information on structural connectivity is captured by the *edges* between nodes, we can define a region/node's *degree* by the number of connections it has to other regions/ nodes.[3] Recent advances in network theory have shown that "hubs," regions with a high degree of connectivity (e.g., one standard deviation above the mean of the graph in question; figure 8.4), are important in regulating the flow and integration of information between regions (region A_N in figure 8.2B and C; Guimerà and Nunes Amaral 2005; Guimerà, Sales-Pardo, and Amaral 2007; "importance" will be further refined in a later section).

Just as the number of connections a region has is important in determining whether it will operate as a hub, so, too, is its structural *topology*. Some regions are best characterized as "*provincial* hubs" because they occupy a central position within a single functional cluster (Guimerà and Nunes Amaral 2005). For example, being highly connected with other visual regions, visual area V4

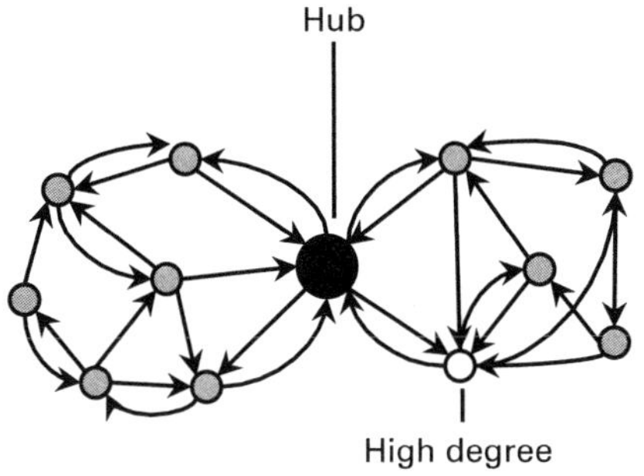

Figure 8.4
Network structure and *hub* nodes. Nodes with unusually high connectivity may be considered "hubs" and are likely to have important roles in determining information flow. The high degree node in white may also have greater influence in processing, though more locally. Adapted with permission from Rubinov and Sporns (2010).

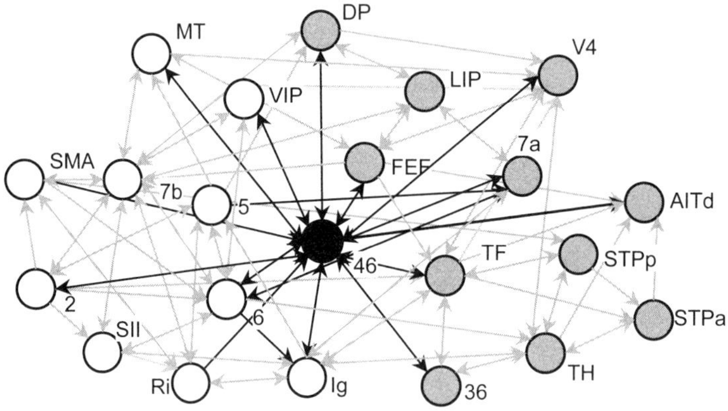

Figure 8.5
Network associated with Brodmann area 46 in dorsal-lateral prefrontal cortex. Given its high connectivity, the area can be considered a hub. AITd, dorsal subregion of anterior inferotemporal area; DP, dorsal posterior nucleus; FEF, frontal eye field; Ig, granular insula; LIP, lateral intraparietal area; MT, medial temporal area (also known as "V5"); Ri, retroinsular area; SII, somatosensory area II; SMA, supplementary motor area; STPa, anterior subregion of superior temporal polysensory area; STPp, posterior subregion of superior temporal polysensory area; TH, thalamus; TF, visual area TF; 2, 5, 9, 7a, 7b, 36, 46, Brodmann areas 2, 5, 9, 7, 7b, 36, 46; V4, visual area 4; VIP, ventral intraparietal area. Reproduced with permission from Sporns, Honey, and Kotter 2007.

can be considered a provincial—and, in this case, visual—hub (Sporns, Honey, and Kotter 2007). Other regions might be better characterized as "*connector* hubs" since they link multiple functional clusters. For example, given that Brodmann area 46 in dorsal-lateral prefrontal cortex is extensively connected to other prefrontal regions as well as to visual area 4, it acts as a connector hub linking prefrontal and visual regions; indeed, it serves as one of the main hubs in the brain (Modha and Singh 2010; Sporns, Honey, and Kotter 2007; figure 8.5; see below for further discussion regarding hub-like functions of lateral prefrontal cortex). Interestingly, we might also think of area V4 as a connector hub itself, linking visual and prefrontal regions (see also Modha and Singh 2010). Thus there is never a single way to "slice" networks, a theme to which I will return later in the chapter.

The above considerations are pertinent to understanding the repercussions of brain lesions on behavior. The topological characterization of brain connectivity suggests that the impact of lesions will be strongly dependent on a region's *structural embedding*: lesions of more peripheral (nonhub) regions will produce relatively specific deficits, whereas lesions of hub regions will have a much broader impact on behavior, one that will be strongly determined by the precise topology of the hub (e.g., provincial vs. connector). In particular, lesions of connector-hub regions will have widespread effects on cognitive and affective behaviors, effects that are sometimes difficult to characterize.

The systematic compilation of anatomical data has revealed massive connectivity between cortical regions (Felleman and Van Essen 1991; Young et al. 1994), between subcortical regions (Petrovich, Canteras, and Swanson 2001; Risold, Thompson, and Swanson 1997), and between subcortical and cortical regions (Young et al. 1994). For instance, Gorica Petrovich, Newton Canteras, and Larry Swanson (2001) reported that on the order of 600 connections of the amygdala to other brain regions were already known at the time of their publication. More recent work has quantified anatomical connectivity in important ways. Particular interest has been focused on characterizing these data and relating them to specific network topologies. Notably, "small-world" networks (Watts and Strogatz 1998), which are ubiquitous in natural, social, and technological systems, combine densely clustered connectivity with an admixture of "random" connections, including *long-range* ones. Each such network preserves a high degree of connectivity within local neighborhoods while allowing *all* its nodes to be linked by surprisingly short paths (i.e., connection steps). Pairs of regions are linked by short paths despite large network size and *sparse* overall connectivity, thus creating a "small world" within the network—as in the famous "six degrees of separation" (Milgram 1967). For example, Olaf Sporns and Jonathan Zwi (2004) have suggested that the cortical connectivity of the macaque exhibits small-world properties.

Drawing on their recent large-scale network analysis of as many as 383 brain regions (figure 8.6A; plate 9) and of cortico-cortical, subcortico-subcortical, and cortico-subcortical connections, Dharmendra Modha and Raghavendra Singh (2010) have found clear evidence of the small-world nature of brain connectivity. They describe a tightly integrated "core circuit" (figure 8.6B; plate 9), spanning parts of premotor, temporal, parietal, and prefrontal cortex, thalamus, "basal brain" (subcortical nuclei at the base of the forebrain, including the amygdala and basal ganglia), cingulate cortex, insula, and visual cortex. The central idea of the core network is that it is strongly connected to all other regions of both the core and the rest of the brain, with several important properties: (1) it is a subnetwork far more tightly integrated than the overall brain network; (2) information likely spreads more swiftly within it than through the overall network; and (3) it serves as the principal hub through which the overall brain network communicates with itself.

Modha and Singh's large-scale network analysis (2010) also revealed that prefrontal cortex contains a disproportionate share of topologically central regions in the brain, as evaluated in terms of measures of efficiency in aggregation

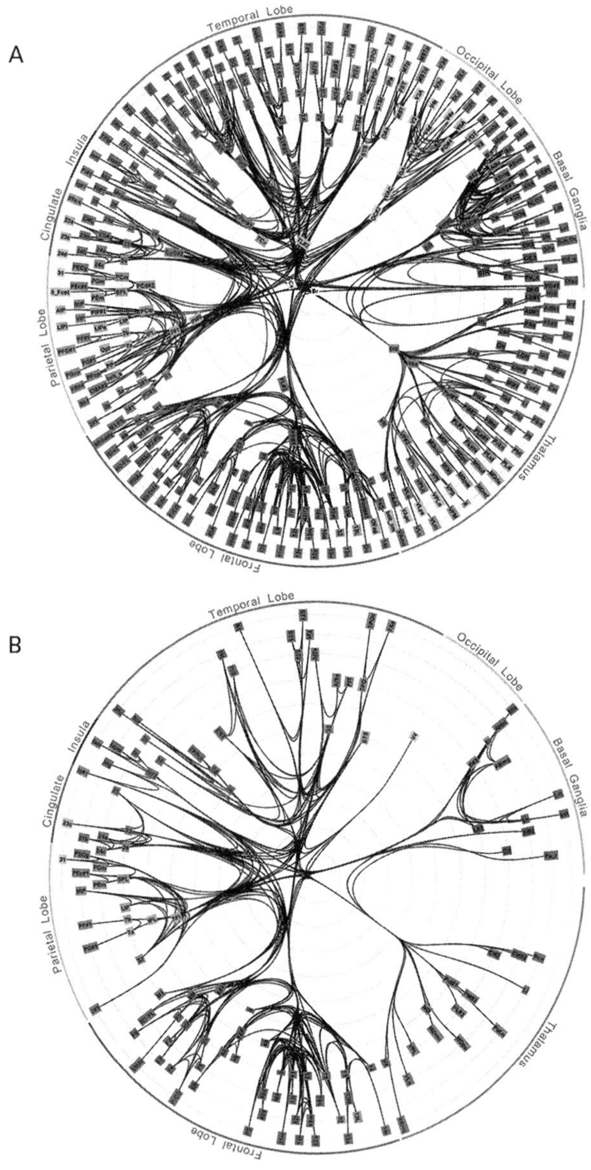

Figure 8.6 (plate 9)

Whole-brain network connectivity structure. (A) In their analysis, Modha and Singh (2010) considered existing anatomical connectivity data of an extensive set of cortical and subcortical regions spanning most major brain sectors. The basal ganglia refer to nuclei at the base of the forebrain, including the amygdala. (B) Innermost "core" circuit. Notably, several amygdala nuclei were included in the inner core. Reproduced with permission from Modha, D. S., and Singh, R. (2010). Network architecture of the long-distance pathways in the macaque brain. *Proceedings of the National Academy of Sciences of the United States of America*, *107*(30), 13485–13490.

and distribution of information, among others. Their findings are consistent with the work of Bruno Averbeck and Moonsang Seo (2008), who decomposed the macaque prefrontal cortex into twenty-five regions and characterized how they are connected to one another, in addition to sixty-eight other brain regions (Averbeck and Seo 2008). They found that prefrontal regions exhibited a very high degree of interconnectivity and that input information from sensory, motor, and limbic (as labeled by the authors) regions could reach anywhere in the prefrontal network within at most two connections. More generally, it appears that the connectivity of macaque cortex tends to minimize the number of "processing steps" between brain regions through long-distance connections (Kaiser and Hilgetag 2006). Although such connections may entail developmental and metabolic costs, they help reduce the overall physical distance between regions. Interestingly, subnetworks in macaque cortex with minimal wiring length, which are in fact metabolically more efficient, have significantly greater distances between regions than the overall network has (Kaiser and Hilgetag 2006).

"Hub nodes" exhibit a high degree of connectivity and are sometimes densely *inter*connected, forming a structure known as a "rich club," namely, a dominant cluster of highly influential nodes (Colizza et al. 2006). Rich clubs determine several important properties of the *entire* network, another reason why hub regions are of particular importance. In one study of structural brain data, Martijn van den Heuvel and Olaf Sporns (2011, 15784) state that "the aggregation of hubs into a rich club suggests that the communication hubs of the brain do not operate as individual entities, but instead act as a strongly interlinked collective" (see also Zamora-Lopez, Zhou, and Kurths 2010).

Functional Connectivity

Anatomy first, and then physiology; but if physiology first, then not without anatomy.
—Attributed to Bernhard von Gudden

Thus far, the discussion has emphasized the role of structural connectivity, which provides the physical linkage for functional connectivity between regions. Although, at first glance, the notion of an architecture anchored in physical connections is clear cut, the boundary between anatomy and function quickly blurs when we consider specific anatomical factors such as the receptor subtypes involved, the laminar profile of the connections (often interpreted in terms of "modulatory" vs. "driving" inputs; see note 2 to chapter 3), the presence and proportion of excitatory and inhibitory interneurons, and the strength of the connections (Lee, Harrison, and Mechelli 2003). The existence of complex circuits with multiple feedforward and feedback connections and diffuse projection systems further complicates the picture.

Thus, to understand how regions and networks contribute to brain function, it helps to identify the way regions are "functionally connected." Devised to characterize how neurons interact, *functional connectivity* was initially defined as the "temporal coherence" (Friston et al. 1997) among the activity of different neurons, as measured by cross-correlating their spike trains (Aertsen et al. 1989; Aertsen and Preissl 1991; Gerstein and Perkel 1969), or, more generally, as the "temporal correlation between neurophysiological (functional) measurements made in different brain areas" (Friston et al. 1993, 9). Hence functional connectivity is essentially a "model-free" description of the *joint state* of multiple brain elements (e.g., neurons, regions). Functional connectivity is at times contrasted with "*effective* connectivity," which attempts to explain the origins of the observed functional relationship and, by definition, assumes a model (Stevenson et al. 2008). For instance, when applied to neurons, "effective connectivity" was initially defined as the simplest neuron-like physical circuit that would produce the same temporal relationship observed experimentally between neurons in a cell assembly.

The relationship between structural and functional connectivity is a complex one. In principle, responses in two regions could be perfectly correlated (barring noise) due to common inputs (figure 8.7B). But these two regions could also be perfectly correlated yet have the effect entirely mediated via an intermediate region (figure 8.7C). More generally, we need to consider at least three types of relationship between structural and functional connectivity. In the first type—the "default condition"—a functional relationship between two regions is likely the result of relatively direct structural connectivity (figure

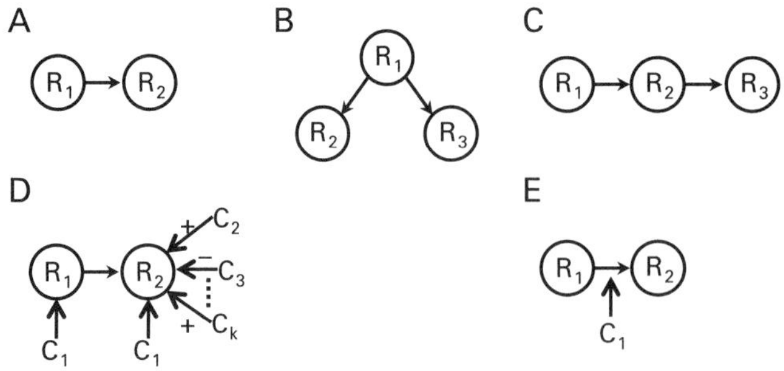

Figure 8.7
Function and structure. The relationship between structure and function can be nuanced and complex. Whereas panel A depicts the "default" structural connection leading to a functional relationship, panels B–E depict more intricate scenarios. See text for further discussion. R, region; C, context.

8.7A). In the second, a functional relationship between two regions is *not* supported in a direct manner by structural connectivity (figure 8.7C). And, in the third, structural, excitatory connectivity between two regions exists in the absence of a detectable functional relationship. In this third, less intuitive type of relationship, many scenarios are conceivable. For instance, in figure 8.7D, R_1 and R_2 are physically connected, but, whereas R_1 is robustly engaged only during context C_1, R_2 is engaged both during context C_1 and during a range of other contexts, some of which may lead to a suppression of baseline levels of activity in the region. Accordingly, unless researchers focus exclusively on context C_1, they are not likely to observe a functional relationship between R_1 and R_2.

A variant of the third type of relationship was illustrated by Biyu He and colleagues (2007), who reported a case in which functional connectivity between regions was disrupted in the *absence* of anatomical damage to those regions or their connections, specifically, interhemispheric functional connectivity of posterior cortex (intraparietal sulcus) in spatial neglect patients (see note 3 to chapter 3). The authors suggested that anatomical connectivity may be necessary but *not sufficient* for normal functional connectivity: excitatory/inhibitory neuronal inputs from other regions may be needed as well. Another illustration is given by a study of patients with multiple sclerosis, a pathology that compromises central white matter in a diffuse manner (Hawellek et al. 2011). Functional connectivity in the patients increased in the face of a concomitant *reduction* of anatomical connectivity.

Another instance of the third type of relationship between structural and functional connectivity, which has been investigated in a number of neuroimaging studies, reveals an important way that structure and function differ from each other (figure 8.7E). Specifically, the functional connectivity between two regions with known (or presumed) structural connections is *context dependent*: it can increase or decrease as a function of several variables without any concomitant modification in structure. These variables include task performance (Rissman, Gazzaley, and D'Esposito 2004), motivation (Padmala and Pessoa 2011), and emotion (Pessoa, McKenna, et al. 2002). In one study, my colleagues and I investigated the functional connectivity pattern between early visual areas in affective and neutral contexts (Damaraju et al. 2009). In the affective context, subjects viewed faces surrounded by a ring whose color signaled the possibility of mild shock (figure 8.8A; plate 10); in the neutral context, they viewed faces surrounded by a ring whose color signaled safety, namely, that a shock would not occur. We found that a measure of functional connectivity was considerably strengthened in the affective versus neutral context (figure 8.8; plate 10). Thus an affective context not only changes the

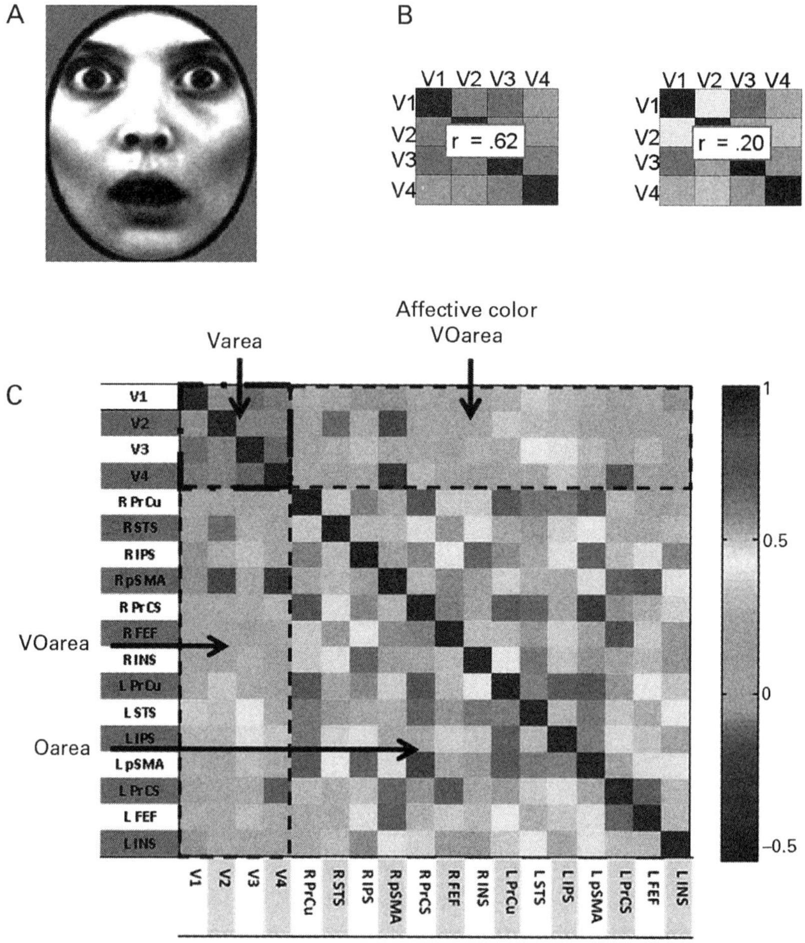

Figure 8.8 (plate 10)
Emotion alters the response pattern across early visual cortex. (A) Both fearful and neutral (not shown) faces were presented with a ring that indicated whether it was a safe or threat (possible mild shock) condition. (B) Functional connectivity between visual areas V1, V2, V3, and V4 during affective (left) and neutral (right) context as indexed using pairwise correlations (average correlation shown). (C) Correlation matrix of differential responses for fearful versus neutral faces in both affective and neutral contexts (ring colors). The upper-left part is the same as in panel B. Of the entire matrix, that was the only part that was robustly altered as a function of context. Varea, correlations within early visual areas; Oarea, correlations within "other areas"; VOarea, correlations between early visual and "other areas"; L, left; R, right; FEF, frontal eye field; INS, insula; IPS, intraparietal sulcus; PrCs, precentral sulcus; PrCu, precuneus; pSMA, presupplementary motor area; STS, superior temporal sulcus. Colors code for correlation values as indicated by the color scale. Adapted with permission from Damaraju et al. 2009.

magnitude of evoked responses as commonly observed, but also altered the response *pattern* of responses across early visual cortex.

The results obtained by my colleagues and I revealed an increase in functional connectivity in early visual areas for fearful versus neutral faces in an affective color context (Damaraju et al. 2009). This suggests that affective salience impacts the *relationship* between responses evoked in different brain regions and that signals are integrated in early, retinotopically organized areas in a manner that depends on affective information. Critically, the impact of context was specific to early visual areas since the correlations observed within "other areas" did not differ as a function of context in the same manner. Combined, the results indicated that during an affectively potent condition, responses across early visual cortex become more "coherent," suggesting that an increase in functional integration accompanies such conditions. In the absence of potential threat, responses across early visual cortex were more varied and possibly less well integrated. The increase in functional integration in response to affective context may be an important way in which affective stimuli modulate visual processing and have an increased impact on perception and behavior (see also figure 8.10; plate 11 and related text).

A growing number of studies are seeking to explore the relationship between functional and structural connectivity more generally. In their review of studies, Jessica Damoiseaux and Michael Greicius (2009) employed both tractography and resting-state connectivity to compare functional and structural connectivity.[4] Although some of their findings supported a direct relationship between the two, several did not. For instance, they detected no fiber tracts between medial prefrontal cortex and the medial temporal lobe, yet observed robust functional connectivity between them. Moreover, Bharat Biswal and colleagues (1995) observed interhemispheric correlation of functional MRI signals in cortical regions having only a few direct connections, such as left and right hand representations in primary motor cortex and spatial representations in left and right primary visual cortex (see also Lowe, Mock, and Sorenson 1998). And, finally, in comparing anatomical and functional connectivity between regions of interest on adjacent cortical gyri, Martin Koch, David Norris, and Margret Hund-Georgiadis (2002) found that, whereas low functional connectivity rarely occurred in combination with high anatomical connectivity (as expected), high functional with low anatomical connectivity often did.

A striking example of structure-function dissociation was reported by J. Michael Tyszka and colleagues (2011), who investigated a population of adults lacking a corpus callosum, the major communication pathway between the brain's two hemispheres. During the resting state, despite their marked structural difference, these adults exhibited a high degree of functional similarity to controls. "The present findings argue that largely normal functional networks can emerge in brains with dramatically altered structural connectivity. . . . Perhaps the most profound aspect of the present findings is the suggestion that the functional organization of the brain subserving cognition can be driven by factors other than direct structural connectivity" (Tyszka et al. 2011, 15161). And, finally, although Dante Mantini and colleagues (2011) found that resting-state functional connectivity in the monkey brain fit well with data on structural connectivity, in some cases, they also observed resting-state connectivity in the *absence* of documented monosynaptic anatomical connectivity. Assuming that their estimates of functional connectivity are valid, the findings by Mantini and colleagues imply that functional connectivity exists between regions that are not monosynaptically connected. If so, *poly*synaptic connections might be supporting network organization during the resting state. Together, the findings of the studies reviewed here further highlight the complex relationship between structural and functional connectivity (see also van den Heuvel et al. 2009).

An additional consideration in understanding functional connectivity pertains to the "dynamics" implemented in particular regions. Thus, as is well documented, lateral prefrontal circuits can implement reverberating activity that can be sustained for several seconds (e.g., Fuster 2008). Such dynamics not only extend the repertoire of lateral PFC computations but also influence the precise form of functional connectivity that lateral prefrontal cortex has with other regions (Honey et al. 2007). Therefore, to elucidate functional interactions, we need to know more not only about structural connectivity properties, but also about how local physiological properties impact both short- and long-range brain interactions—yet another example of the complex relationship between structural and functional connectivity.

Functional Networks

Just as structural connectivity can be used to determine brain networks, so can functional connectivity, too. As mentioned previously, based on patterns of coactivation across a large number of studies, Toro, Fox, and Paus (2008) identified three functional networks: a fronto-parietal attention network, a resting-state network, and a motor network. Using a similar approach, Ronald Postuma and Alain Dagher (2006) identified patterns of coactivation between cortex and the striatum that appear to reflect the anatomical model of parallel cortico-striatal loops proposed by Garrett Alexander, Mahlong DeLong, and Peter Strick (1986).[5] *Large-scale* networks have also been investigated based on functional data. For example, in their study of resting-state activity,

Raymond Salvador and colleagues (2005) submitted the pattern of coactivations between ninety regions (involving major cortical gyri and subcortical nuclei) to hierarchical clustering analysis; they identified six major systems corresponding roughly to four cortical lobes, a medial temporal cluster, and a cluster of subcortical nuclei. As in the analysis of structural connectivity, analysis of *functional* path length between regions was consistent with a small-world topology.

In the past decade, graph-theoretic analysis of functional neuroimaging data has focused almost exclusively on characterizing the large-scale properties of resting-state datasets (Bullmore and Sporns 2009; Wang, Zuo, and He 2010). In a recent study, however, my colleagues and I sought instead to understand the network properties of a focused set of brain regions during task conditions engaging them (Kinnison et al. 2012). Using graph-theoretic network analysis to characterize how emotional and motivational "mini-states" potentially alter functional connectivity, we investigated two separate tasks in which emotional or motivational cues separately preceded the execution of a response-conflict task. In the emotion task (Choi, Padmala, and Pessoa 2012), subjects viewed an initial cue that determined whether they were in a threat or safe trial (see figure 5.7 for the task structure); in the motivation task (Padmala and Pessoa 2011), they viewed an initial cue that indicated whether they were in a reward or control trial (see figure 6.4 for the task structure).

The effects of emotional and motivational cues exhibited several similarities. At the network level, *global efficiency* (a measure of integration) increased, and *decomposability* (a measure of how easily a network can be divided into smaller subnetworks or "communities") decreased.[6] In other words, the network became less segregated when the context was signaled by the cue (possible shock in one experiment, or possible reward in the other), revealing that one way in which both emotional and motivational processing affect brain responses is by increasing functional connections across brain regions.

The dual competition model (chapter 7) proposes that the effects of reward during perception and cognition depend in part on interactions between valuation regions and fronto-parietal regions important for attention and executive control. Such interactions lead to the upregulation of control and improve behavioral performance during challenging task conditions with higher likelihood of reward. The increased functional connectivity between the two communities that we detected in the motivation dataset (figure 8.9A) is consistent with these ideas and also suggests that the increases in connectivity can be quite broad. For example, the caudate (figure 8.9B) and the nucleus accumbens showed increases in functional connectivity to nearly all cortical regions that were systematically reward driven. This finding considerably extends the

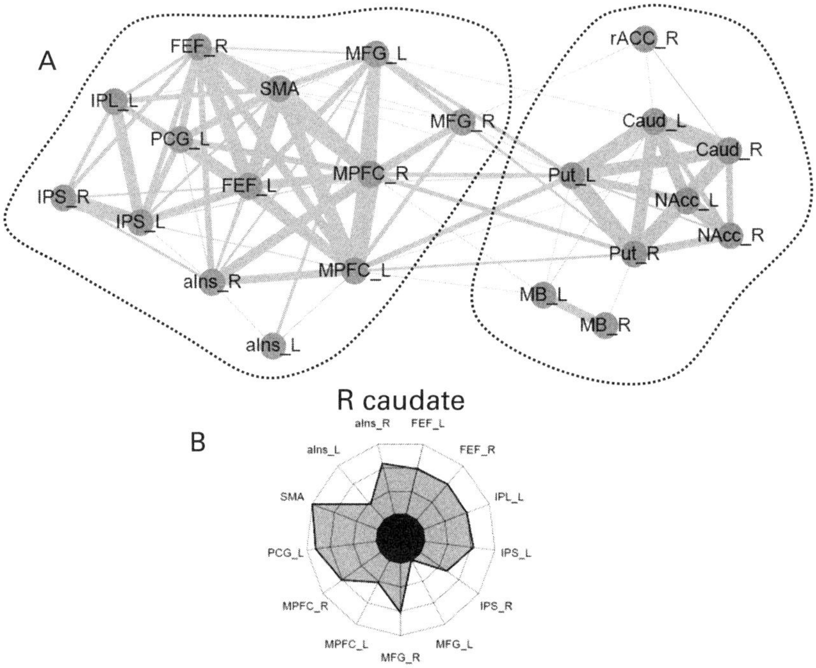

Figure 8.9
Network structure and reward. (A) Community detection was applied to the set of brain regions that responded more strongly to reward than to no-reward context at the cue phase (see figure 6.4). Two communities were detected. _L, left; _R, right; Caud, caudate; FEF, frontal eye field. (B) Comparison of the pattern of connectivity between reward and no-reward contexts revealed increases during the former, mostly *between* the two communities, reflecting increased integration with reward. The polar plot shows increases in functional connectivity of the right caudate with nearly all regions belonging to the "other" community. Line width represents the relative strength of the functional connectivity between regions. IPL, inferior parietal lobe; aIns, anterior insula; IPS, intraparietal sulcus; PCG, precentral gyrus; MB, midbrain; MFG, middle frontal gyrus; MPFC, medial prefrontal cortex; NAcc, nucleus accumbens; Put, putamen; rACC, rostral anterior cingulate cortex.

inferences that can be made from using standard functional connectivity analysis (including "psychophysiological" analysis; Friston et al. 1997).

We observed a related pattern with the emotional manipulation (Kinnison et al. 2012). In this case, functional integration increased between subcortical regions, such as the bed nucleus of the stria terminalis and thalamus, and cortical regions, including the insula and medial prefrontal cortex. A notable difference between the results of the two datasets concerned the way functional connectivity changed within the communities involving cortical areas. In the case of reward, functional connectivity increased in the cortex, whereas, in the case of threat, functional connectivity decreased between several pairs of corti-

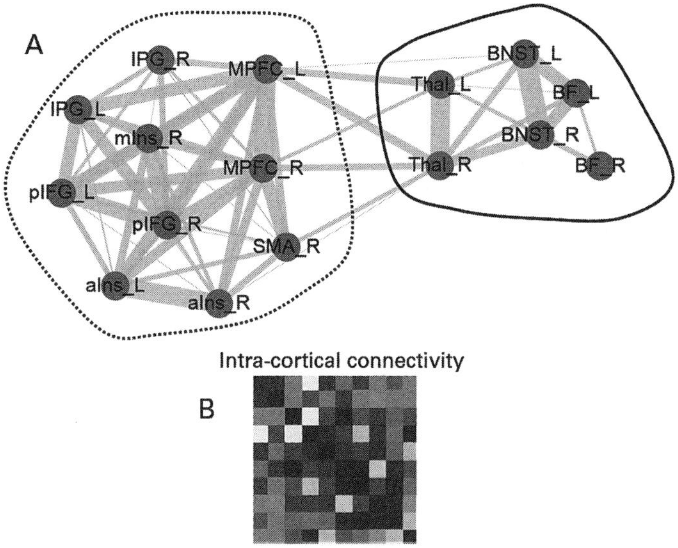

Figure 8.10 (plate 11)
Network structure and threat. (A) Community detection was applied to the set of brain regions that responded more strongly to threat than to safe context at the cue phase (see figure 5.7). Two communities were detected. _L, left; _R, right; BF, basal forebrain; BNST, bed nucleus of stria terminalis; mIns, midinsula; MPFC, medial prefrontal cortex; IPG, inferior parietal gyrus; pIFG, posterior inferior frontal gyrus; SMA, supplementary motor area; Thal, thalamus. (B) Changes in threat versus safe connectivity for all pairs of regions within the community on the left in panel A. Dark colors indicate no change; warm colors indicate threat connectivity greater than safe; cool colors indicate threat connectivity smaller than safe. One of the effects of threat may have been to "disconnect" cortical regions from each other, possibly leading to performance impairments.

cal regions (figure 8.10; plate 11). How should we interpret these findings? The dual competition model proposes that emotional processing interferes with executive function by diverting needed resources, thus impairing cognitive performance. Reduced functional connectivity among some of the cortical areas may therefore have reflected the interference effect that the threat of shock exerted on subsequent cognitive performance (the conflict task)—responses across some of the cortical regions may have become less coherent with threat.

Our findings (Kinnison et al. 2012) revealed several ways in which both emotional and motivational processing altered functional connectivity, including increased global efficiency and reduced decomposability. An analysis of magnetoencephalography data has suggested that greater *cognitive* effort is associated with the emergence of a less modular network topology (Kitzbichler et al. 2011). Because MEG is less sensitive to deeper brain signals, it is likely that the changes observed were more directly tied to cortical processing. The results described here show that the processing of emotional and motivational stimuli may have a similar impact on network organization, but that it may emphasize enhanced vertical, *cortico-subcortical* functional integration in a behaviorally appropriate manner. Potential reward may contribute to improved task performance (and reward attainment), and potential threat may redirect mental resources in the service of mobilizing the body toward safeguarding the organism against harm.

"Importance": Structural and Functional Embedding

Neurons whose responses are systematically related to a task are typically assumed to play a role in the underlying computations performed by the brain when processing that task. In practice, some form of statistical test is applied to establish the significance of the association. Iannis Vlachos, Ad Aersten, and Arvind Kumar (2012) provocatively suggest, however, that we need to go "beyond statistical significance" to consider also the implications of *network structure* with respect to neuronal activity. Noting that "not all observed activity modulations of neurons in a task are relevant for the specific task itself," they assert that "statistical significance of recorded neural events is only a *necessary* but not *sufficient* condition for making inferences regarding the functional importance of these events for the computations performed by the investigated brain area" (Vlachos, Aersten, and Kumar 2012, 4).

How should we go about determining the *importance* of particular neurons to a given computation? One strategy is to consider their degree of *structural embeddedness* and *functional embeddedness* (or "effective embeddedness," as termed in Vlachos, Aersten, and Kumar 2012). "Structural embeddedness" refers to the way neurons are physically embedded in their surrounding network; "functional embeddedness" refers to the influence neurons have on the activity of the surrounding network, which depends on structural embeddedness, in addition to other synaptic and cellular properties, ongoing neuronal activity, neuromodulators, and the like. Thus the importance of task-related neurons is governed by their relative position in the topological space of the network.

This reasoning can be applied to brain areas and not just particular neurons. Take the case of structural embeddedness. Regions (i.e., nodes) with a high degree of connectivity have the potential to be influential, particularly if they function like connector hubs, as described earlier. Counterintuitively, however, in some circumstances, the strongest influence does not correspond to the most highly connected nodes of a network (Kitsak et al. 2010; see also Liu et al. 2011). Instead, the most influential nodes are those located within the *core* of the network. In other words, they belong to a *topologically central* subnetwork.[7] Hence one way to measure embeddedness is to determine nodes that exhibit the property of *centrality* (Newman 2010).

In general, no single measure will perfectly capture *influence* or *importance* because different measures will convey different aspects of network structure. Indeed, many different measures of centrality have been proposed (Rubinov and Sporns 2010), and, not surprisingly, these generate different results (Zuo et al. 2012). Therefore a combination of different metrics will provide a better measure of embeddedness and, more important, of how a node affects network properties (Vlachos, Aersten, and Kumar 2012; Modha and Singh 2010; see also Liu et al. 2011).

Networks: Further Issues

Although networks have increased our understanding of structure-function relationships in the brain, and even though network theory research has advanced considerably since the publication of seminal papers in the late 1990s, several important questions remain regarding the application of existing techniques.

Community Assignment and Overlap

A key issue of network science is the determination of densely interconnected groups of elements known as "communities."[8] For instance, figure 8.6 displays a decomposition of brain regions in terms of frontal, parietal, cingulate, and the like communities. Most partitioning schemes of brain networks group individual elements (brain regions) into unique communities; for example, Brodmann area 46 belongs to the "frontal community." This widely used approach is inadequate, however, for many complex systems, whose elements naturally belong to multiple communities (Ahn, Bagrow, and Lehmann 2010; Palla et al. 2005). Thus, in social networks, individuals may be members of several different communities at the same time, with each community characterized by a different type of relationship (family, work, sports).

Network scientists determine communities by dividing a network into groups of nodes, often by maximizing the number of *within*-group links, and minimizing the number of *between*-group links (Girvan and Newman 2002). In other words, it is typically accepted that a community should have more internal than external connections, a property that is aligned with the notion of a dense group of elements (e.g., regions, people). But consider, for instance, the communities around the word "Newton" in a network of commonly associated English words (Ahn, Bagrow, and Lehmann 2010). "Newton" is part of many intersecting communities (scientists, physics, gravity). Critically, allowing

"Newton" to participate in several communities provides a better description of its relationship to other words than forcing it to belong to a single one. Thus the multiple relationships of "Newton" are simultaneously captured, instead of focusing on a (possibly) dominant one—one for which "Newton" has the greatest strength of membership, say. Highly overlapping communities can have more *external* than internal connections (Ahn, Bagrow, and Lehmann 2010). In fact, conceptualizing networks in terms of overlapping communities prevents the clear-cut attribution of individual nodes to specific communities.[9]

As in many applications of network theory, the investigation of brain networks has focused on establishing unique partitions—that is, *non*overlapping communities of regions. For instance, in the study by Dharmendra Modha and Raghavendra Singh (2010), as noted above, regions in the frontal lobe constituted a frontal community, regions in parietal lobe, a parietal community, and so on. Earlier work that did not rely on algorithmic methods of dividing brain networks into subnetworks also favored describing nonoverlapping communities. For example, Mesulam (1998) identified at least five major networks in the human brain, which he characterized as relatively separable.[10] And Bressler and Menon (2010) identified three *non*overlapping networks: a "default mode" (resting-state) network, a salience network, and a central-executive network. Although these are valuable proposals of how multiregion coalitions carry out mental functions, it is unclear whether decomposing the brain into nonoverlapping networks reflects the best way to group its cortical and subcortical regions. Moreover, it is important to note that nonoverlapping partitioning schemes poorly characterize brain organization (see Sadaghiani et al. 2010; Ferrarini et al. 2009). One telling illustration of their inadequacy is the network overlap observed at connector-hub regions, which crucially combine diverse sources of information to produce novel signals (Zamora-Lopez, Zhou, and Kurths 2010). Consider, for example, a connector-hub region as shown in figure 8.4. Whereas a traditional, nonoverlapping decomposition would place this region in one of the two network communities, the hub clearly belongs to both. Put more generally, the characterization of community overlap and how this depends on context is an exciting direction for future research. Indeed, community overlap has vital implications for understanding structure-function relationships because region overlap generates several of the difficulties reviewed during the discussion of region-to-function mapping (figure 8.1)—overlap, by definition, compounds the one-to-many problem.

Optimal Partitions

A second key issue of network science is that complex networks contain structure at *multiple scales* simultaneously. Such networks are frequently encountered in the brain. For example, the "visual network" can be decomposed into ventral and dorsal subnetworks (Ungerleider and Mishkin 1982), and prefrontal cortical networks involved in executive control can be decomposed into "rapid adaptive control" and "stable set control" subnetworks (Dosenbach et al. 2006). Multiscale information actually suggests that we may need to go beyond establishing communities; instead, it is important to explicitly account for information at multiple scales in a network *simultaneously*. For example, Averbeck and Seo (2008) detected not only prefrontal communities, but also a possible hierarchical structure within them.

The findings of Averbeck and Seo (2008) prompt the discussion of the "decomposition problem," a methodological issue central to understanding community detection methods. Although, in computing network partitions, one typically chooses the "best" among a larger set of candidate partitions, in complex networks, many other decompositions exhibit "scores" quite similar to the "best" candidate partition (see figure 9 of Good, de Montjoye, and Clauset 2010), which suggests that the "best" is not inherently superior to other decompositions. Likewise, most real-world networks have many plausible representations of roughly equal likelihood (Clauset, Moore, and Newman 2008; Good, de Montjoye, and Clauset 2010). Indeed, as summarized by Benjamin Good, Yves-Alexandre de Montjoye, and Aaron Clauset (2010, 046106–046110), "there are typically an exponential number of structurally diverse alternative partitions with modularities very close to the optimum." Averbeck and Seo (2008) proposed an ingenious strategy to tackle this problem with respect to prefrontal networks. They built a "consensus" prefrontal tree generated from the fifty top, most likely trees, which revealed many features not observed in the "best" tree—although many notable features were commonly found in the top trees.

More generally, I contend that there is no single inherently "best" solution to the decomposition problem and that many different representations of a network's structure may convey important information. It is thus better to see the representation of a network as a *family* of possible decompositions. Unlike laboratory conditions, real-world situations are *not* devoid of noise; here the problem of multiple viable solutions to the decomposition problem is considerably more acute. Thus treating specific decompositions as inherently superior is unwarranted absent compelling evidence to the contrary.

In a related matter, although hierarchical decompositions are frequently attempted for brain data, many may, in fact, be *indeterminate*. This is strikingly illustrated in the analysis of the connectivity pattern of the visual cortex. For some time, the cortical lamination pattern between the visual thalamus and area V1 has been used as a guide that is extrapolated to other connections, which are then labeled as "feedforward," (i.e., originating in superficial cortical layers and terminating in layer IV) or "feedback" (i.e., originating in inferior layers and terminating in layers I and VI while avoiding layer IV). Based on these assumptions, Claus-C. Hilgetag, Mark O'Neill, and Malcolm Young (1996) investigated the organization of the visual system and concluded that "the information in the anatomical constraints cannot be expressed satisfactorily by any single hierarchical ordering." In this sense, the visual hierarchy might be considered *indeterminate*, such that "no single hierarchy can represent satisfactorily the number and variety of hierarchical orderings that are implied by the anatomical constraints" (777).[11]

A similar conclusion was reached by Dwight Kravitz and colleagues (2013) in their recent, authoritative description of the ventral visual pathway. The authors reject the concept of a sequential, multistage architecture and suggest, instead, that the "ventral visual pathway is actually a complex recurrent network" (29). Their framework also speaks to the necessity of understanding brain regions in terms of overlapping communities as discussed above. Thus, in their view, many regions in ventral visual cortex (along occipital and temporal cortex) are simultaneously part of at least six circuits: circuits involving the amygdala and striatum, in addition to circuits involving the medial temporal lobe, ventral-lateral prefrontal cortex, and orbitofrontal cortex.

Multirelational Relationships and Communicability

Thus far, we have considered networks that are unirelational in the sense that the connection between two nodes represents only one kind of information, such as whether a connection is present or how strong it is. However, networks are often *multi*relational (also called "multiplex"), as when several types of information define the interactions between two individuals in social networks (Szell, Lambiotte, and Thurner 2010). In the case of brain data, structural connectivity information such as the strength, laminar profile, and type of neurotransmitter for anatomical connections may be available, as may functional connectivity information. Algorithms of community detection for multirelational networks (Mucha 2010) offer exciting new avenues to explore network structure.

A final issue to be discussed here concerns *communicability* in complex brain networks (Estrada and Hatano 2008). Many important measures that characterize networks are based on the shortest paths connecting two nodes, including the determination of communities. Counterintuitively, as described by Ernesto Estrada and Naomichi Hatano (2008), "information" can in fact spread along other than the shortest paths (see also Borgatti 2005; Newman 2005). This has implications for the understanding of brain networks because direct anatomical connectivity is frequently emphasized as the chief mode of communication between brain regions. The notion that communicability does not necessarily rely on shortest paths reminds us of the need to obtain network-level properties in describing the flow of signals in neural networks. It also highlights the need to characterize functional connectivity between regions, which does uniquely depend on direct anatomical connections.

Summary

The issues discussed in this section, as well as those described in "'Importance': Structural and Functional Embedding," point to the complexities in trying to characterize and understand brain networks. In reality the problem is even more daunting given that in this chapter the temporal dimension was barely discussed. At a minimum, investigators utilizing graph-theoretic analysis of brain data should, first, discuss the properties of multiple community decompositions with similar "optimization scores," especially when different classes of solutions offer comparable scores. Second, both overlapping (where a region can belong to more than one community) and nonoverlapping decompositions should be investigated. In particular, the "default" strategy of reporting simple nonoverlapping decompositions should be justified (and used sparingly).

Understanding a Region's Function via Functional Fingerprints

The availability of databases containing the results of thousands of neuroimaging studies allows us to investigate human brain function in novel ways (Yarkoni 2010). A popular strategy is to perform meta-analyses that address questions such as "Is the amygdala engaged selectively for fear?" And, given the wealth of information in these databases, we can investigate a host of other questions as well.

If brain regions are engaged in many processes based on the networks they are aligned with in particular contexts, they should be activated by a range of tasks in neuroimaging experiments. Michael Anderson, Josh Kinnison, and I employed a data-driven approach to investigate the functional repertoire of brain regions based on a large set of functional MRI studies (Anderson and Pessoa 2011; Anderson, Kinnison, and Pessoa 2013). We characterized the function of brain regions in a multidimensional manner via their *functional fingerprint* (see Passingham, Stephan, and Kotter 2002), classifying activations in terms of twenty task *domains* chosen to represent a range of mental processes, including perception, action, emotion, and cognition, as developed in the BrainMap database (Laird, Lancaster, and Fox 2005). The functional fingerprint

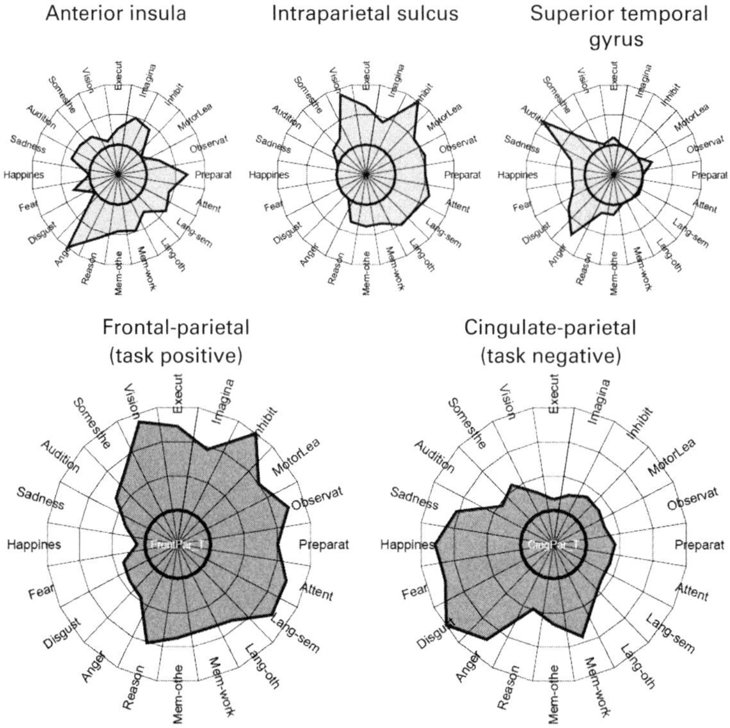

Figure 8.11
Functional fingerprints of regions and networks. (Top) Polar plots illustrate the fingerprints of three brain regions. Each vertex corresponds to one of the domains investigated. Both the left anterior insula and the left intraparietal sulcus exhibited diverse functional profiles. The superior temporal gyrus in the vicinity of auditory cortex was less diverse, though the fingerprint revealed its involvement in emotional processing, in addition to audition. (Bottom) Polar plots illustrate the fingerprints of two brain networks, which were defined by Toro, Fox, and Paus (2008) based on a meta-analysis of task activation data. The frontal-parietal "attention" network was a task-*positive* network generated by "seeding" the left intraparietal sulcus. The cingulate-parietal "resting-state" network was a task-*negative* network generated by "seeding" ventral-anterior medial prefrontal cortex. Although both networks are quite diverse, the analysis revealed that they are fairly complementary to one another.

for a given region thus represented both the set of domains that systematically engaged the given region and the relative degree of engagement (for functional fingerprints of sample regions see figure 8.11, top). The concept of functional fingerprint can be extended to include networks, thus helping describe the functional repertoire of coalitions of regions (figure 8.11, bottom).

From the brain region fingerprints, we calculated a *diversity index* to further characterize the degree of *functional diversity* (Anderson and Pessoa 2011; Anderson, Kinnison, and Pessoa 2013). A high-diversity brain region would be engaged by tasks in many domains, whereas a low-diversity region would be engaged by only a few domains. The scholarly literature is replete with measures of diversity, particularly in biology and economics (e.g., Magurran 2004). The Shannon (1948) diversity (or entropy), H, of a fingerprint was defined as

$$H = -\sum_{i=1}^{S} p_i \ln p_i$$

where $S = 20$ was the number of task domains and p_i corresponded to the ith domain proportion (for an improved Shannon index, see Chao and Shen 2003). The diversity we observed varied considerably across cortex (figure 8.12; plate 12), with "hot spots" apparent in dorsal-medial and dorsal-lateral prefrontal cortex and anterior insula, among others, and with "cool spots" apparent in lateral temporal cortex, parts of posterior medial frontal/parietal cortex, and ventral-medial prefrontal/orbitofrontal cortex, among others.

To probe whether there are principles that govern the organization of functional fingerprints in the brain, we examined how fingerprint similarity varied as a function of *task*-based coactivation strength, which allowed us to determine whether regions that tend to coactivate have more similar profiles. We computed task-based coactivation by choosing a specific "seed" region and interrogating the database to determine the likelihood that, if the seed region was active, another specific voxel was active (for every voxel). To establish functional similarity, we compared the functional fingerprint of the seed region to that of all voxels in the brain. Figure 8.13 (plate 13) shows results from a

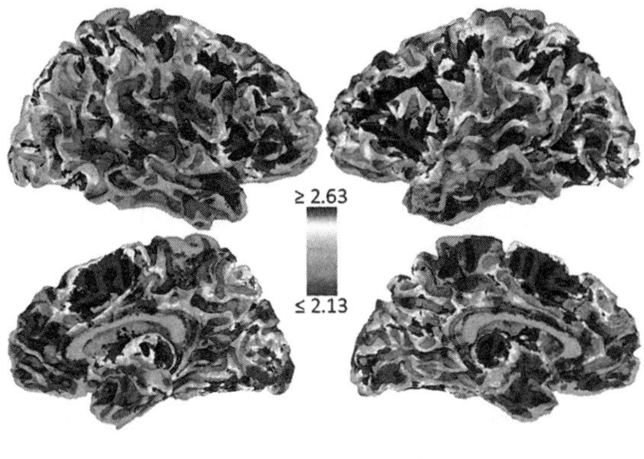

Figure 8.12 (plate 12)
Functional diversity map. Areas of higher functional diversity are shown in warm colors, and areas of lower diversity are shown in cool colors (color bar represents diversity H values). Locations without colors did not have sufficient findings for the estimation of diversity. Adapted with permission from Anderson, Kinnison, and Pessoa 2013.

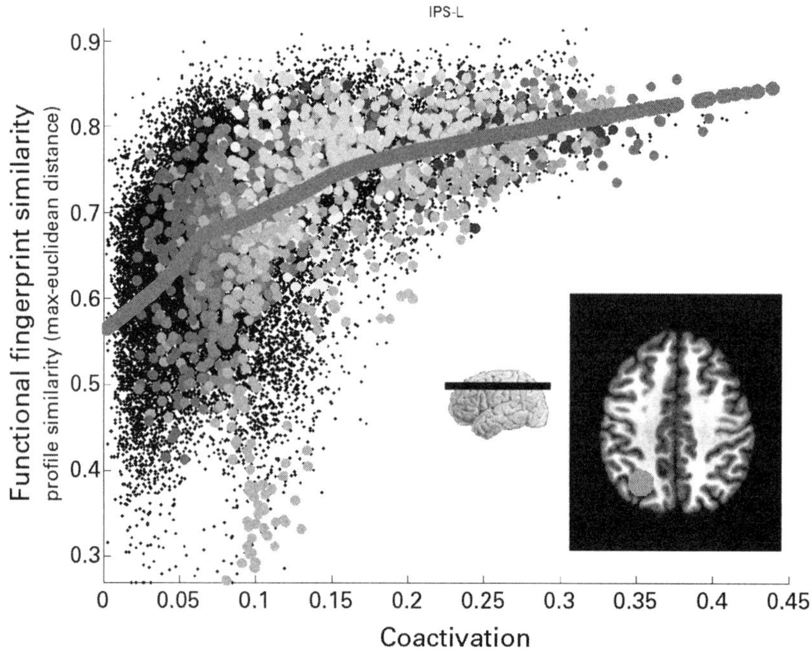

Figure 8.13 (plate 13)
Task coactivation and diversity. Task coactivation is plotted against functional fingerprint similarity. The seed region was the left intraparietal sulcus (red circle in inset). Each point represents one voxel in the brain. Spearman's rho = .53; $p < .001$. Data fit via standard locally weighted scatter plot smoothing ("lowess"). The colors refer to coordinates of regions from specific networks published in the literature. From Toro, Fox, and Paus 2008: red, frontal-parietal "attention" network (general task-positive); blue, cingulate-parietal "resting-state" network (task-negative). From Dosenbach et al. 2007: orange: frontal-parietal "rapid adaptive control" network; cyan: cingular-opercular "stable set control" network. From Yeo et al. 2011: dark green: "dorsal attention" network; yellow: "ventral attention" network; pink: "control network"; light purple: "default" network; light green: heterogeneous set of regions including visual and motor regions.

seed in the left intraparietal sulcus (each point represents a voxel). The scatter plot reveals that coactivation was significantly related to profile similarity: the higher the coactivation strength, the more similar the fingerprints between two sites.

Overall, we found that functional fingerprint similarity varied directly with (1) task-based coactivation strength and (2) resting-state functional connectivity. The two properties can be viewed as indicating general network *assortativity*, a preference for a node to link to others that have similar characteristics. For example, a friendship network may be highly assortative if it connects individuals who have similar tastes (Newman 2003). Our analysis revealed that brain regions that coactivate during tasks tended to have more similar profiles; regions that are functionally connected at rest also tended to exhibit more similar profiles. Our next objective was to investigate the relationship of the functional repertoires of regions belonging to a given network. In other words, how homogeneous are the set of fingerprints of a specific network? Are fingerprints from network *X* more similar to each other than to those from network *Y*? Answering this question may help advance our understanding of brain networks and possibly reveal common principles of organization. Thus we studied properties of well-established networks, such as the "dorsal attention," "ventral attention," and "executive control" networks (Corbetta, Patel, and Shulman 2008; Dosenbach et al. 2008). In addition, we studied "task-negative" networks defined either via coactivation data (Toro, Fox, and Paus, 2008) or resting-state data (Yeo et al. 2011).

To address whether a region is more likely than not to display a functional fingerprint more similar to those of regions of the same network than to fingerprints of other networks, we set about "contrasting" two networks (Anderson, Kinnison, and Pessoa 2013). We compared them to each other by using a multivariate test based on the "statistical energy" (Aslan and Zech 2005) of two samples (each comprising a set of functional fingerprints, one set per network) to evaluate whether the networks were drawn from the same parent distribution. This strategy revealed that, for instance, the dorsal and ventral attention networks (Corbetta and Shulman 2002) are indeed fairly distinct based on their functional fingerprints. This simple strategy can be easily generalized to help understand the relationship of the functional profiles of regions across a given network. Thus a positively assortative network would be one in which the average functional distance between region pairs within the network are smaller than distances between regions of that network and all regions outside it (on average).

We analyzed a group of eight networks previously described in the literature (Anderson, Kinnison, and Pessoa 2013). For all "task-positive" networks, we found that *within*-network fingerprint differences tended to be smaller than *between*-network differences, consistent with assortativity. In contrast, a "default" network, which was based on resting-state functional data (see Yeo et al. 2011), exhibited near-zero assortativity, and the task-negative network of Toro, Fox, and Paus (2008), which was based on coactivation data (shown in figure 8.11, bottom), was *dis*assortative. These results are consistent with the notion that task-negative networks are relatively *heterogeneous*. Along these lines, recent work has suggested that task-negative networks in fact can be decomposed in a number of ways (Andrews-Hanna et al. 2010).

Together, our findings suggest that brain regions are functionally highly diverse, in line with inferences drawn by Russell Poldrack (2006, 2011). The findings also indicate that the operations of a given brain region can be understood in terms of its *functional fingerprint*, namely, the task domains that systematically engage the region. Beyond its descriptive aspects, the functional fingerprint approach outlines a framework in which a region's function is viewed as inherently multidimensional: a vector defines the fingerprint of a region in the context of a specific domain structure. Although the domain that we employed used a task classification scheme from an existing database, it was not the only one possible. How should one define the domain structure? One hope is that cognitive ontologies can be defined that meaningfully carve the "mental" into stable categories (Bilder et al. 2009; Price and Friston 2005). I contend, however, that *no* single ontology will be sufficient. Instead, in characterizing brain function and behavior, it is better to conceive of several task domains that are useful and complementary. Thus a region's functional fingerprint needs to be understood in terms of a *family* of (possibly related) domains. Finally, the framework can be extended to networks to elucidate, for instance, the degree to which they are assortative.

Conclusions

In this chapter, I discussed several aspects that inform the mapping between structure and function in the brain. In it, I argued (as others before me) that a network perspective should supplant the common strategy of understanding the brain in terms of individual regions. Although this perspective is needed for a fuller characterization of the mind-brain, it should not be viewed as a panacea. For one thing, it does not dissolve the challenges posed by the many-to-many mapping between regions and functions. Clearly, however useful the network approach may be, we should *not* anticipate a *one*-to-*one* mapping when it is adopted. Furthermore, decomposition of the brain network into meaningful clusters of regions, such as those generated by community-finding algorithms, does not by itself reveal "true" subnetworks. Given the complex and multirelational relationship between regions, multiple decompositions will offer different "slices" of a broader landscape of networks within the brain. And, finally, I described how the function of brain regions can be characterized in a multidimensional manner using diversity profiles, which can also be used to describe the way different brain regions participate in networks.

Notes

1. With respect to pluripotentiality and degeneracy, see Pessoa 2008 for examples involving emotion and cognition; Cisek 2010 for examples involving perception and action; and Schultz 2006 for examples involving dopamine function.

2. The operculum is the part of cerebral cortex that covers the cortex within the lateral sulcus (also called the "Sylvian fissure"), which includes the insula.

3. This definition of "degree" applies to "binary" networks in which edges are either present or not. "Degree" can also be defined in weighted networks, where edge strength varies continuously.

4. "Tractography" refers to computational techniques that estimate major white matter fiber tracts based on diffusion-weighted imaging (Basser et al. 2000); "resting-state connectivity" refers to methods that evaluate functional connectivity across brain areas during conditions that do not involve overt tasks, that is, during "rest" (Fox and Raichle 2007).

5. The "striatum" refers to the caudate and putamen, structures that are part of the basal ganglia. It is also useful to distinguish the "*dorsal* striatum" (i.e., caudate and putamen) from the "*ventral* striatum," which includes the nucleus accumbens. Intriguingly, the "head" of the caudate, like the prefrontal cortex areas that project to it, may have evolved in primates (Passingham and Wise 2012).

6. Although the technical term in network science is "modularity," which is used in a semantically neutral fashion, "modularity" in neuroscience is a tremendously loaded term.

7. One way to evaluate the topologically central subnetwork is to use k-shell decomposition (Pittel, Spencer, and Wormald 1996). Intuitively, core decomposition recursively "peels off" the least connected nodes to reveal progressively more closely connected subnetworks.

8. Although these communities are at times also called "modules," the term should not be interpreted as a module in the cognitive science sense.

9. More technically, the decomposition problem is related to the existence of structure at multiple scales simultaneously (Ahn, Bagrow, and Lehmann 2010; see also "Optimal Partitions").

10. Though Mesulam (1990) also discussed network overlap.

11. These ideas are also aligned with the multiple waves model described in chapter 3. From the model's perspective, the notion of hierarchy is *not* well defined.

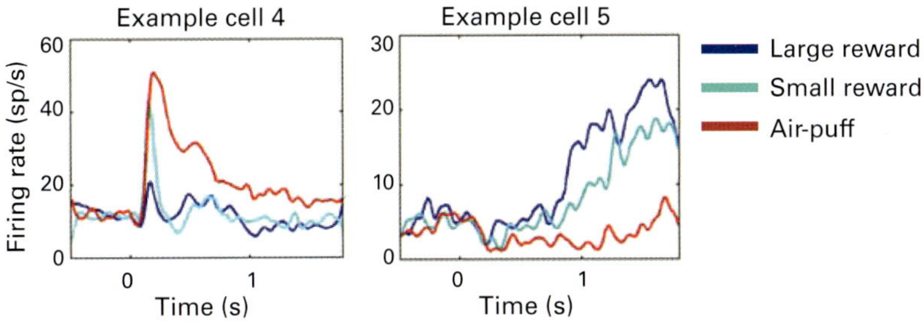

Plate 1 (figure 2.9) Graded representation of value. (Left) Cell responses during presentation of the visual conditioned stimulus from a cell that encoded negative value. (Right) Cell responses during the "trace interval" (interval between the visual stimulus offset and outcome delivery) from a cell that encoded positive value. Reproduced with permission from Belova, Paton, and Salzman 2008.

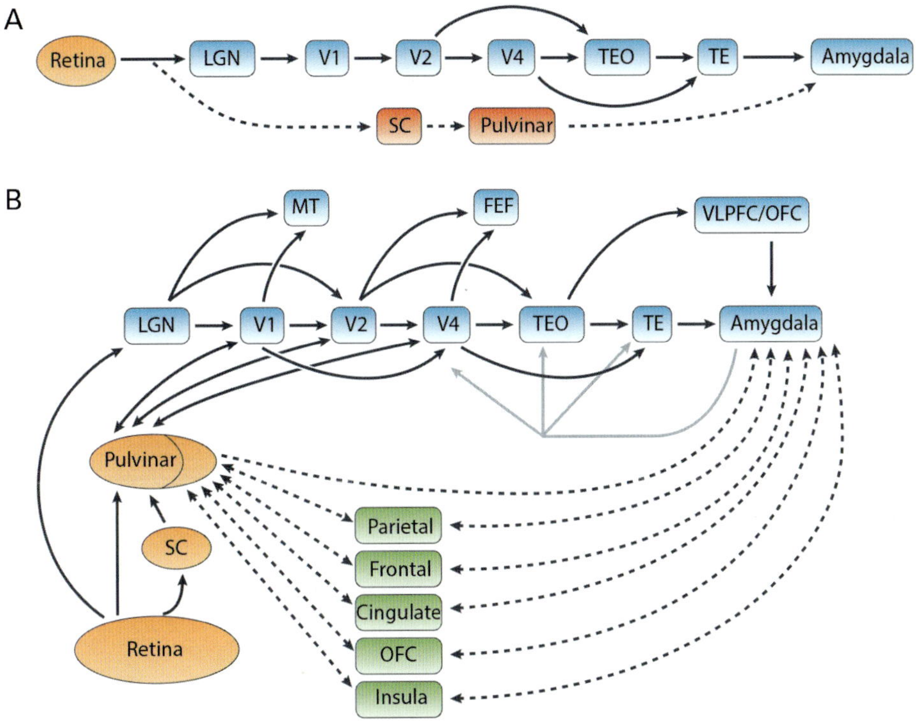

Plate 2 (figure 3.1) Visual pathways. (A) Traditional flowchart of visual processing typically emphasizes the lateral geniculate nucleus–V1–V2–V4–TEO–TE pathway, although the scheme is not strictly hierarchical. According to the "standard hypothesis," a subcortical pathway involving the superior colliculus and the pulvinar nucleus of the thalamus provides fast and automatic access to the amygdala. (B) Alternative flowchart of visual signals via multiple pathways, including "shortcuts." The "multiple waves" of activation initiate and refine cell responses. For simplicity, feedback pathways, which are extensive, have been omitted. The existence of such feedback pathways dictates, however, that a complex ebb and flow of activation sculpts the neuronal profile both of activation throughout visual cortex and of amygdala responses. Some of the connections between the pulvinar and visual cortex and between the pulvinar and "associational" areas are also indicated. The curved line in the pulvinar schematically separates the medial pulvinar (to the right of the line) from the rest of the structure. FEF, frontal eye field; LGN, lateral geniculate nucleus; MT, medial temporal area (also known as "area V5"); OFC, orbitofrontal cortex; SC, superior colliculus; TE, TEO, inferior temporal areas TE, TEO; VLPFC, ventral-lateral prefrontal cortex; V1, V2, V4, visual areas 1, 2, 4. Reproduced with permission from Pessoa and Adolphs 2010.

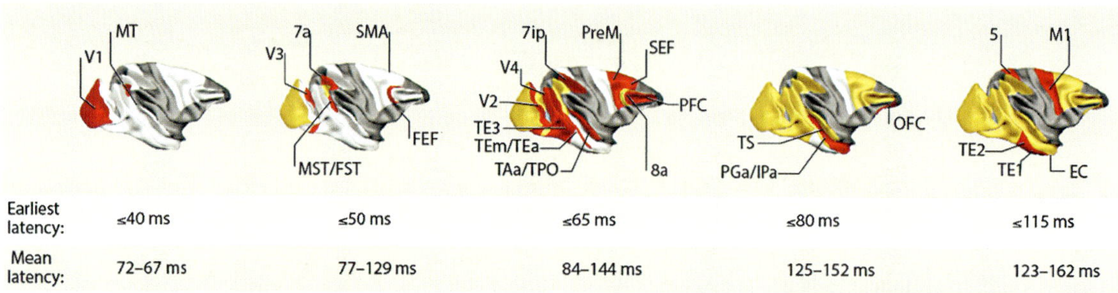

Plate 3 (figure 3.2) Response latencies to visual stimulation in macaque cortex. The earliest latencies are remarkably short, and even mean response latencies reveal very fast cortical processing. Areas that became active at the given latency after visual stimulation are shown in red, those activated earlier in yellow and those not yet activated in white. Areas for which no information was available are shown in dark gray. EC, entorhinal cortex; FEF, frontal eye field; FST, fundus of superior temporal cortex; 5, 7a, 7ip, 8a, areas 5, 7a, 7 intraparietal, 8a; IPa, superior temporal area IPa; M1, primary motor cortex; MST, medial superior temporal cortex; MT, medial temporal area (also known as "V5"); OFC, orbitofrontal cortex; PFC, prefrontal cortex; PGa, superior temporal area PGa; PreM, premotor cortex; SEF, supplementary eye field; TAa, anterior subregion of superior temporal area TA; TE1–TE3, inferior temporal areas TE1–TE3; Tem/TEa, medial and anterior subregions of inferior temporal area TE; TPO, superior temporal area TPO; TS, superior temporal sulcus; V1–V4, visual areas 1–4. Reproduced with permission from Pessoa and Adolphs 2010 and adapted from Lamme and Roelfsema 2000.

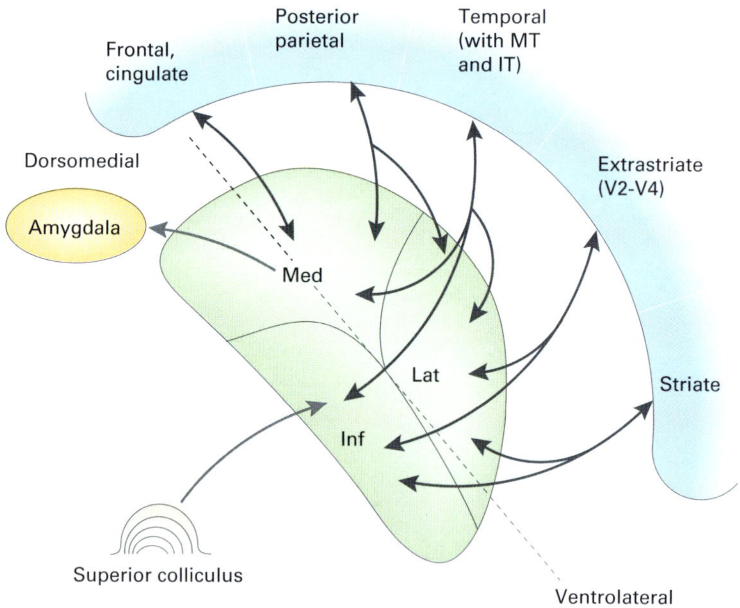

Plate 4 (figure 3.6) Schematic diagram of pulvinar connectivity. Most pulvinar nuclei and subnuclei (including those not shown here) are involved in thalamo-cortical loops that target different cortical territories. The inferior nucleus is reciprocally connected to striate and extrastriate cortex, the lateral nucleus is connected to association cortex in temporal and parietal lobes (as well as extrastriate cortex), and the medial nucleus is connected to higher-order association cortex in parietal, frontal, orbital (not shown), cingulate and insular regions (not shown), in addition to the amygdala. Thus the medial nucleus is not only connected with the amygdala but is also part of multiple thalamo-cortical loops (note, however, that the connection to the amygdala does not seem to be bidirectional). The superior colliculus is a layered structure whose superficial layers are visual in nature and project to the inferior nucleus. Its intermediate and deeper layers are multimodal and involved in motor preparation, including for eye movements, and project to the medial nucleus. Inf, inferior; IT, inferior temporal cortex; Lat, lateral; Med, medial; MT, medial temporal area (also known as "V5"). Reproduced with permission from Pessoa and Adolphs 2010 and adapted from Stepniewska 2004.

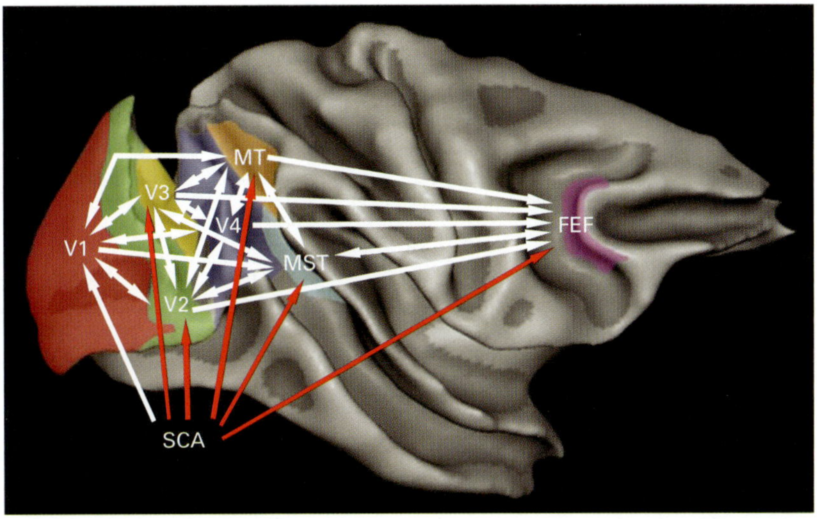

Plate 5 (figure 3.10) Visual cortex connectivity model. Connectivity was determined by taking structural data into account and estimating the "best-fit" pathways based on cell response timing. Including subcortical regions and their connections (red arrows) provided the best model. FEF, frontal eye field; MST, middle superior temporal area; MT, middle temporal area; SCA, subcortical area; V1–V4, visual areas 1–4. Reproduced with permission from Capalbo, Postma, and Goebel 2008.

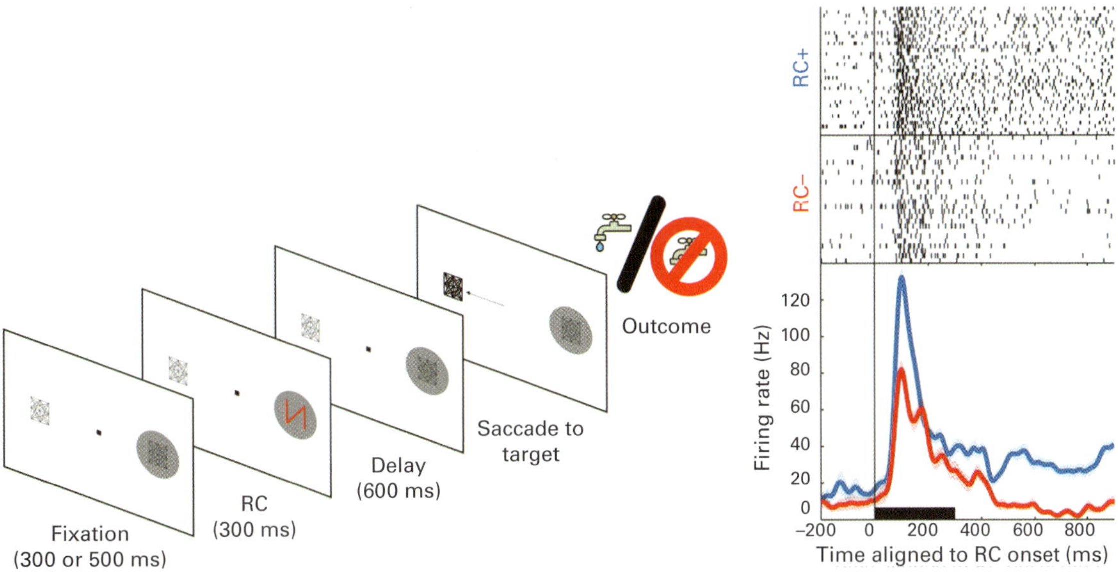

Plate 6 (figure 6.2) Behavioral task and cell responses in the lateral intraparietal area. (Left) Two patterns, one to the right and one to the left of fixation, were shown throughout the trial. After a fixation period, a cue appeared (inverted "N"), followed by a delay period. At the end of the delay, one of the patterns was brightened (indicated by arrow), indicating the saccade target location. A reward was administered on correct reward trials. The gray circle indicates the area of the receptive field of the cell being recorded. (Right) Response of a representative neuron. Trial onset is aligned with the presentation of the cue stimulus. Graph shows average firing rates during reward (blue) and no-reward (red) conditions. The black horizontal bar denotes the cue duration. Reproduced with permission from Peck et al. 2009 and kindly provided by Christopher Peck.

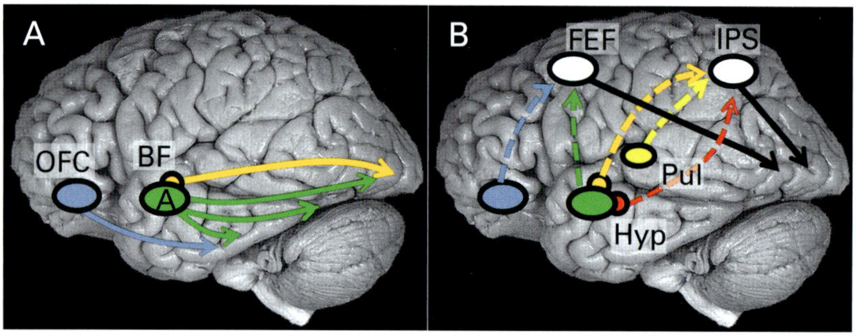

Plate 7 (figure 7.2) Interactions between emotion and perception in the brain. (A) Visual processing is modulated by affective value via several mechanisms, including those involving projections from the amygdala, basal forebrain, orbitofrontal cortex, and possibly hypothalamus (not shown). (B) Other modulatory mechanisms rely on interactions between evaluative sites (colored ellipses and circles) and attentional control sites (white ellipses). Located in fronto-parietal cortex, the latter sites are known to exert top-down influences on visual processing. Dotted lines indicate possible indirect connections. All locations are approximate, and the positions of the amygdala, basal forebrain, hypothalamus, and pulvinar are shown on the lateral surface for schematic purposes. Labels from panel A were omitted to reduce clutter. A, amygdala; BF, basal forebrain; FEF, frontal eye field; Hyp, hypothalamus; IPS, intraparietal sulcus; OFC, orbitofrontal cortex; Pul, pulvinar. Reproduced with permission from Pessoa 2010a.

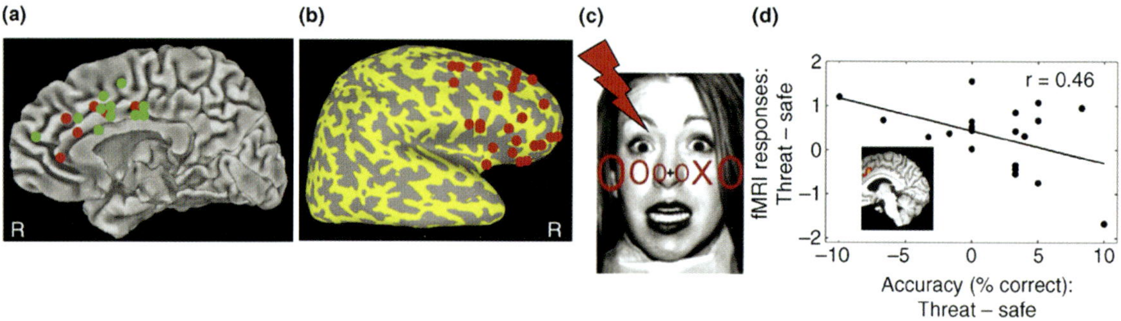

Plate 8 (figure 7.5) Processing resources and threat processing. Summary of results from thirty-four PET and functional MRI studies of conditioning from 1995 to 2008, illustrating the coordinates provided for the contrast of threat (CS+) versus safe (CS−) conditions. (a) Peak activation sites observed in medial prefrontal cortex are shown for the left (green) and right (red) hemispheres (coordinates were projected onto a midline view for display purposes). (b) Peak activation sites observed for the right lateral surface are shown on an "inflated" surface, including the middle frontal gyrus, inferior frontal gyrus, and anterior insula (note that the surface inflation "pushed up" some of the activation sites relative to their standard anatomical positions). (c) Subjects viewed an array of letters superimposed on task-irrelevant faces and were asked to report whether the target letter "X" was present. In the low-demand condition shown here, the target appeared among a uniform array of distractors (pop-out condition). In the high-demand condition (not shown), a nonuniform array of letters was employed (search condition). In the threat condition, faces were previously paired with mild electrical shock. (d) Differential responses to task-irrelevant faces (threat–safe) were inversely related to behavioral performance, suggesting that the processing of threat captured processing resources needed for task execution; and the greater the differential response magnitude, the worse the performance. Results are shown for a region of interest in dorsal-medial prefrontal cortex that was defined in terms of a separate contrast of high versus low attentional demand (shown in the inset). Data reanalyzed from Lim, Padmala, and Pessoa 2008. Reproduced with permission from Pessoa 2009.

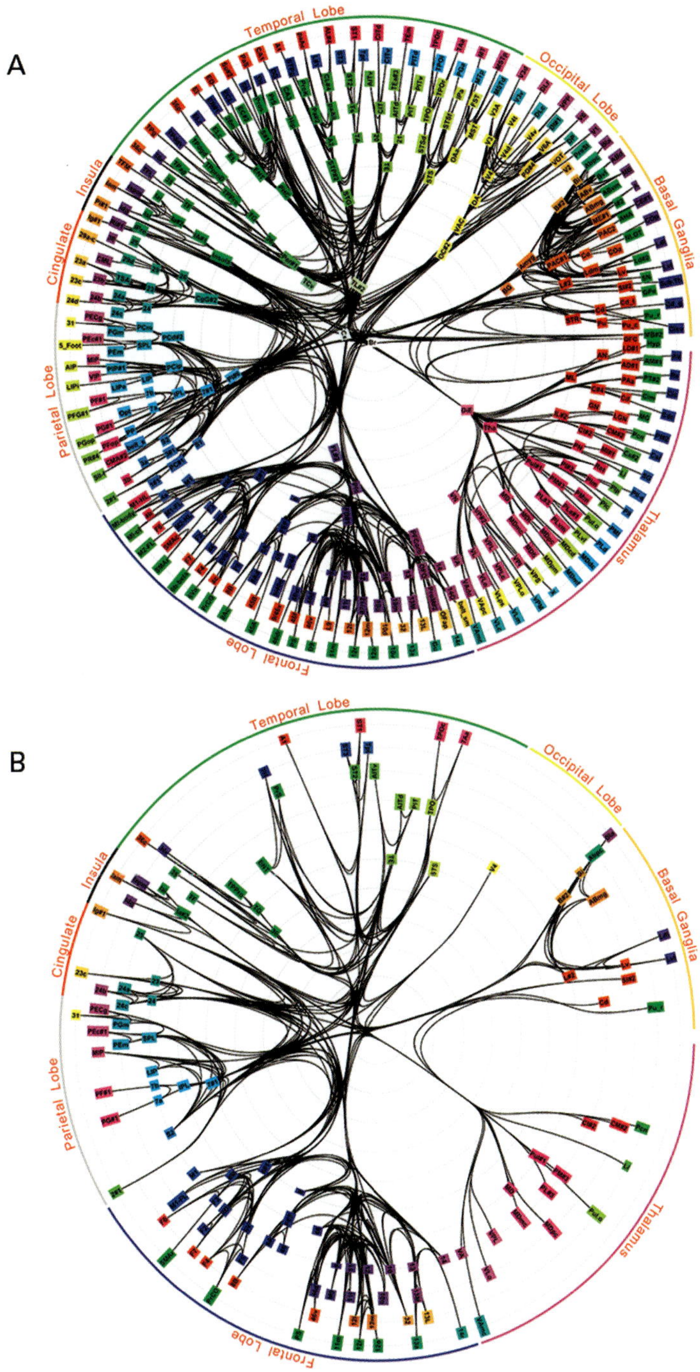

Plate 9 (figure 8.6) Whole-brain network connectivity structure. (A) In their analysis, Modha and Singh (2010) considered existing anatomical connectivity data of an extensive set of cortical and subcortical regions spanning most major brain sectors. The basal ganglia refer to nuclei at the base of the forebrain, including the amygdala. (B) Innermost "core" circuit. Notably, several amygdala nuclei were included in the inner core. Reproduced with permission from Modha, D. S., and Singh, R. (2010). Network architecture of the long-distance pathways in the macaque brain. *Proceedings of the National Academy of Sciences of the United States of America, 107*(30), 13485–13490.

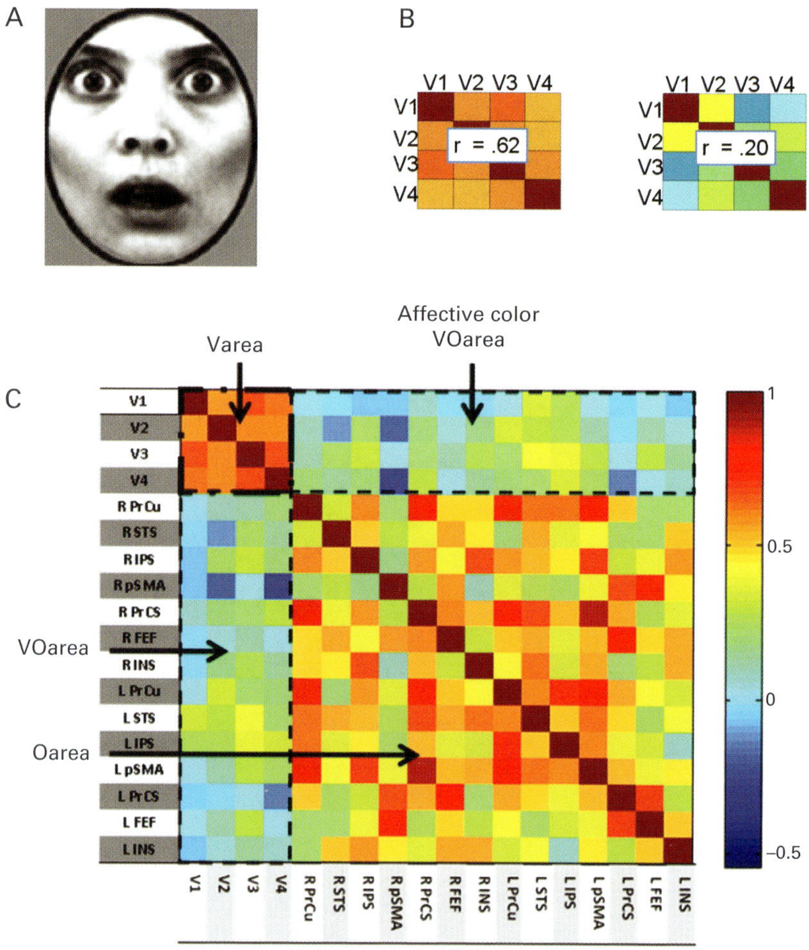

Plate 10 (figure 8.8) Emotion alters the response pattern across early visual cortex. (A) Both fearful and neutral (not shown) faces were presented with a ring that indicated whether it was a safe or threat (possible mild shock) condition. (B) Functional connectivity between visual areas V1, V2, V3, and V4 during affective (left) and neutral (right) context as indexed using pairwise correlations (average correlation shown). (C) Correlation matrix of differential responses for fearful versus neutral faces in both affective and neutral contexts (ring colors). The upper-left part is the same as in panel B. Of the entire matrix, that was the only part that was robustly altered as a function of context. Varea, correlations within early visual areas; Oarea, correlations within "other areas"; VOarea, correlations between early visual and "other areas"; L, left; R, right; FEF, frontal eye field; INS, insula; IPS, intraparietal sulcus; PrCs, precentral sulcus; PrCu, precuneus; pSMA, presupplementary motor area; STS, superior temporal sulcus. Colors code for correlation values as indicated by the color scale. Adapted with permission from Damaraju et al. 2009.

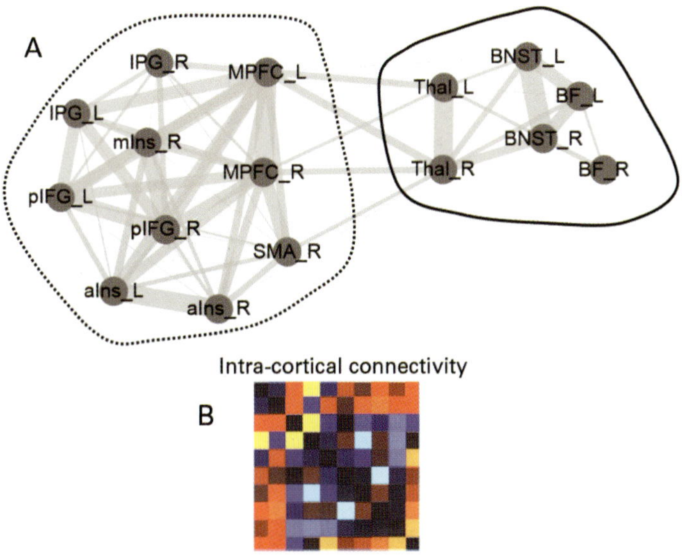

Plate 11 (figure 8.10) Network structure and threat. (A) Community detection was applied to the set of brain regions that responded more strongly to threat than to safe context at the cue phase (see figure 5.7). Two communities were detected. _L, left; _R, right; BF, basal forebrain; BNST, bed nucleus of stria terminalis; mIns, midinsula; MPFC, medial prefrontal cortex; IPG, inferior parietal gyrus; pIFG, posterior inferior frontal gyrus; SMA, supplementary motor area; Thal, thalamus. (B) Changes in threat versus safe connectivity for all pairs of regions within the community on the left in panel A. Dark colors indicate no change; warm colors indicate threat connectivity greater than safe; cool colors indicate threat connectivity smaller than safe. One of the effects of threat may have been to "disconnect" cortical regions from each other, possibly leading to performance impairments.

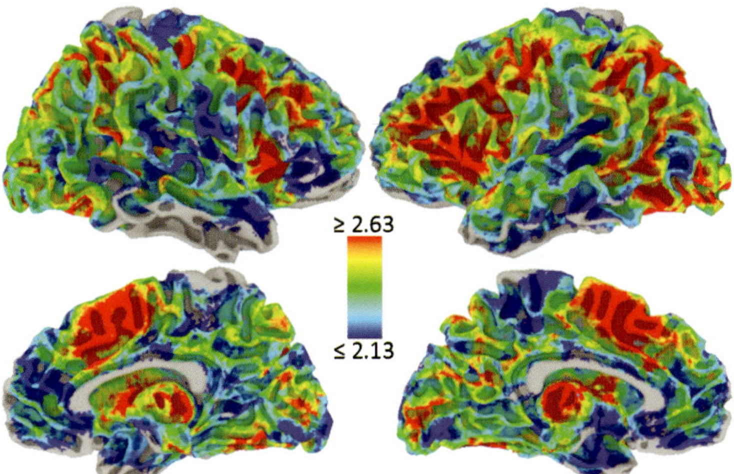

Plate 12 (figure 8.12) Functional diversity map. Areas of higher functional diversity are shown in warm colors, and areas of lower diversity are shown in cool colors (color bar represents diversity H values). Locations without colors did not have sufficient findings for the estimation of diversity. Adapted with permission from Anderson, Kinnison, and Pessoa 2013.

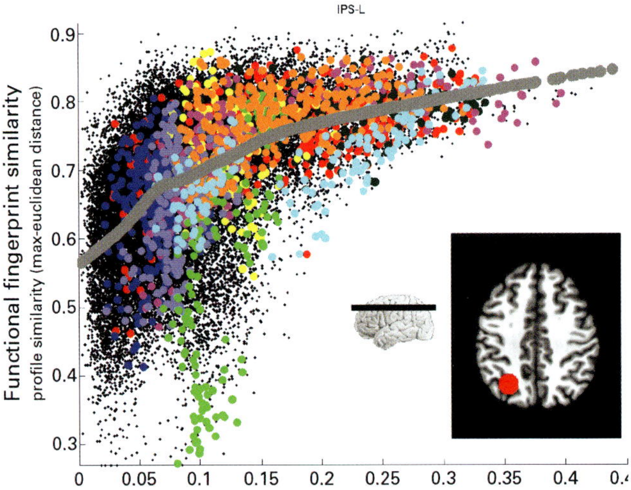

Plate 13 (figure 8.13) Task coactivation and diversity. Task coactivation is plotted against functional fingerprint similarity. The seed region was the left frontal eye field (red circle in inset). Each point represents one voxel in the brain. Spearman's rho = .53; $p < .001$. Data fit via standard locally weighted scatter plot smoothing ("lowess"). The colors refer to coordinates of regions from specific networks published in the literature. From Toro, Fox, and Paus 2008: red, frontal-parietal "attention" network (general task-positive); blue, cingulate-parietal "resting-state" network (task-negative). From Dosenbach et al. 2007: orange: frontal-parietal "rapid adaptive control" network; cyan: cingular-opercular "stable set control" network. From Yeo et al. 2011: dark green: "dorsal attention" network; yellow: "ventral attention" network; pink: "control network"; light purple: "default" network; light green: heterogeneous set of regions including visual and motor regions.

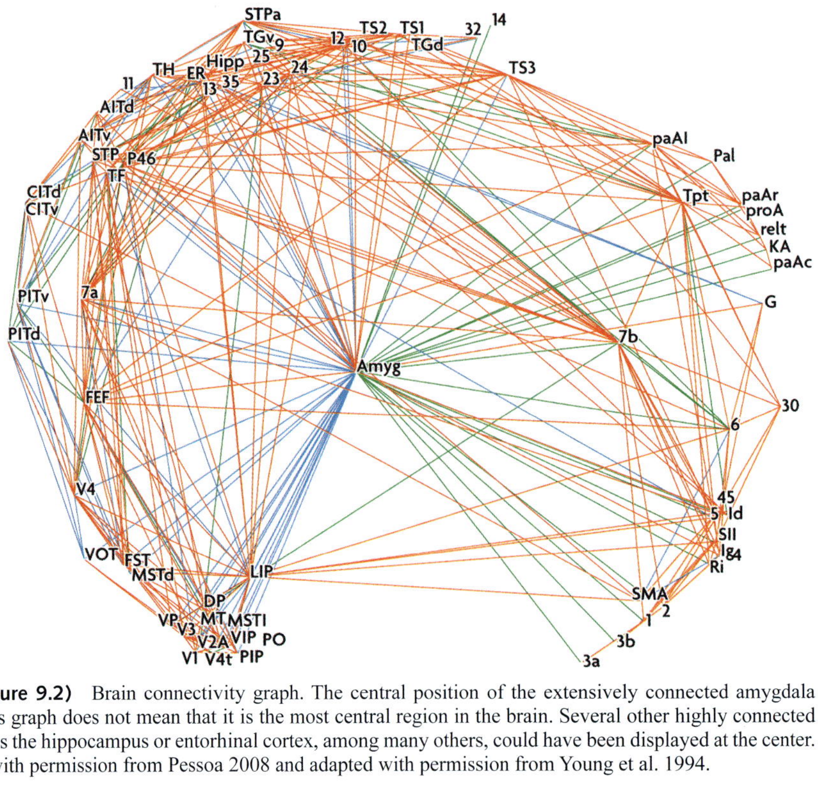

Plate 14 (figure 9.2) Brain connectivity graph. The central position of the extensively connected amygdala (Amyg) in this graph does not mean that it is the most central region in the brain. Several other highly connected regions such as the hippocampus or entorhinal cortex, among many others, could have been displayed at the center. Reproduced with permission from Pessoa 2008 and adapted with permission from Young et al. 1994.

9 Anatomical Substrates for Cognitive-Emotional Interactions

Chapter 8 described a network perspective on brain function in general terms. The present chapter will illustrate how the brain's architecture exhibits extensive avenues for information interaction and integration, and how its underlying structure provides the substrate for the coordinated flow of information that characterizes complex behaviors. The focus will be on architectural features of several brain regions that have been linked to emotion, including two subcortical regions, the hypothalamus and the amygdala. The chapter will also discuss features of the basal forebrain, a system historically linked with arousal processes, but suggested to be an important region for cognitive-emotional communication. Finally, it will review anatomical properties of prefrontal cortex and closely related areas, including medial and lateral prefrontal cortex, orbitofrontal cortex, and the anterior insula. In illustrating structural connectivity features, by necessity, the present chapter will enumerate many brain regions. Although this enumeration makes for somewhat tedious material for those less familiar with brain anatomy (who may choose to skip to the conclusion of the chapter), I believe it is a valuable exercise since it underscores the principle that regions involved in emotional circuits are among the most widely connected in the brain, suggesting that they may play at times a "quasi-global" communication role. Indeed, I propose that an important feature of emotional processing is its ability to engage a vast gamut of territories that are linked with multiple perceptual, motor, and cognitive operations—in addition to the more frequently emphasized bodily and homeostatic ones. In this manner, emotional processing is concerned with pervasive mechanisms that mobilize both mental and bodily resources. And, though the discussion to follow uses the word "emotion," by and large, the considerations here apply to emotion and motivation alike, given their close relationship.[1]

Hypothalamus

The human hypothalamus is a pearl-sized structure containing a number of nuclei. As its name implies, it is located just below the thalamus—and thus just above the brainstem. The importance of the hypothalamus in certain aspects of emotion is well known, as highlighted by the early work of Philip Bard and Walter Cannon, who showed in their "decortication" experiments that coordinated emotional expressions were abolished when the hypothalamus was excised, but not when only cortex was compromised (Bard 1928, 1929, 1934; Cannon 1929; Cannon and Britton 1925). Indeed, in 1929, Harvey Cushing—considered by many to be the father of modern neurosurgery—expressed his views this way: "Here in this well-concealed spot, almost to be covered with a thumb nail, lies the very mainspring of primitive existence—vegetative, emotional, reproductive—on which, with more or less success, man has come to superimpose a cortex of inhibitions" (as quoted by Card, Swanson, and Moore 2008, 795). Since the 1920s and 1930s, our knowledge of hypothalamic function has been greatly extended and refined, and current understanding concurs with the earlier notion that the hypothalamus is involved in several important "basic" operations. For example, it coordinates many complex homeostatic mechanisms, such as hormonal and behavioral circadian rhythms, in addition to neuroendocrine outputs. To execute these functions, the hypothalamus works in concert with a multitude of other sites, several of which are located in the brainstem and spinal cord.

Historically, the hypothalamus has been conceptualized in terms of "descending" systems, a view that is summarized by its designation as the "head ganglion" of the autonomic nervous system. However important the hypothalamus may be for descending control, though, a significant recent insight is that mammalian cerebral cortex and the hypothalamus share massive *bidirectional* connections. In the rat, the best-studied mammal species, there are four major routes from the hypothalamus to cerebral cortex (Risold, Thompson, and Swanson 1997; figure 9.1). These include a robust direct projection to *all* parts of the cortical mantle, and three indirect routes by way of the thalamus, basal nuclei (specifically, basal forebrain and amygdala), and brainstem.

The hypothalamus appears to be the largest source of nonthalamic direct input to cortex, as reviewed in detail by Larry Swanson and colleagues (Risold, Thompson, and Swanson 1997; Swanson 2000). In the rat, some notable targets of hypothalamic input include regions in medial prefrontal cortex (infralimbic, prelimbic, and anterior cingulate cortex) and insular cortex. Interestingly, less prominent projections of the hypothalamus to lateral prefrontal cortex and even to primary sensory areas have been found. An important

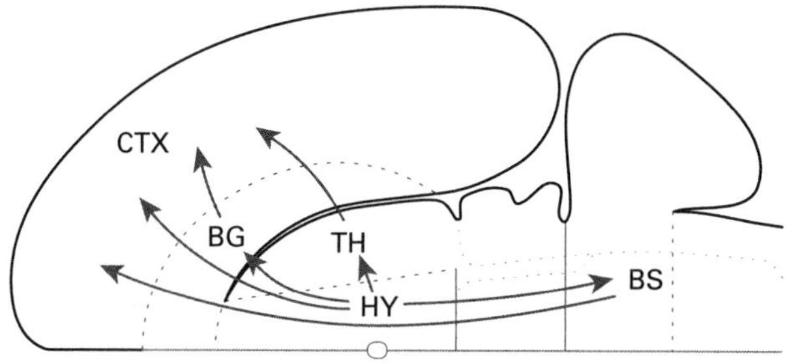

Figure 9.1
Hypothalamic ascending connectivity. Summary of the four major pathways from the hypothalamus to cerebral cortex schematized on a flattened representation of the rat brain. The basal ganglia here refer to the basal forebrain and the amygdala complex. Note that one of the indirect connections first descends to the brainstem. BG, basal ganglia; BS, brainstem; CTX, cortex; HY, hypothalamus; TH, thalamus. Reproduced with permission from Risold, Thompson, and Swanson 1997.

indirect system connects the hypothalamus to cortex via the basal forebrain. Another noteworthy route to cortex involves several amygdala nuclei, including projections via the basolateral nucleus that reach visual, cingulate, and motor areas. The connections between prefrontal cortex and the hypothalamus have also been investigated in nonhuman primates (Rempel-Clower and Barbas 1998), where they were found to closely resemble those observed in rats. Notably, as stated by Nancy Rempel-Clower and Helen Barbas (1998, 413; italics added), "the hypothalamus has widespread projections to *all* sectors of the prefrontal cortex. Retrogradely labeled neurons in the hypothalamus were found for every tracer injection on the orbital, medial, and *lateral* surfaces." In addition to the systems linking the hypothalamus to cortex, major telencephalic projections link cortex to the hypothalamus, including those from the hippocampal formation, amygdala, insular cortex, and prefrontal cortex (mostly orbital and medial prefrontal cortex).

In sum, whereas the hypothalamus is involved in a host of basic control functions, it is part of an extensive bidirectional connective system with cortex and many other subcortical structures in a manner that allows for integration of wide-ranging signals. Critically, the hypothalamus is linked to other structures that have themselves broad connectivity, including the basal forebrain and the amygdala, further expanding its potential for influencing information processing.

Basal Forebrain

The basal forebrain comprises a heterogeneous set of structures close to the medial and ventral surfaces of the cerebral hemispheres. As discussed in chapter 2, the basal forebrain originates "ascending" cholinergic and GABA-ergic projections that innervate extensively throughout the *entirety* of cortex. Major projections reach visual, parietal, cingulate, frontal (including lateral and orbital regions), and insular cortex (Mesulam 2000); substantial connections have also been found to both the hippocampus and amygdala (Mesulam et al.1992).

Cerebral cortex, though sharing a basic organization, is heterogeneous and certain parts have more or less refined patterns of lamination. Areas of cortex made up of six clearly defined layers are referred to as "isocortex"; areas with less differentiation, as "noniscortex." The basal forebrain projection system exhibits a connectivity gradient, such that the *densest* projections to cortex are found toward *less* differentiated areas, such as cingulate cortex and nonisocortical sectors of temporal, insular, and orbitofrontal cortex (Mesulam et al. 1992).[2] The pattern is consistent with connections to both the hippocampus and the amygdala (two regions with simplified cytoarchitecture), which are stronger than those to cortex.

Whereas the output pathways of the basal forebrain are organized in a distributed fashion, its inputs originate from a more circumscribed, though still broad, set of regions. Connections from cortex originate largely from nonisocortical sites, for instance, sites in orbitofrontal, temporal, insular, and cingulate cortex (Mesulam et al. 1992; Zaborszky et al. 1999). The major subcortical projections to the basal forebrain are from the amygdala, hypothalamus, and various brainstem cell groups, including midbrain dopaminergic cells and the locus coeruleus (Zaborszky et al. 1999). The last two are particularly pertinent given their roles in processing unexpected or salient stimuli (Aston-Jones and Cohen 2005; chapter 6 discusses dopamine in the context of motivation). For instance, Laszlo Zaborszky and colleagues (1999) suggest that the locus coeruleus sends a "warning signal" to the forebrain via its input to the basal forebrain.

Given its overall connectivity pattern, the basal forebrain is in a position to influence processing across the brain. As discussed in other chapters, some of this structure's effects include increased alertness and enhanced selective attention, disturbances of which have been proposed to contribute to neuropsychiatric disorders (Sarter and Bruno 1999). The basal forebrain also plays a key role in cortical plasticity (Dykes 1997; Kilgard and Merzenich 1998; Weinberger 2003). Given that its projections are so general, the impact of the basal forebrain is frequently presented as "unspecific," although it can also be selective. Thus, in one study (Goard and Dan 2009), direct stimulation of the basal forebrain enhanced the coding of natural scenes in visual cortex by markedly improving the reliability of cell responses.

Although less understood, how the basal forebrain participates in several *specific* circuits may be at least as important as its extensive effects on cortical function. The cortical regions that receive the largest number of basal forebrain projections are the same ones that have pathways to the basal forebrain. In this manner, basal forebrain–cortex–basal forebrain circuits can be established (Zaborszky 2002). Circuits involving the basal forebrain and subcortical regions also exist. For instance, basal forebrain-amygdala interactions support multiple aspects of attentional processing (Holland 2007; Maddux et al. 2007).

Amygdala

A remarkable property of the primate amygdala is its massive interconnection with cortex. Based on the available data at the time, analysis of amygdala pathways by Malcolm Young and colleagues (1994) revealed that, of the 72 cortical areas included in their study, it was connected to all but eight (see also Barbas 1995; Swanson 2003; figure 9.2; plate 14). Although the connectivity illustrated in figure 9.2 (plate 14) is remarkable, it is probably considerably *underestimated*, given that as many as 1,000 separate cortical and subcortical pathways may exist (Petrovich, Canteras, and Swanson 2001). The connectivity is all the more notable given that it involves all cortical lobes as well as subcortex.[3] Combined, these properties indicate that the amygdala is an extensively interconnected *connector hub*. Furthermore, in the network analysis by Dharmendra Modha and Raghavendra Singh (2010) discussed previously, several amygdala nuclei (e.g., lateral nucleus, accessory basal nucleus) were identified as part of a *core* brain circuit, all of whose regions have extremely high connectivity. Thus the amygdala is proposed to participate in the brain's "rich club" as far as anatomical information is concerned (see chapter 8). Together, these findings reveal that the amygdala has exceptional potential for information integration.

When whole-brain connectivity data are analyzed, not surprisingly, prefrontal regions are among those receiving some of the most highly processed and integrated sensory information (Young et al. 1994). This structural feature is thought to be significant because it provides prefrontal cortex with relative insulation from unprocessed incoming sensory stimuli. Indeed, M.-Marsel Mesulam (2000) has proposed that this key anatomical feature of prefrontal cortex may confer the primate brain with a greater degree of flexibility than the brains of "less complex" species. This view is problematic, however, because it does not take into account the widespread bypass circuits discussed in chapter 3 that can rapidly convey information across the brain in the case of visual processing. Nevertheless, the availability of highly elaborated information

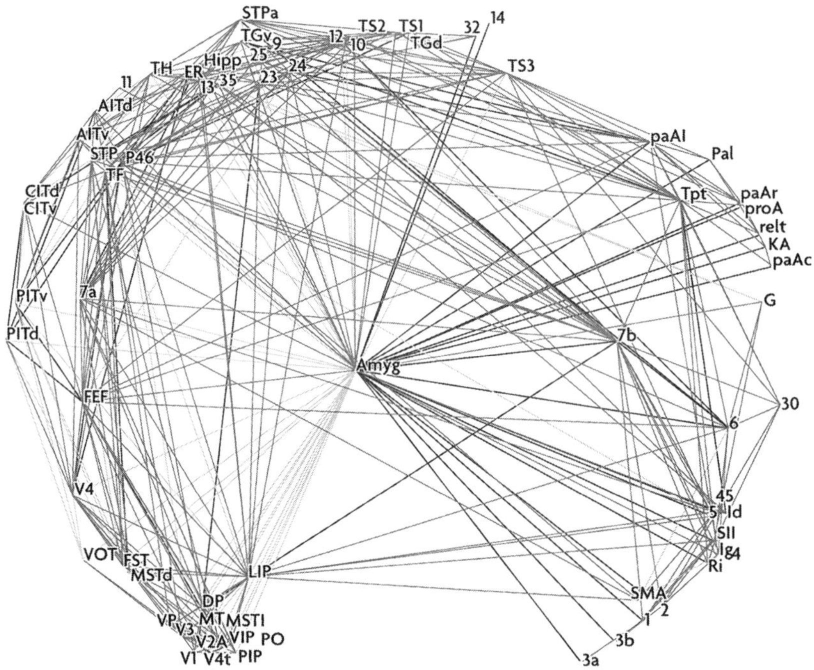

Figure 9.2 (plate 14)
Brain connectivity graph. The central position of the extensively connected amygdala (Amyg) in this graph does *not* mean that it is the most central region in the brain. Several other highly connected regions such as the hippocampus or entorhinal cortex, among many others, could have been displayed at the center. Reproduced with permission from Pessoa 2008 and adapted with permission from Young et al. 1994.

is likely valuable in supporting more abstract computations required during cognitive tasks. It is thus noteworthy that the amygdala (and other regions, such as the hippocampus and entorhinal cortex) also receives highly processed information (Amaral et al.1992; Young et al. 1994), although this probably does not play the same role in the amygdala that it does in prefrontal cortex, where such information is important for solving abstract problems (e.g., fluid intelligence problems; Duncan et al.1996). I suggest that highly processed information is important for *value* representation (see chapter 2) since value may often depend not on superficial sensory properties, but on those showing a fair degree of “invariance.” For example, a dominant male’s face is biologically significant irrespective of viewing angle, illumination, and the like. Also relevant, the amygdala is the recipient of signals from all sensory modalities (Sah et al. 2003), a property especially useful for the representation of value. Thus, to take the example above a step further, the significance of a dominant male's vocalizations should be evident by recognizing that they were uttered by him and not by another male further down the hierarchy.

The pattern of connectivity between the amygdala and prefrontal cortex (Amaral and Price 1984) is of particular interest given the latter's role in cognitive functions. In addition to substantial connections between the amygdala and both medial and orbital aspects of prefrontal cortex, recent findings indicate that *bidirectional* connectivity is present also along the lateral surface (Ghashghaei, Hilgetag, and Barbas 2007)—although the connections are relatively weak. More generally, to understand how amygdala signals are potentially broadcast to all sectors of prefrontal cortex, it is important to consider PFC connectivity itself. In one study, although the amygdala was estimated to be directly connected to approximately 40 percent of prefrontal regions, approximately 90 percent of prefrontal cortex was deemed capable of receiving amygdala signals after a single additional connection *within* prefrontal cortex (Averbeck and Seo 2008; figure 9.3). This "one-step" property seriously undermines the notion that affective signals are confined to orbital and medial prefrontal territories, as often assumed in the literature—here equating "affective" with amygdala signals. Furthermore, arguments that amygdala outputs have limited influence on responses in lateral prefrontal cortex have less traction once we consider these architectural features. Thus in a functional MRI study by Adam Messinger and colleagues, microstimulation of monkey amygdala revealed responses in lateral prefrontal cortex, including dorsal sites (Messinger et al. 2011). Although the downstream effects of microstimulation

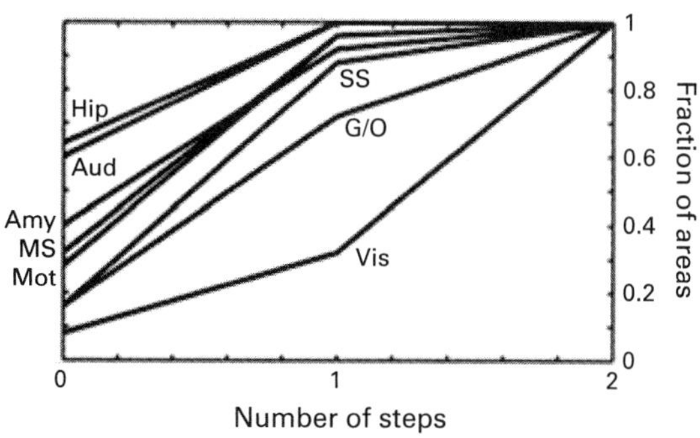

Figure 9.3
Prefrontal cortex connectivity. Fraction of frontal areas that receive signals from each modality as a function of the number of connectivity "steps" within frontal cortex. Zero indicates the areas that receive a direct projection from the amygdala, hippocampus, or the indicated sensory or motor modality, and one indicates the fraction of areas that would receive the signal after a single step within frontal cortex. Amy, amygdala; Aud, auditory; G/O, gustatory/olfactory; Hip, hippocampus; Mot, motor; MS, multisensory; SS, somatosensory; Vis, visual. Reproduced with permission from Averbeck and Seo 2008.

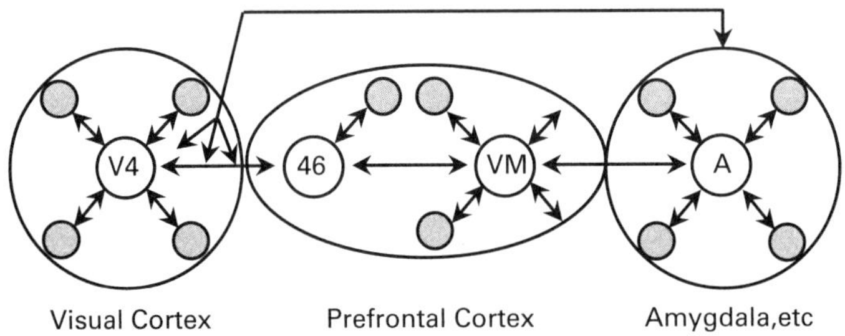

Figure 9.4
Connector hubs and clusters of "visual," "prefrontal cortex," and "amygdala" communities. The central regions are area V4 in the visual cluster, Brodmann area 46 and ventral-medial cortex (VM) in the prefrontal cortex cluster, and the amygdala (A) in the amygdala cluster. Connectivity between the visual and amygdala hubs is highly asymmetrical, with the amygdala receiving projections from regions in anterior temporal cortex, but sending them to nearly all levels of the ventral visual system in temporal and occipital cortex (see multiple arrows within "Visual Cortex" circle).

may involve both mono- and polysynaptic connections, this finding demonstrates that the amygdala has the ability to influence lateral prefrontal cortex.

In summary, the connectivity pattern between the amygdala and prefrontal cortex reveals ample opportunities for signal communications (figure 9.4); and, through couplings of varying strengths, amygdala–prefrontal cortex interactions enable cognitive-emotional interactions and integration.

Other notable aspects of amygdala connectivity not discussed in this chapter include interactions between the amygdala and the basal forebrain that are important for attentional functions (see chapter 2) and substantial projections from the amygdala to visual cortex that influence competition in visual cortex, thus contributing, again, to attention (see chapter 7).

Prefrontal Monitoring and Control of Visceral and Other Bodily Functions

Although prefrontal cortex is critical for abstract reasoning, planning, and other complex mental processes, it is also involved in processes that are considerably more "elementary." Indeed, "as early as 1869, the British neurologist John Hughlings Jackson . . . proposed that within the convolutions of the cerebrum were represented not only voluntary movements of the whole body but also the involuntary movements of the blood vessels and viscera" (Cechetto and Saper 1990, 208). And the notion that prefrontal cortex itself participates in the control of the autonomic nervous system dates back to at least the start of the twentieth century (for a historical account of this notion, see Neafsey 1990).

Walle Nauta (1971, 182; italics added) eloquently summarized some of these ideas: "The frontal lobe is characterized so distinctly by its multiple associations with the limbic system, and in particular by its direct connections with the hypothalamus, that it would seem justified to view the frontal cortex as the *major—although not the only—neocortical representative of the limbic system.* The reciprocity in the anatomical relationship suggests that the frontal cortex *both* monitors and modulates limbic mechanisms."

More recently, the tight interrelation between prefrontal cortex and bodily functions was central to the somatic marker hypothesis of Antoine Bechara, Hanna and Antonio Damasio, and colleagues, especially with respect to orbitofrontal and ventral-medial prefrontal cortex (Bechara, Damasio, and Damasio 2000). Likewise, the notion that the anterior insula, a region discussed here in conjunction with prefrontal sites, is also involved in sophisticated representations of the body has gained considerable visibility (Craig 2002, 2009; Damasio 1999, see also Morrison, Bjornsdottir, and Olausson 2011).

Cingulate Cortex

The functions of cingulate cortex, which surrounds the entire extent of the corpus callosum and may comprise as many as forty subareas, are quite complex (Vogt 2008). The anterior sector of cingulate gyrus is involved in a broad range of processes, including willed action, executive function, and emotion.[4] A remarkable property of this cortical tissue is that its *descending* projection system is probably more extensive than that of any other cortical region (Vogt and Vogt 2009) and includes major projections to autonomic regulatory structures—notably, lateral hypothalamus, periaqueductal gray, parabrachial nucleus, and the nucleus of the solitary tract (Vogt and Derbyshire 2009). This connectivity is consistent with stimulation studies that have documented effects of cingulate electrical stimulation on virtually all autonomic and many endocrine mechanisms (Vogt 2008).

Conversely, cingulate cortex is the recipient of *ascending* signals from the brainstem (Vogt, Aston-Jones, and Vogt 2009). The most notable of these is perhaps the one from the nucleus of the solitary tract, the major visceral sensory cell group in the brain. Several nociceptive circuits also reach anterior and midcingulate areas indirectly via the thalamus. These findings clearly establish that posterior as well as more anterior regions of cingulate gyrus are involved in *bidirectional* signal communication. Thus, contrary to some accounts, cingulate cortex participates in more than "motor" autonomic functions (see also discussion on the anterior insula below).

Interestingly, several regions within cingulate cortex are part of the core circuit proposed by Modha and Singh (2010). Brodmann areas 24 and 32 are especially noteworthy because of their extensive connectivity within the core circuit itself. These cingulate regions ranked among the "top 10" brain regions according to several key metrics of network "centrality," which capture capability of signal integration, distribution, and intermediation (see table 1 of Modha and Singh 2010). In all, given the well-known roles of cingulate cortex in cognitive functions, the bidirectional architecture described above provides exceptional opportunities for cognitive-emotional interaction and integration.

Orbitofrontal Cortex

Orbitofrontal cortex is the portion of prefrontal cortex that sits just above the orbits of the eyes and extends posteriorly several centimeters to form the frontal "base" of the brain. Descriptively, locations within anterior prefrontal cortex can be expressed in terms of medial and lateral orbitofrontal cortex. A better way to view this cortex is based on its connectivity pattern, which reveals *orbital* and *medial* "networks" (Carmichael and Price 1996). The orbital network receives wide-ranging sensory information and appears to integrate it, particularly in relation to the assessment of food and reward. The medial network is distinctively and heavily connected with regions of the medial wall of the brain, including those of cingulate cortex and surrounding areas. In contrast to the orbital network, the medial network receives few sensory inputs. Importantly, the *medial* orbitofrontal cortex projects to the hypothalamus and other visceral-control areas, leading Dost Öngür and Joseph Price (2000) to suggest that it is involved in the "visceral modulation of emotion." Indeed, via the hypothalamus, descending medial orbitofrontal influence appears to extend as far as autonomic centers in the spinal cord (Barbas et al. 2003). In contrast, there are relatively few projections to the hypothalamus from the orbital network. Therefore, at least through its medial network, the orbitofrontal cortex interfaces with autonomic brain regions. Notably, the same network is strongly interconnected with cingulate cortex.

Anterior Insula

Orbitofrontal cortex is closely associated with the anterior insula.[5] Insular cortex is the part of cortex folded within the lateral sulcus (also called the "Sylvian fissure") between the frontal and temporal lobes—it is thus not visible from the lateral surface of the brain (cortex covering it is referred to as the "operculum," meaning "lid"). In neuroimaging studies, researchers have found that the anterior insula is engaged by a remarkable array of tasks spanning perceptual, cognitive, and emotional ones (e.g., Craig 2009). Indeed, in a recent large-scale analysis of neuroimaging data, the anterior insula was among the most functionally diverse regions of the brain (Anderson, Kinnison, and Pessoa 2013; chapter 8).

Containing as it does precise representations of the internal state of the body, the anterior insula is strongly involved in the processing of bodily signals (Craig 2002, 2009). Both sympathetic and parasympathetic bodily signals are at first conveyed to the posterior insula, and then to anterior insula via the midinsula. A. D. (Bud) Craig (2002) has proposed that the anterior insula participates in the "afferent representation of 'feelings' from the body" (see also Damasio 1999), whereas the cingulate is involved in the initiation of behaviors and thus is more "motor" in function. Although this proposal may hold at a broad level of description, as noted above, cingulate cortex also receives visceral, ascending signals from important autonomic nuclei, including the nucleus of the solitary tract. At the same time, descending connections from the insula to the brainstem also exist, although ascending ones are often emphasized. And, even though descending connections are mostly found in the posterior insula (Yasui et al. 1991), they probably exist in the anterior insula as well. Thus Yukihiko Yasui and colleagues (1991) observed inhibition and increases of gastric mobility primarily in distinct parts of anterior insular cortex upon microstimulation in the rat. Thus, too, projections from the insula to the lateral hypothalamus and the parabrachial nucleus (Saper and Loewy 1980) have been documented. Moreover, autonomic responses have been observed upon stimulation of several insular sites (Cechetto and Chen 1990), consistent with a visceral motor role (Augustine 1996). Clearly, then, in both cingulate cortex and the insula, *bidirectional circuits* are present—although they are likely asymmetrical in their efficacies.

Lateral Prefrontal Cortex

The role of lateral prefrontal cortex in a range of cognitive operations is well documented. Indeed, this region is perhaps *the* paradigmatic cognitive brain region. Yet the connectivity pattern of lateral prefrontal cortex is quite extensive, indicating that its function should be thought of more broadly. Brodmann area 46 in dorsal-lateral prefrontal cortex is not only an important hub (Sporns, Honey, and Kotter 2007); it also has major connectivity with other regions that are topologically central themselves (see discussion of "rich club" in chapter 8), which considerably expands the reach of the dorsal-lateral prefrontal cortex in influencing information flow. In several studies, Brodmann area 46 scored very highly (often as the "top" brain region) based on network-based metrics of "centrality" (e.g., Sporns, Honey, and Kotter 2007). Being robustly connected to medial and orbital prefrontal areas, as well as to the insula, the region has access to afferent signals from the body and can influence efferent signals that reach the body. More generally, I suggest that its connectivity pattern allows lateral prefrontal cortex to play a significant role in cognitive-emotional integration (see also Averbeck and Seo 2008).

Conclusions

Although subcortical structures such as the hypothalamus and the amygdala have historically been implicated in emotion, it is becoming increasingly clear that their connectivity affords them great potential to interact with both cortical and subcortical structures involved in cognition. As noted in the prescient words of David Amaral and Joseph Price (1984, 493; see quotation at the beginning of chapter 2), the amygdala "should be included among the structures which are responsible for the elaboration of higher cognitive functions."

Our knowledge of hypothalamic connectivity has considerably expanded, too. The traditional view of the hypothalamus emphasized its "descending" functions. Thus Jeffrey Gray (1990, 276) asserted that "neither the hypothalamus nor the central grey has, to my knowledge, been implicated in cognitive processing." Recent findings, however, have made it clear that, through extensive ascending projections, the hypothalamus provides direct input to the *entire* cortical mantle (Risold, Thompson, and Swanson 1997), with the potential to influence both perception and cognition.

These findings strongly support the proposal by Josef Parvizi (2009, 358) that subcortical regions play much more prominent roles in complex behaviors:

> So-called higher functions of the brain might in fact depend on signals from subcortical to cortical structures rather than the other way around. . . . Currently we do not have sufficient knowledge about the mode of subcortical involvement in cognition and behavioral regulation. In fact, we know very little about the role of subcortical structures in these "higher" functions, precisely because a significant proportion of current research does not see beyond the cerebral cortex.

Parvizi goes on to cite the vast reciprocal relationships between the frontal lobe and subcortical structures.

Our understanding of prefrontal cortex connectivity has also evolved. Large prefrontal sectors are strongly connected with brainstem nuclei that are responsible for controlling autonomic and endocrine mechanisms in the service of supporting survival and bodily integrity through homeostasis. The prefrontal and related sectors comprising cingulate, orbitofrontal, and insular cortex are strongly interconnected not only with each other but also with the amygdala. In all, prefrontal cortex participates in descending and ascending circuits, highlighting its role in the *vertical* integration of information. This perspective on prefrontal cortex function complements the more common emphasis on *horizontal* communication via cortico-cortical pathways, which relies on the robust bidirectional connections between parietal and frontal cortex, for instance. Walle Nauta (1972, 138) made similar points more than thirty years ago:

> In past decades, a direct fronto-hypothalamic connection would likely have been interpreted as evidence of a cortical "control" over hypothalamic effector mechanisms. Within the frame of more contemporary notions, however, a somewhat different interpretation would seem possible in which the relation between frontal lobe and hypothalamus is expressed in terms of information rather than control of "higher" over "lower" levels. It could be asked, for example, if loss of the frontal cortex could perhaps entail the loss of a channel whereby neural codes representing perceptual or ideational processes in the cerebral cortex are conveyed to the hypothalamus, and by which the latter, in turn, can report to the frontal lobe any changes in its functional state, whether resulting from such cortical afflux or from neural and humoral signals emanating from the organism's viscero-endocrine periphery, or from both.

The brain regions reviewed here exhibit extremely high connectivity. Indeed, with the exception of the hypothalamus, Modha and Singh (2010) found them all to be part of a core brain circuit, indicating that they communicate widely with one another.[6] Finally, lateral prefrontal cortex, often thought to be more directly involved in cognitive operations, is both richly and directly connected to areas in medial, orbital, and insular cortex—regions replete with bodily signals.

Chapter 8 described broad conceptual issues related to a network perspective on brain function. The present chapter has briefly described architectural features of brain connectivity that demonstrate not only extensive communication involving diverse types of signals but also the exceptional potential for cognitive-emotional interactions.

Notes

1. One might very roughly distinguish the two by saying that emotion is aligned with sensory (or input) aspects of behavior and motivation is aligned with motor (or output) aspects.

2. Cingulate cortex is at times described as "paralimbic" (Mesulam 2000). Lamination is well defined, but less so than in isocortex. Technically, cingulate cortex is agranular, that is, it lacks a conspicuous layer IV. But note that, like almost everything in biology, agranular versus granular is relative (hence "dysgranular" cytoarchitecture).

3. Connections to parietal cortex are sparse and target the intraparietal sulcus (Amaral and Price 1984).

4. The anterior sector of the cingulate gyrus includes anterior and midcingulate cortex as defined by Vogt (2008).

5. By some criteria, there is a substantial extension of the insula onto the posterior orbital surface of cortex (Price 2006).

6. In their analysis, Modha and Singh (2010) considered only a very small fraction of hypothalamic pathways, some fifty connections in all (see their table S1). However, the hypothalamus has at least fifty distinct cell groups (Petrovich, Canteras, and Swanson 2001), each with its own distinct pattern of connectivity.

10 Conclusion

It would be naïve, therefore, to ascribe the neural substrata of behavioral motivation to any one particular level of the central nervous organization.
—Walle Nauta (1972)

The preceding chapters have reviewed how emotion and motivation interact with perception and cognition. This conclusion summarizes some of the main ideas presented in the book and discusses some implications and extensions of these ideas.

The Amygdala and Its Role in Brain Processing

Characterizations of amygdala function have evolved considerably in the past two decades and go well beyond the notion of a "fear center" and the handling of negative-only information. Notable steps in this direction include the suggestions by Paul Whalen (1998) that the amygdala plays important roles in vigilance and ambiguity processing and by David Sander, Jordan Grafman, and Tiziana Zalla (2003) that it operates as "an evolved system for relevance detection." The role of the amygdala in Pavlovian processes and how they influence several facets of decision making was elaborated by Ben Seymour and Ray Dolan (2008); Ralph Adolphs (2003, 2010) and Leslie Brothers (1997) described the amygdala's role when social stimuli are concerned.

The idea advanced in this book is that a key function of the amygdala is to shape *selective information processing*. Selection of information for further analysis is, of course, a central problem that needs to be solved for effective behavior. The amygdala is a core structure in a system involved in "What is it?" processing and thus contributes to highlighting what is of significance to the organism. By and large, this function can be called "attention," which, as William James emphasized (1890, 404), "implies withdrawal from some things in order to deal effectively with others." I have argued that the functions of the amygdala extend beyond those surrounding "What is it?" and involve "What's to be done?" A key reason for this is that the amygdala participates in the representation of value (including positive value) and in decision making, as revealed in both nonhuman and human studies. For example, amygdala lesions impair behavior on the Iowa Gambling Task in humans and alter delay-based decision making in rats (e.g., they become more impulsive).

The amygdala thus takes part in an impressive array of processes that far exceed some of its proposed functions, such as vigilance, arousal, salience detection, novelty detection, and relevance detection. "Information gathering" (Whalen 1998) better captures several of its functions but comes short, too. In the end, it is better simply to refrain from summarizing its functional repertoire so as to better appreciate the wide scope of the amygdala's contributions to brain mechanisms and behavior.

Subcortical "Low Road" Pathway and Emotional Processing

A purported division of labor between cortical and subcortical regions has been present from the time of the earliest *circuit models* of emotion. For example, in the circuit proposed by James Papez (1937), the hypothalamus constituted the output station in control of bodily responses, and a pathway through the thalamus and hypothalamus coexisted with a longer cortical pathway via both sensory and cingulate cortex, which eventually reached the hypothalamus after coursing through the hippocampus. Emotional experiences (or feelings) depended on the integration of signals by cingulate cortex. However, given that the hypothalamus was responsible for "emotional coloring," bodily responses to emotional stimuli could be generated by engaging the thalamic pathway in a more direct manner: "Emotion may arise in two ways: as a result of psychic [cortical] activity and as a consequence of hypothalamic activity" (Papez 1937, 728). Many versions of this type of dual processing model exist, including some variants that have captured the popular imagination, such as the "triune brain" model of Paul MacLean (1970, 1990).[1]

In the case of vision, it has been suggested that a subcortical pathway from retina to superior colliculus to pulvinar to amygdala that entirely bypasses cortex enables the processing of emotion-laden visual stimuli to be fast, automatic, and nonconscious. Chapter 2 argued against this notion on several general grounds: (1) affective visual information is not handled qualitatively faster than other visual information; (2) processing of affective visual stimuli involves both low– and high–spatial frequency information; and (3) the amygdala is not required for rapid, nonconscious detection of affective information. It also raised several additional points of a physiological and anatomical character: (1) pulvinar physiology demonstrates that its responses are closer to those of visual cortex than to those of the superior colliculus; (2) pulvinar physiology indicates that cell responses are sensitive to behavioral relevance, attention, and awareness; and (3) pulvinar anatomy is not consistent with an uninterrupted colliculo-pulvino-amygdalar pathway.

For these and many other reasons, Ralph Adolphs and I proposed the "multiple waves" model as an alternative to the low road pathway scheme (Pessoa and Adolphs 2010). In broad terms, one of the goals of our model is to shift the debate away from whether there is a unique subcortical pathway to whether a processing architecture exists that is capable of rapidly transmitting information via multiple pathways. For example, the pulvinar provides inputs to visual areas V1, V2, and V4—in the case of V4, the proportion of labeled neurons from the pulvinar is comparable to the that from the lateral geniculate nucleus to V1, for instance (Markov et al. 2011). From V4, signals can rapidly reach many areas (via mono- or polysynaptic routes), including the amygdala and ventral-lateral prefrontal/orbitofrontal cortex. The past few years have been particularly exciting in demonstrating visual pathways that either bypass the lateral geniculate nucleus (in the thalamus) or involve this nucleus but bypass early visual cortex.

The multiple waves model emphasizes the role of the pulvinar in coordinating and regulating the flow of multimodal information, which is accomplished via a series of thalamo-cortical loops. In this role, the pulvinar moves from being a passive relay station of the "standard hypothesis" to being an active element of information processing. The medial pulvinar, in particular, is bidirectionally connected to several areas that assess biological significance, including cingulate, insular, and orbitofrontal cortex. And, as noted in chapter 3, there is increasing empirical support for the pulvinar's role in regulating information transfer between cortical regions (i.e., cortico-thalamo-cortical communication).

The multiple waves architecture also has implications for how we characterize the processing of neutral information. "Nonstandard" subcortical pathways via the superior colliculus (see also Takaura, Yoshida, and Isa 2011), pulvinar, and lateral geniculate nucleus have long been a problem for more hierarchical views of visual processing. For example, David Van Essen (1985, 292) stated that

> the current version of the cortical hierarchy contains no reference to the massive subcortical connections that visual cortex has. . . . A great deal of visual and nonvisual information flows into and out of each cortical area via these pathways, but the relationship of this traffic to the direct corticocortical pathways remains largely obscure. . . . Structures such as the superior colliculus and pulvinar cannot be fit into the current

anatomical hierarchy using the same criteria as are used for corticocortical connections.

Although we still do not understand the precise contributions of nonstandard pathways to visual perception, there is growing appreciation for their relevance and, hence, for the role of nonhierarchical processing.

New Evidence for the Subcortical Pathway and Implications

With the refinement of MRI methods for studying fiber tracts in humans, several research groups have started to investigate the connectivity of the superior colliculus, pulvinar, and amygdala. In particular, Marco Tamietto and colleagues (2012) have reported tracking a direct colliculo-pulvino-amygdalar pathway. Here I must note, however, that, even though MRI–based tract-finding algorithms have grown in sophistication in recent years, they have several key limitations. Two are worth mentioning. First, MRI tractography may suggest the existence of a connection between two structures, but it does not permit us to infer its directionality. Second, its algorithms do not allow us to determine uninterrupted connections between multiple structures and, more critically, those in a *functional* neural circuit. Indeed, to determine whether the neurons in the pulvinar that receive superior colliculus inputs do in fact project to the amygdala, researchers need to employ traditional anatomical and electrophysiological methods. Which is what Rebecca Berman and Robert Wurtz (2010) did in describing a subcortical circuit that processes moving stimuli and links the superior colliculus to visual area MT via the pulvinar. Thus, however valuable a tool for investigating brain connectivity in humans, current tract-finding methods are not well suited to the study of multiregion *circuits*.

Even assuming for the moment that new findings provided evidence for the purported subcortical pathway, I contend they would not substantially alter the conclusions of chapter 2, for the following reasons. First, the pulvinar is a *higher-order* thalamic nucleus, not a relay. It projects to and receives projections from essentially the entire cortical mantle. The medial pulvinar in particular, which projects to the amygdala, receives inputs from parietal, orbitofrontal, cingular, and insular cortex. Thus pulvinar responses are expected to reflect signals from all these regions. Second, and critically, the responses of the amygdala—end recipient of the pathway—are increasingly understood to depend on attention, context, and awareness. Indeed, recent studies have challenged the "necessary status" of the amygdala for many of its "special" functions. Third, response latencies in the amygdala are not unusually fast, indicating that, routinely, signals are *not* being conveyed by the subcortical pathway. And, finally, even if a subcortical pathway via the pulvinar exists, other potential connections also carry related signals so that, at a minimum, research needs to be expanded to elucidate their contributions and to contrast them to those of the purported subcortical route.

What Kind of Unawareness Matters?

The research literature is replete with paradigms such as backward masking and attentional blink that challenge the visual system so that awareness can be studied. At times, much is made about the neuroimaging responses observed in the amygdala for very brief stimuli (e.g., 15–30 ms). In such cases, subjects may report not seeing them ("subjective unawareness"). Although this type of unawareness is certainly interesting, I would argue that it is not the one most relevant to the impact of affective content on behavior and on clinical conditions such as anxiety.

Multiple meanings of the terms "aware" and "unaware" are conflated in the research literature. In addition to the sense of "unaware" linked to "subliminal," we need to consider a second sense associated with the idea of *unintentional* processing. This sense, which has been the object of much research in social psychology, may indeed prove to be more important to the understanding of human behavior. As aptly summarized by John Bargh and Ezequiel Morsella (2008, 78):

> We propose an alternative perspective, in which unconscious processes are defined in terms of their unintentional nature and the inherent lack of awareness is of the influence and effect of the triggering stimuli and not of the triggering stimuli (because nearly all naturally occurring stimuli are supraliminal). By this definition of the unconscious, which is the original and historic one, contemporary social cognition research on priming and automaticity effects have shown the existence of sophisticated, flexible, and adaptive unconscious behavior guidance systems. These would seem to be of high functional value, especially as default behavioral tendencies when the conscious mind, as is its wont, travels away from the present environment into the past or the future.

Whether the unintentional unconscious is so sophisticated and flexible is a matter of debate. There can be no doubt, however, that it is *qualitatively* different from the type of subliminal unconscious sometimes emphasized in the emotion literature (which, incidentally, may be quite "dumb"; Loftus and Klinger 1992—if at all present; Pessoa 2005).

In sum, if nonconscious processing is significant for the impact of affective content on behaviors, it would be in the sense of the "unintentional unconscious." The contributions of the amygdala and, in fact, many other evaluative brain regions, such as orbitofrontal, medial prefrontal, and insular cortex, to unintentional processing need to be unraveled for a clearer understanding of the underlying mechanisms.

Why Is the Amygdala Important?

In the broader neuroscience literature, the amygdala is viewed as a central node in emotional processing in part due the "low-level" properties ascribed to the subcortical pathway. Defects in the amygdala system are said to underlie phobias, mood disorders, and posttraumatic stress syndrome, and variability in its functioning, to reflect individual differences at the genotypic and personality level.

Although chapter 3 challenged many of the properties typically ascribed to the subcortical pathway, the amygdala is indeed important for behavior and mental health. First, together with the hypothalamus and medial prefrontal cortex, the amygdala has extensive projections to downstream regions in the brainstem that are capable of mobilizing the body; indeed, its central nucleus is at times described as a "controller of the brainstem." The autonomic and neuroendocrine connections of these brain regions are part of sympathetic and parasympathetic networks that coordinate bodily responses in the face of challenges to the organism.

Second, the amygdala, hypothalamus, medial prefrontal cortex, and related regions, being among the most extensively connected parts of the brain, are optimally positioned to influence information processing. As hubs through which evaluative signals are communicated, they are thought to have widespread effects on mental function and to play a significant role in affective and cognitive impairments observed in mood disorders. Metaphorically speaking, as one of these hubs, the amygdala is strategically positioned to "ignite" both body and brain.

Processing of Emotion-Laden Information and Automaticity

Richard Shiffrin and Walter Schneider (1977, 155–156) defined an "automatic process . . . as a sequence of nodes that nearly always becomes active in response to a particular input configuration, where the inputs may be externally or internally generated and include the general situational context, and where the sequence is activated without the necessity of active control or attention by the subject." The concept of automatic process—or "automaticity"—is clearly aligned with introspection: certain sensations and mental operations seem to occur "automatically," with no effort. Others, however, are qualitatively different and best viewed as "*controlled* processes." These require "active attention by the subject [and] are therefore tightly capacity-limited, [although] the cost of capacity limitations is balanced by the benefits deriving from the ease with which such processes may be set up, altered, and applied in novel situations for which automatic sequences have never been learned" (Shiffrin and Schneider 1977, 156–157).

Because automatic and controlled processes appear to be qualitatively opposed, it is natural to dichotomize mental phenomena into these two classes. But such a dichotomy has simply not held up in the face of new data: reports of automaticity have invariably been countered by reports of capacity limitation; behavioral effects assumed to operate automatically are affected in ways that belie that assumption.[2]

The argument made here is that a better framework is one where performance is always considered capacity limited and described as a performance-resource function. Recall from chapter 4 how behavior is contingent on the resources available to perform a given task. Some behaviors will exhibit shallower performance-resource relationships, where performance only rises slowly based on the mental effort exerted—these behaviors are thus "controlled." Other behaviors exhibit steeper such relationships, and ceiling performance is reached even when conditions are degraded (e.g., under short exposure)—these behaviors are thus "efficient." Although the performance-resource function may seem to be an abstract construct when little is known about the task at hand, it forces researchers to consider a spectrum of scenarios when studying how a behavior depends on the multiple factors that influence performance.

Why is a continuous framework better than a dichotomous one? For one thing, it fits the empirical data better: researchers have repeatedly found capacity limitations for "automatic" phenomena (e.g., Pashler 1998). For another, the dichotomous framework is plagued by serious conceptual issues. As reported by Agnes Moors and Jan De Houwer (2006, 321): "Of the various methods that we reviewed for testing the presence of automaticity features, not a single test gave airtight guarantees." Schneider and colleagues (1984, 20–21) examined twelve criteria and concluded that none was necessary or sufficient for the distinction between automatic and nonautomatic processes. Shiffrin (1988) evaluated several criteria for automaticity, ultimately retaining not one as generally applying to all automatic processes. He eventually came to favor a gradualist view.

Another reason why a continuous framework is better is that we still have an incipient understanding of *competition*—and, hence, of whether or not interference will result when multiple items are involved. In broad terms, the notion of competition, as well understood and accepted by most researchers, goes roughly as follows. Because processing capacity is limited, competition (among visual items in the case of vision) is proposed to "select" the most relevant information at any given time (Desimone and Duncan 1995; Grossberg

1980). When resources are not fully consumed, spare capacity is used to process task-irrelevant items (Lavie 1995). The problem is that we do not always know whether interference will occur in any given situation. Generally, multiple factors determine how information competes in visual cortex and beyond, including task difficulty, set size, spatial arrangement, cueing, "dilution," and the like. Thus, although considerable progress has been made in describing how attention influences visual responses (Reynolds and Heeger 2009), much remains to be worked out.

Finally, a continuous framework demystifies the processing of certain complex features. For example, processes such as reading and the perception of elaborate emotional images (e.g., those in the International Affective Picture System database) are at times depicted as "automatic" in a sense that is almost magical (for a cogent in-depth discussion, see Pourtois, Schettino, and Vuillemier 2012).[3] Indeed, the underlying mechanisms of abilities such as proficient reading and the perception of emotional scenes are remarkably fast. That we do not understand why they are so fast, however, simply means we are still quite some way from a better mechanistic description of these processes.

Dual Process Models

The above discussion of automatic versus controlled processes is also pertinent to *dual process* models. Although exploring specific models is beyond the scope of this book, common to all of them is the strong assumption of the existence of two *qualitatively* different mental systems, for instance, "intuition" and "reasoning" (for a lucid discussion, see Keren and Schul 2009). A popular trend is to call the two components "system 1" and "system 2," where the first is automatic/heuristic/reflexive and the second is controlled/analytic/reflective (Evans 2008). But as others have expressed in the past, the idea of a dual system model is both slippery and conceptually unclear (see Keren and Schul 2009). For one thing, nearly all dual process models have as a central component the automatic versus controlled dichotomy, which as discussed above is not a viable distinction.

I would argue that, as with the question of automatic versus controlled processing of emotion-laden visual stimuli in chapter 4, the question of whether there are two systems in dual process models is not an entirely empirical one. This is because no single critical experiment can provide a final, definitive answer. In the end, however irresistible dichotomies are to the human mind (Kelso and Engstrøm 2006; Newell 1973), dichotomizing implies *oversimplifying* (Keren and Schul 2009; Kruglanski et al. 2006). A continuous framework is better, albeit more complex (Kruglanski et al. 2006).

Interactions between Cognition and Emotion in Prefrontal Cortex: An Integration Model Is Better Than a Segregation One

Cognition and emotion have been traditionally described as mutually antagonistic, the one interfering with the other and vice versa. As reviewed in chapter 5, some evidence from functional MRI studies is consistent with this notion. At the same time, many interactions between emotion and cognition do not fit into a simple push-pull relationship. Instead, cognitive and emotional signals are combined in complex ways in prefrontal cortex, such that both contribute to observed activity—and ensuing behavior. That said, is the mapping of cognition and emotion onto the brain better characterized by a *segregation* or an *integration* model?

Chapter 5 rejected the segregation model and advocated for the integration model. But if emotion and cognition are not antagonistically related, how do they interact in prefrontal cortex? As illustrated in chapter 5, cognitive-emotional interactions take many forms. Notably, lateral prefrontal cortex, which has well-documented roles in cognition, is also an important convergence site for cognitive and emotional signal interactions. Thus, in many cases, the "cognitive" or "emotional" nature of mental processes is blurred in a way that highlights the integration of these domains in the brain. Thus, too, it is both conceptually and empirically unproductive to carve the brain into cognitive and emotional realms.

Another point concerns the ability of emotional information to influence task performance. In the perceptual literature, emotional items are described as powerful enough to modify ongoing performance when they are task irrelevant and outside the focus of attention. In the laboratory, however, the impact of emotion on cognitive tasks is often relatively modest and can sometimes be absent even when stronger manipulations are administered. Two studies serve to illustrate this point. In a recent behavioral experiment, my colleagues I observed that the threat of shock during a working memory task did not lead to impaired performance (trials actually containing shock were not included in our analysis; Choi, Padmala, and Pessoa, unpublished observations). In another study (Qin et al. 2009), a stress induction prior to an N-back working memory task was associated with changes in activation in frontal cortex, but a behavioral effect was not detected (see also Hu et al. 2011; Robinson et al. 2011). Such findings indicate that the demands of many cognitive tasks can *counteract* the effect of emotional information (see chapter 5). Although this is in line with the view of an antagonistic relationship between cognition and emotion, it also challenges the notion that emotion is more powerful than cognition—in many cases, it is the other way around.[4]

Motivation and Reward Have Specific Effects on Perception and Cognition

Renewed interest in the effects of motivation on perception and cognition has resulted in a growing number of studies involving behavioral, neuroimaging, and monkey physiology techniques. The findings of these studies have led to a reconceptualization of how motivation shapes perception and cognition.

Early perception studies manipulated reward payoffs, but focused on understanding changes in the response criterion. More recent research has attempted to elucidate how motivation influences performance in visual tasks. Does it enhance attention or does it have less specific effects, such as speeding up reaction time? Are the effects of motivation on attention related to changes in sensory mechanisms, or are they related to more decision-related processes? As reviewed in chapter 6, in contrast to earlier findings, *selective* effects have recently been found. Examples include the increase in visual sensitivity (d-prime) in both endogenous and exogenous attention tasks, enhanced stimulus coding, and target/distractor effects. The impact of motivation on visual performance has been found in reward studies both when reward was contingent on performance and when it was not. Indeed, *value learning* studies showed that the effects of reward are manifested not only when subjects are encouraged to perform better, but also via learning mechanisms. Stimuli consistently associated with reward behave as if they had increased *salience*.

Traditional accounts described motivation's effect on cognition in terms of a "general activation" function that was deemed independent of control demands. In other words, motivation was viewed as a blunt process. Findings from the studies reviewed in chapter 6, however, underscore the striking ability of motivation to shape cognitive behavioral performance selectively. These findings include reduction of response conflict, reduction in task-switch costs, selective effects on working memory, and improved long-term memory.

Emotion and Motivation Influence Competition at Perceptual and Executive Levels

Emotion and motivation alter how information is processed in the brain. The dual competition model describes these effects in terms of competition at the perceptual and executive levels. In the context of perception, a common effect of emotion and motivation is to confer enhanced *salience* to visual objects. Thus task-relevant items are detected better, whereas task-irrelevant items act as more potent distractors. In the case of emotion, one mechanism has been repeatedly emphasized, namely, the modulation of visual areas by the amygdala. In chapter 7, however, I described five other circuits that may influence visual processing, although little is known about them. Unraveling their contributions is important, especially given growing evidence that the effect of emotion-laden visual stimuli is maintained in patients with amygdala lesions. In the case of motivation, less is known about the mechanisms involved and no structure appears to exert a role similar to that of the amygdala and its connections to visual cortex. A study by Jacqueline Gottlieb and colleagues (Peck et al. 2009) showed that attentional signals in parietal cortex incorporate reward value but did not describe how parietal signals are endowed with value. In this case, interactions between structures in the striatum and parietal cortex, such as those observed by Srikanth Padmala and me (Padmala and Pessoa 2011) and by Helga Harsay and colleagues (2011), are suggested to be involved.

The effect of emotion and motivation on executive function can be viewed in terms of *resources*. In the case of emotion, resources needed for cognition are thought to be consumed by emotional processing when arousal is sufficiently high, thereby impairing performance. In the case of motivation, processing is prioritized so that additional resources are allocated to reward-relevant items (or mechanisms). Alternatively, the resources may be used in a more efficient manner, as conveyed by the idea of *sharpening*.

A problem with this formulation is that we do not have a mechanistic account of "resource," which is treated metaphorically in a way that connects with a *physical* sense of the word—as if it were a fuel of some sort, for example, that can be used up and replenished. Although the metaphor may be useful in some contexts, we clearly need a more formal and precise description of the term. Without it, the epistemological status of the notion of resources will remain problematic.

Mara Mather and Matthew Sutherland (2011) have attempted to account for both perception and memory findings in emotion studies through "arousal-biased competition," where arousal, whether elicited by external stimuli, internal thoughts, or stress hormones, modulates the strength of competing mental representations, enhancing memory for items that dominate the contest for selective attention. In a nutshell, arousal enhances *biased competition* in the sense articulated by Robert Desimone and John Duncan (1995). In particular, Mather and Sutherland (2011) propose that if the arousing stimulus is not in direct competition with the task-relevant stimuli, then the processing of neutral, goal-relevant stimuli will be enhanced (e.g., Chajut and Algom 2003; Cornsweet 1969), and that of less relevant stimuli will be reduced (e.g., Callaway and Thompson 1953; Nobata, Hakoda, and Ninose 2010).

An additional issue centers around the direction of the effects of emotion on prefrontal brain responses. To understand the direction (discussed at greater length in chapter 7), we need to consider the relationship between behavior and PFC responses during regular processing (i.e., with neutral items). Suppose that regular behavior is correlated with *increased* responses in a specific area of prefrontal cortex. Then, if emotion impairs performance, we would expect decreased activity and, conversely, increased activity if emotion improves performance. In contrast, where regular behavior is correlated with *decreased* responses in a given area, the predicted patterns are reversed. Although this is a simple point, it helps clarify why both types of patterns are observed.

Ideas of how emotion and motivation influence perceptual and cognitive processes have a long history, of course. Starting in the 1970s, Stephen Grossberg and his collaborators proposed formal neural network models detailing interactions between these diverse domains. From the outset, one of their goals was to establish mechanisms of emotional and motivational processing in a way that was integrated with perception and cognition (see Grossberg 1971). In fact, cross-fertilization between advances in perceptual and cognitive architectures, on the one hand, and those in motivational and emotional architectures, on the other, has been a recurring theme of this work (see Grossberg 1980). The strict *integration* of diverse signals was already hinted at during some of Grossberg's earliest work: "If strong coupling mechanisms exist in every nucleus, then the embeddings representing a given external input are not localized. Rather, they are spread over all nuclei" (Grossberg 1964, 382). In other words, because of the high degree of interactivity between brain regions, representations of external stimuli are distributed across the brain.

Structure-Function Mappings and Network Perspective on Emotion and Cognition

If we are to consider how concepts like emotion and cognition are mapped onto the brain, we must consider the issue of structure-function mapping more broadly. Chapter 8 provided a general treatment of issues that inform the understanding of brain-behavior relationships. If the mapping from brain areas to functions is both one-to-*many* and *many*-to-one, namely, *many*-to-*many*, trying to come up with the *main function* that a region implements will be unproductive. Indeed, because regions carry out functions only as parts of networks with other regions, speaking of a "region's function" is not entirely meaningful. As I have proposed, a better way to characterize a region is in terms of its *functional profile*, namely, the range of processes it is involved in as it interacts with different groups of brain regions. For example, the functional profile of the amygdala is quite diverse and involves evaluation of affective significance, reward value, attentional processes, and engagement of bodily responses, among other functions. In each case, the amygdala accomplishes the function through interactions with other regions, in these cases, with orbitofrontal cortex, ventral striatum, visual cortex, and brainstem, respectively. Thus labels such as "perception," "cognition," and "emotion" are useful linguistic categories, but only in the limited sense of providing placeholders for descriptive purposes—they do *not* map onto behavior *or* the brain.

Although a network perspective is needed for a richer characterization of the mind-brain, it should not be viewed as a panacea. As discussed previously, we should not anticipate a one-to-one mapping when the network approach is adopted (contrary to the recent trend of labeling networks with specific functions; see examples in Bressler and Menon 2010). Critically, decomposition of brain regions into "communities" (also called "modules") does not by itself reveal "true" subnetworks. Instead, multiple decompositions offer different viewpoints of existing structure. A network perspective, though valuable and indeed necessary, needs to be employed in a nuanced manner: its application is not straightforward.

The network perspective has significant implications for appreciating the roles of structural and functional connectivity. Great emphasis is often put on the existence of robust, direct structural connectivity when trying to determine the function of brain regions. However, anatomical connectivity is *not sufficient* for inferring functional connectivity. Areas having no strong or direct structural connections may still have close functional ones. Generally, the intricate way in which multiple regions are interconnected, including the existence of inhibitory connections, greatly complicates the interpretation of how structure will support function.

Once we shift our focus from brain regions to brain *networks*, the question arises, how best to understand them? As reviewed in chapter 8, modern network/graph theory offers several tools that are helpful in describing brain networks and how they are linked to behaviors. For example, brain architecture exhibits *small-world* properties. Specifically, a high degree of local connectivity is mixed with long-range connections that allow all regions of the brain to be linked by very short paths (typically two or less). This has a profound influence on the way we should view the organization of the brain.

It is interesting to contrast small-world architecture with the model proposed by M.-Marsel Mesulam (1998). Although very much a distributed processing model, it emphasizes the stepwise nature of brain connectivity.[5] For example, prefrontal regions are among those most distant from the sensory periphery—based on the average number of connections (Young et al. 1994). Thus the prefrontal cortex receives highly processed and integrated sensory information. This structural feature is thought to be important because it provides prefrontal cortex with relative *insulation* from the sensory periphery. However, long-range connections—and small-world properties—break this pattern in a way that permits prefrontal cortex to *also* operate on much less processed data. Consider here the notion of a tightly integrated *inner core circuit* as proposed by Dharmendra Modha and Raghavendra Singh (2010) and discussed in chapter 8. Thus, even though prefrontal cortex is capable of operating on signals that have undergone considerable refinement by virtue of being processed by several brain regions, it is also susceptible to "less abstract" influences. This suggests that prefrontal cortex can be rapidly influenced by many types of information, including biologically significant information. These properties are exactly the ones we would expect from a multiple waves architecture.

Long-range connections and multiple pathways operate in the "reverse" direction as well, namely, from the "rest of the brain" *to* visual cortex. Indeed, all levels of the visual cortex are influenced by affective and motivational significance, a view that runs counter to a position taken by Mesulam (and a number of other researchers):

> Early sensory processing . . . is relatively protected from value-based modulations. This arrangement [anatomical connectivity] allows internal states to influence sensory-fugal [that is, emanating from sensory cortex] processing only after the essential features of sensory events have been encoded. In fact, there is little adaptive sense in altering responses to shapes, motion or colour, whereas it makes a lot of sense to modulate responses to specific faces, objects or locations. (Mesulam 1998, 1036)

Anatomical Potential for Cognitive-Emotional Interactions

Chapter 9 briefly reviewed how the architecture of the brain includes multiple avenues for communicating information. The substrates for information interaction and integration are indeed plentiful. One intriguing example is that of the hypothalamus, whose descending control of bodily functions is well documented, but whose contributions to *ascending* processing are poorly understood. In particular, the hypothalamus has widespread projections to *all* sectors of prefrontal cortex (Rempel-Clower and Barbas 1998). Given the role of the hypothalamus as a critical component of the central autonomic nervous system, this pattern of connectivity implies that the hypothalamus has the ability to influence processing throughout prefrontal cortex. Notably, this includes lateral prefrontal cortex, which is important for cognitive function. Intriguingly, lateral prefrontal regions appear to have few, if any, *descending* projections to the hypothalamus (Rempel-Clower and Barbas 1998), whereas medial and orbital regions have many (Rempel-Clower and Barbas 1998). Overall, this is suggestive of an asymmetrical arrangement: whereas communication from prefrontal cortex to hypothalamus is quite restricted, communication from hypothalamus to prefrontal cortex is quite broad. As crisply summarized by Nauta (1972, 137):

> The unique feature of the neural circuitry outlined in the foregoing account is that it places the frontal cortex in a reciprocal relationship with two great functional realms, namely: (i) parietal and temporal regions of the cerebral cortex involved in the processing of visual, auditory and somatic sensory information, and (ii) the telencephalic limbic system and its subcortical correspondents, in particular the hypothalamus. . . . Because of its reciprocal connections with the limbico-hypothalamic axis the frontal cortex could be viewed at once as a modulator of hypothalamic mechanisms and as a cortical region receiving information concerning the functional state of the limbic system and hypothalamus. . . . Judged by anatomical data, a re-representation of both external and internal environment would seem to be a major characteristic of the frontal cortex.

Implications of Network Perspective for Emotion and Cognition

In reflecting on the network perspective described in chapter 8, five significant implications come readily to mind. First, given the extensive interactions between brain territories, with this perspective, our emphasis shifts from attempting to understand the brain a region at a time to characterizing how coalitions of regions support the mind-brain. And because brain regions are not the unit of interest, they should not be viewed as "cognitive" or "emotional." This stands in sharp contrast to the traditional view, which sees regions whose function involves homeostatic processes or bodily representations as emotional, and those less aligned with such operations as cognitive.

Second, considered from the network perspective, the architectural features of the brain provide *massive* opportunity for cognitive-emotional interactions, encompassing *all* brain territories. For example, extensive communication between the amygdala and visual cortex exists. Thus visual processing takes place within a context that is defined by signals occurring in the amygdala and related brain regions (e.g., orbitofrontal cortex) whose processing is linked to affective significance. In this sense, vision is never *pure*, but only *affective*, vision.[6] A similar point can be made for other sensory modalities. Cognitive-emotional interactions also abound in prefrontal cortex, which is commonly described with reference to abstract processes. More generally, given interregion interactivity, and the fact that networks intermingle signals of diverse origin, although a characterization of brain function in terms of networks is still needed, the networks themselves are best thought of as neither "cognitive" nor "emotional."

Third, regions traditionally viewed as central for affective processing appear to be extremely well connected, which suggests that they have at times important "quasi-global" roles and that this may be a central feature of their class (but see Achard et al. 2006).[7] Nor are they the only regions with high connectivity: we encounter highly connected regions throughout the brain, including occipital, temporal, parietal, and frontal lobes, in addition to insula, cingulate, thalamus, and other regions at the base of the brain.

Fourth, the network perspective reminds us that emphasizing only interactions between brain regions that are connected by direct, robust structural connections is misleading. The strength of functional connectivity is equally important and at times will deviate from the strength of the structural connection. Architectural features guarantee the rapid integration of information even when strong structural connections are not present—and support functional interactions that will vary based on context. A case in point is illustrated by the "one-step" property of amygdala–prefrontal cortex connectivity (i.e., amygdala signals reach nearly all prefrontal regions within a single connectivity step in prefrontal cortex), which allows the amygdala to engage in functional interactions with lateral PFC regions not heavily connected to it.

And, fifth, the insights gained from adopting a network perspective suggest that the mind-brain is *not* decomposable in terms of emotion and cognition. In other words, the neural basis of emotion and cognition should be seen as governed less by properties that are intrinsic to specific sites and more by interactions among multiple brain regions. In this sense, emotion and cognition are *functionally integrated systems*, namely, they more or less continuously impact each other's operations: "The problem is then not one of isolating the localized mechanisms, but of exhibiting the organization and the constituent functions. . . . An explanation in terms of organization supplants direct localization." (Bechtel and Richardson 2010, 151).

Concluding Thoughts

Emotion feels different from cognition. Thus our feeling of outrage when a careless driver dangerously cuts in front of us, forcing us to rapidly swerve away to avoid being hit seems *qualitatively* different from our calmly considering the various driving routes we might take on a cross-country summer trip with the family. And these two mental states and associated processes also appear to be subserved by fairly independent brain networks. Yet, when we consider the available neuroscientific data, attempts to characterize regions as either "emotional" or "cognitive" quickly break down. An architecture of rich interconnectivity leads to a structure-function mapping that is both one to many and many to one. Ultimately, looking at the brain from the perspective of one brain region at a time is bound to produce a highly distorted and, more critically, impoverished description. The development of neural network theory (Grossberg 1980; Rumelhart and McClelland 1986) and modern graph-theoretic methods illustrate alternative frameworks that focus on distributed computations. As these methods evolve into newer theoretical frameworks, we can anticipate that a truly dynamic network view of the brain will become more widely adopted. In such dynamic frameworks, "emotion" and "cognition" may be used as labels in the context of certain behaviors, but will not map cleanly into compartmentalized pieces of the brain.

Notes

1. The triune brain is, of course, a "triple" model, not a "double" one. Carl Sagan, who did much to widen the appeal of science in general, may have contributed to the popular appeal of the triune brain idea in his book *The Dragons of Eden* (1977). I thank Brenton McMenamin for reminding me of this fact.

2. If one relaxes the definition of "automaticity," it could be argued that several of these behavioral effects are still "automatic." See chapter 3 for further discussion on "strong" and "weak" automaticity.

3. At times, some descriptions of "automatic processes" are reminiscent of past claims that light travels instantaneously. To be sure, the speed of light is incredibly fast—but it is still *finite*.

4. Of course, the cases where cognition is more powerful than emotion do not involve threats to one's survival.

5. Mesulam (1998) also called attention to the importance of heteromodal regions of cortex in connecting diverse brain systems.

6. In the rat, the hypothalamus has connections with occipital cortex, including primary and secondary visual areas, although these are less prominent than other pathways (Risold, Thompson, and Swanson 1997).

7. Barbas and Pandya (1989) also suggest that poorly laminated limbic (their term) regions of prefrontal cortex (which are frequently implicated in affective processing) have a more global role in neural processing.

Glossary

At the most basic level, the brain is composed of lobes, as shown in figure G.1. Orientation in the brain is based on three orthogonal axes: front to back (anterior/rostral to posterior/caudal), top to bottom (superior/dorsal to inferior/ventral), and left to right. See figure G.2 for lateral and medial views of the brain along the left-to-right axis, where "lateral" refers to "outside" and "medial" to "inside," and figure G.3 for views along the other two reference axes.

acetylcholine Neurotransmitter that has excitatory, inhibitory, and modulatory functions, acting in both the peripheral and central nervous system, and that, with associated neurons, makes up the cholinergic system.

aversive conditioning See **fear conditioning.**

basal forebrain Heterogeneous set of structures close to the ventral and medial surfaces of the cerebral hemispheres (see figure 2.4).

basal ganglia Group of nuclei lying deep in the subcortical white matter of the frontal lobes, which include caudate, putamen, globus pallidus, substantia nigra, and subthalamic nucleus.

bed nucleus of the stria terminalis (BNST) Structure bordering both the nucleus accumbens and ventral caudate, having several components and a complex topology, and thought to be linked with "sustained fear" (or "anxiety") mechanisms, although growing evidence links it with appetitive functions as well.

blindsight Ability to respond to visual stimuli without consciously perceiving them, which may result from lesions to primary visual cortex.

BOLD signal Blood oxygen level dependent signals, basis of brain oxygenation signal measure of functional magnetic resonance imaging; believed to generally reflect the underlying neuronal activation.

bottom-up Said of factors that are stimulus-based (e.g., contrast or motion); often strictly opposed to "top-down," although the opposition is increasingly challenged.

brainstem Portion of the brain located between the diencephalon (thalamus and hypothalamus) and the spinal cord, comprising the midbrain, pons, and medulla (see figure G.2).

cholinergic system See **acetylcholine**.

cingulate cortex Sector of cortex along the middle part of the brain that surrounds the entire corpus callosum (see figure G.2).

contralateral Referring to anatomical structure on the opposite side of the body to another.

diffusion tensor imaging Magnetic resonance imaging technique that measures the diffusion of water molecules in the brain to estimate the main direction of fiber tracts, hence structural parameters, as opposed to functional activation.

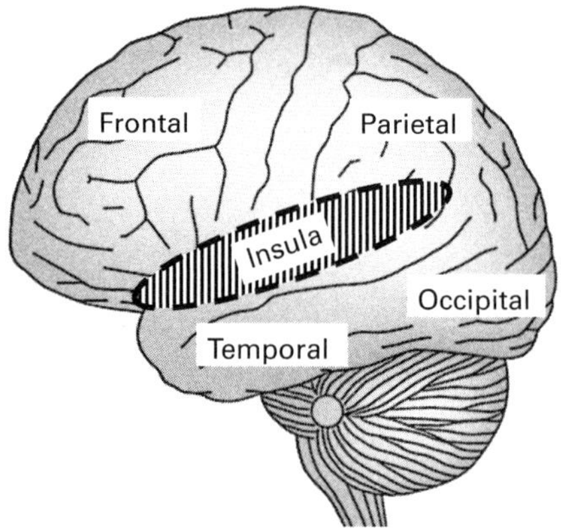

Figure G.1
Lateral view of brain, showing principal lobes.

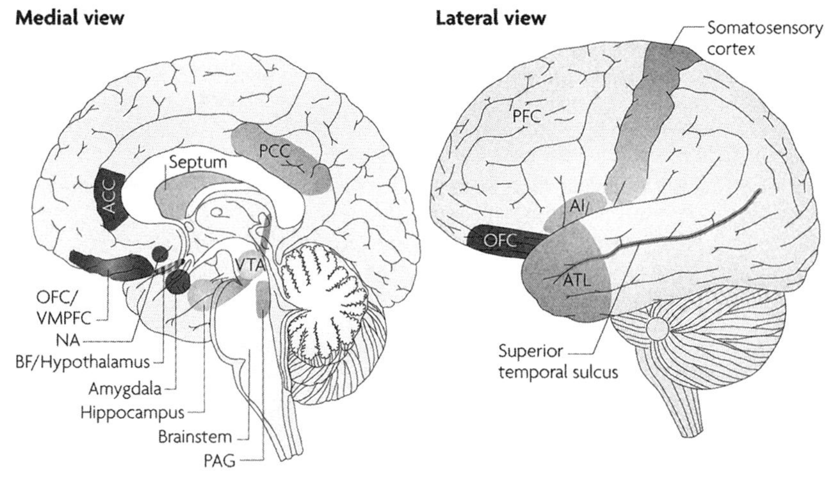

Figure G.2
Medial and lateral views of brain, showing specific regions. ACC, anterior cingulate cortex (term I have avoided using as much as possible because of the inconsistency with which this label is applied); AI, anterior insula; ATL, anterior temporal lobe; BF, basal forebrain; NA, nucleus accumbens; OFC, orbitofrontal cortex; PAG, periaqueductal gray; PCC, posterior cingulate cortex; PFC, prefrontal cortex; VTA, ventral tegmental area; VMPFC, ventral-medial prefrontal cortex.

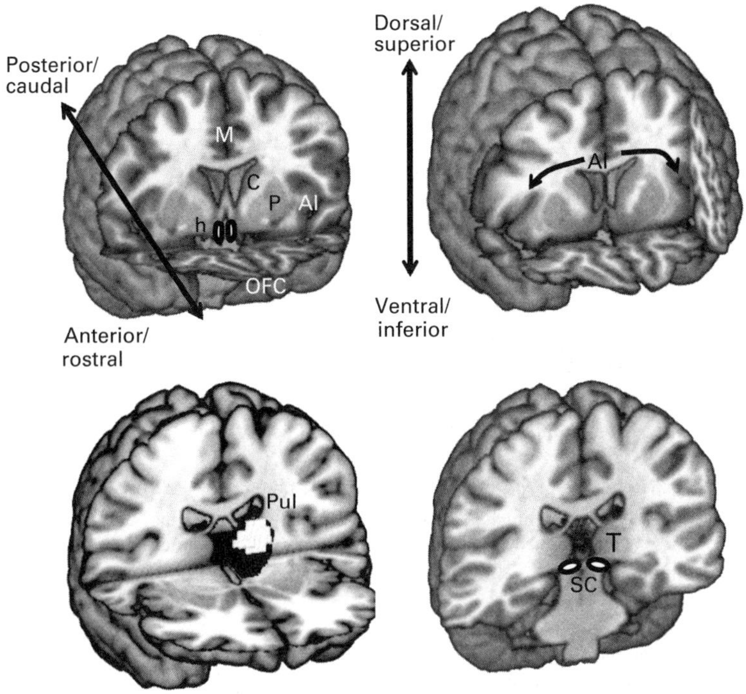

Figure G.3
Brain as viewed along front-to-back and top-to-bottom axes, with approximate locations on anatomical magnetic resonance images. The pulvinar is schematically shown in the bottom left panel as the white part of the thalamus (shown in black). AI, anterior insula; C, caudate; h, hypothalamus; M, medial prefrontal cortex; OFC, orbitofrontal cortex; P, putamen; Pul, pulvinar; SC, superior colliculus; T, thalamus.

dopamine Neurotransmitter that has both inhibitory and excitatory functions; two important dopaminergic sites are the ventral tegmental area and the substantia nigra.

dorsal striatum Caudate and putamen (see figure G.3).

driving versus modulatory inputs Neurons that define the essential patterns of neural activity versus neurons that alter responses without determining the general pattern of response, although this distinction by researchers studying neuron-to-neuron communication is, in general, neither clear cut nor easy to establish.

electroencephalography (EEG) Technique that measures electrical brain signals as picked up by sensors on the scalp.

entorhinal cortex Sector of cortex located in the medial temporal lobe of the brain and closely associated with the hippocampus.

event-related potential (ERP) Average EEG response associated with a specific event, determined by averaging EEG signals during a specific temporal window that starts with the occurrence of the event (e.g., 0–500 ms).

extrastriate Visual cortex outside of primary or striate visual cortex.

fear conditioning Experimental procedure in which a neutral item acquires negative valence by being consistently paired with an inherently negative stimulus; also called "aversive conditioning."

functional connectivity Measure of the strength of association between signals from two brain elements (neurons, regions); the measure is independent of a direct physical connection between the elements.

fusiform gyrus Ridge of ventral temporal cortex thought to be important in processing visual face stimuli or stimuli related to perceptual expertise.

GABA Gamma-aminobutyric acid, neurotransmitter that typically has an inhibitory function; plays a role in regulating neuronal excitatbility throughout the nervous system.

gamma activity Brain signals in the gamma band (30–60 Hz frequency range), thought to be important during many effortful tasks.

gyrus (plural, gyri) Ridge of cerebral cortex typically surrounded by one or more sulci.

heterarchical Replete with overlap, multiplicity, or divergent-but-coexistent patterns of relation, as opposed to "hierarchical."

hypothalamus Subcortical group of small nuclei located just below the thalamus and above the brainstem (see figure G.3).

incentive value Motivational value.

inferior temporal cortex Sector of temporal cortex comprising several visual as well as "associational" areas.

ipsilateral Referring to anatomical structure on the same side of the body as another.

lateral geniculate nucleus (LGN) Nucleus in the thalamus that receives visual projections from the retina and projects to primary visual cortex (V1), as well as to other visual areas.

limbic system Term used in too many different ways and contexts—best avoided to prevent confusion.

magnetoencephalography (MEG) Technique that measures magnetic brain signals picked up by sensors placed around the head; magnetic counterpart of electroencephalography.

magnocellular Referring to cells in the retina, lateral geniculate nucleus, or cortex (e.g., area MT/V5) that are particularly sensitive to moving stimuli and have short response latencies.

medial geniculate nucleus (MGN) Nucleus in the thalamus involved in auditory processing.

medial wall Internal surface of the brain.

medulla Lower part of the brainstem; also called "medulla oblongata."

midbrain Topmost part of the brainstem, just above the pons and below the hypothalamus, containing the tegmentum (including the ventral tegmental area) and the substantia nigra; located between the forebrain, comprising the cerebral hemispheres, thalamus, and hypothalamus, and the hindbrain, comprising the pons and medulla.

modulatory inputs See **driving** versus **modulatory inputs**.

neglect Disabling condition following brain damage in which patients fail to be aware of items to one side of space (e.g., patients with right-sided brain damage may fail to be aware of objects to their left).

neuroaxis Imaginary axis passing through the spinal cord and top of the brain; typically connoting superior-to-inferior (or vice versa) progression along the brain.

norepinephrine Neurotransmitter that originates in brainstem structures such as the locus coeruleus; in broad terms it is involved in arousal and novelty.

nucleus accumbens Subcortical structure located in the ventral striatum and contiguous with ventral sectors of the caudate (see figure G.2).

nucleus of the solitary tract Major visceral sensory cell group in the brain, located in the brainstem.

parabrachial nucleus Region in the pons related to the ascending reticular activating system essential for general arousal.

parvocellular Referring to cells in the retina, lateral geniculate nucleus, or cortex (e.g., area V4) that are particularly sensitive to form information and have longer response latencies than their magnocellular counterparts.

Pavlovian fear conditioning See **fear conditioning.**

periaqueductal gray (PAG) Set of nuclei in the brainstem surrounding the midbrain aqueduct, classically associated with defensive reactions, suppression of pain (analgesia), and autonomic regulation (see figure G.2).

pituitary gland "Master" endocrine gland, attached to the hypothalamus by the pituitary stalk, which secretes hormones directly into the bloodstream.

pons One of the three components of the brainstem, lying below the midbrain and above the medulla.

pop-out search Experimental paradigm in which an item "pops out" solely based on its physical characteristics (e.g., single red circle among large array of green circles).

prefrontal cortex (PFC) Sector of cortex in front of the motor and premotor areas (see figure G.1).

pulvinar Set of nuclei in the thalamus (see figure G.3); involved in part in visual processing as well as related "associational" functions.

receptive field Region of space where a stimulus will influence the firing of a visual cell.

reinforcer Stimulus that strengthens or weakens the behavior that produced it.

reticular formation Network of neurons and axons that occupies the core of the brainstem, giving it a reticulated appearance in myelin-stained material; major functions include control of respiration and heart rate, posture, and states of consciousness.

retinotopic Referring to orderly arrangement of visual receptive fields as external stimuli are gradually moved in space; retinotopically organized areas are said to contain "maps" of visual space.

spatial frequency Measure of how often stimulus intensity changes across space; more technically, a measure of how often sinusoidal components of the signal as determined by the Fourier transform repeat per unit of distance.

speed-accuracy trade-off Pattern of behavior in which subjects emphasize speed at the expense of accuracy or vice versa.

steady-state visual evoked potentials EEG signals that are captured in response to visual stimulation at specific temporal frequencies.

sulcus Depression or fissure on the cortex of the brain that typically surrounds the gyri.

superior colliculus Laminated structure that forms part of the roof of the midbrain (see figure G.3).

telencephalon Part of the brain derived from the anterior embryonic forebrain vesicle that includes the cerebral hemispheres.

top-down Said of processes that rely on endogenous factors, such as internal goals; often described in strict opposition to "bottom-up" processes, although the distinction is increasingly challenged.

ventral striatum Structure comprising the nucleus accumbens and olfactory tubercle; term sometimes refers to the nucleus accumbens only or to ventral portions of the caudate and putamen (see figures G.2 and G3).

ventral tegmental area (VTA) Structure located in the midbrain and an important dopaminergic site; said to originate both "mesolimbic pathways" targeting the nucleus accumbens, amygdala, and hippocampus and "mesocortical pathways" targeting prefrontal cortex and insula (see figure G.2).

ventral visual system Set of regions along the ventral (inferior) surface of occipital and temporal cortex important for form vision.

References

Aarts, E., Roelofs, A., Franke, B., Rijpkema, M., Fernandez, G., Helmich, R. C., et al. (2010). Striatal dopamine mediates the interface between motivational and cognitive control in humans: Evidence from genetic imaging. *Neuropsychopharmacology*, *35*(9), 1943–1951. doi:10.1038/npp.2010.68.

Aarts, E., van Holstein, M., and Cools, R. (2011). Striatal dopamine and the interface between motivation and cognition. *Frontiers in Psychology*, *2*, 163. doi:10.3389/fpsyg.2011.00163.

Achard, S., Salvador, R., Whitcher, B., Suckling, J., and Bullmore, E. (2006). A resilient, low-frequency, small-world human brain functional network with highly connected association cortical hubs. *Journal of Neuroscience*, *26*(1), 63–72.

Adams, W. J., Gray, K. L., Garner, M., and Graf, E. W. (2011). On the "special" status of emotional faces . . . Comment on Yang, Hong, and Blake (2010). *Journal of Vision*, *11*(3): 10, 1–4. doi:10.1167/11.3.10.

Adcock, R. A., Thangavel, A., Whitfield-Gabrieli, S., Knutson, B., and Gabrieli, J. D. (2006). Reward-motivated learning: Mesolimbic activation precedes memory formation. *Neuron*, *50*(3), 507–517. doi:10.1016/j.neuron.2006.03.036.

Adolphs, R. (2003). Cognitive neuroscience of human social behaviour. *Nature Reviews. Neuroscience*, *4*(3), 165–178.

Adolphs, R. (2008). Fear, faces, and the human amygdala. *Current Opinion in Neurobiology*, *18*(2), 166–172.

Adolphs, R. (2010). What does the amygdala contribute to social cognition? *Annals of the New York Academy of Sciences*, *1191*(1), 42–61.

Adolphs, R., Gosselin, F., Buchanan, T. W., Tranel, D., Schyns, P., and Damasio, A. R. (2005). A mechanism for impaired fear recognition after amygdala damage. *Nature*, *433*(7021), 68–72.

Aertsen, A., Gerstein, G. L., Habib, M. K., and Palm, G. (1989). Dynamics of neuronal firing correlation: Modulation of "effective connectivity." *Journal of Neurophysiology*, *61*(5), 900–917.

Aertsen, A., and Preissl, H. (1991). Dynamics of activity and connectivity in physiological neuronal networks. In H. G. Schuster, ed., *Nonlinear Dynamics and Neuronal Networks*, 281–302. New York: VCH.

Aggleton, J. P., ed. (1992). *The Amygdala: Neurobiological Aspects of Emotion, Memory, and Mental Dysfunction*. New York: Wiley.

Aggleton, J. P., ed. (2000). *The Amygdala: A Functional Analysis*. Oxford: Oxford University Press.

Aggleton, J. P., Burton, M. J., and Passingham, R. E. (1980). Cortical and subcortical afferents to the amygdala of the rhesus monkey (Macaca mulatta). *Brain Research*, *190*(2), 347–368.

Aggleton, J. P., and Saunders, R. C. (2000). The amygdala: What's happened in the last decade. In J. P. Aggleton, ed., *The Amygdala: A Functional Analysis*, 1–30. Oxford: Oxford University Press.

Ahn, Y. Y., Bagrow, J. P., and Lehmann, S. (2010). Link communities reveal multiscale complexity in networks. *Nature*, *466*(7307), 761–764.

Alexander, G. E., DeLong, M. R., and Strick, P. L. (1986). Parallel organization of functionally segregated circuits linking basal ganglia and cortex. *Annual Review of Neuroscience*, *9*, 357–381. doi:10.1146/annurev.ne.09.030186.002041.

Alexander, W. H., and Brown, J. W. (2011). Medial prefrontal cortex as an action-outcome predictor. *Nature Neuroscience*, *14*(10), 1338–1344. doi:10.1038/nn.2921.

Alkire, M. T., Hudetz, A. G., and Tononi, G. (2008). Consciousness and anesthesia. *Science*, *322*(5903), 876–880. doi:10.1126/science.1149213.

Alpers, G. W., Gerdes, A. B., Lagarie, B., Tabbert, K., Vaitl, D., and Stark, R. (2009). Attention and amygdala activity: An fMRI study with spider pictures in spider phobia. *Journal of Neural Transmission*, *116*(6), 747–757. doi:10.1007/s00702-008-0106-8.

Amaral, D. G., Behniea, H., and Kelly, J. L. (2003). Topographic organization of projections from the amygdala to the visual cortex in the macaque monkey. *Neuroscience*, *118*(4), 1099–1120.

Amaral, D. G., and Price, J. L. (1984). Amygdalo-cortical projections in the monkey (Macaca fascicularis). *Journal of Comparative Neurology*, *230*(4), 465–496.

Amaral, D. G., Price, J. L., Pitkanen, A., and Carmichael, S. T. (1992). Anatomical organization of the primate amygdaloid complex. In J. Aggleton, ed., *The Amygdala: Neurobiological Aspects of Emotion, Memory, and Mental Dysfunction*, 1–66. New York: Wiley-Liss.

Amiez, C., Joseph, J. P., and Procyk, E. (2006). Reward encoding in the monkey anterior cingulate cortex. *Cerebral Cortex*, *16*(7), 1040–1055.

Anderson, A. K. (2005). Affective influences on the attentional dynamics supporting awareness. *Journal of Experimental Psychology. General*, *134*(2), 258–281.

Anderson, A. K., Christoff, K., Panitz, D., De Rosa, E., and Gabrieli, J. D. (2003). Neural correlates of the automatic processing of threat facial signals. *Journal of Neuroscience*, *23*(13), 5627–5633.

Anderson, A. K., and Phelps, E. A. (2001). Lesions of the human amygdala impair enhanced perception of emotionally salient events. *Nature*, *411*(6835), 305–309.

Anderson, B. A., Laurent, P. A., and Yantis, S. (2011). Value-driven attentional capture. *Proceedings of the National Academy of Sciences of the United States of America*, *108*(25), 10367–10371. doi:10.1073/pnas.1104047108.

Anderson, M. L., Kinnison, J., and Pessoa, L. (2013). Describing functional diversity of brain regions and brain networks. *NeuroImage*, *x*(x), xx–xx. epub ahead of print.

Anderson, M. L., and Pessoa, L. (2011). Quantifying the diversity of neural activations in individual brain regions. Paper presented at the CogSci 2011, Boston.

Andino, S. L., Menendez, R. G., Khateb, A., Landis, T., and Pegna, A. J. (2009). Electrophysiological correlates of affective blindsight. *NeuroImage*, *44*(2), 581–589.

Andrews-Hanna, J. R., Reidler, J. S., Sepulcre, J., Poulin, R., and Buckner, R. L. (2010). Functional-anatomic fractionation of the brain's default network. *Neuron*, *65*(4), 550–562. doi:10.1016/j.neuron.2010.02.005.

Angelucci, A., and Bressloff, P. C. (2006). Contribution of feedforward, lateral and feedback connections to the classical receptive field center and extra-classical receptive field surround of primate V1 neurons. *Progress in Brain Research*, *154*, 93–120. doi:10.1016/s0079-6123(06)54005-1.

Anticevic, A., Repovs, G., and Barch, D. M. (2010). Resisting emotional interference: Brain regions facilitating working memory performance during negative distraction. *Cognitive, Affective and Behavioral Neuroscience*, *10*(2), 159–173. doi:10.3758/CABN.10.2.159.

Arend, I., Machado, L., Ward, R., McGrath, M., Ro, T., and Rafal, R. D. (2008). The role of the human pulvinar in visual attention and action: Evidence from temporal-order judgment, saccade decision, and antisaccade tasks. *Progress in Brain Research*, *171*, 475–483.

Arnold, M. B. (1960). *Emotion and Personality*. New York: Columbia University Press.

Arnsten, A. F. (2009). Stress signalling pathways that impair prefrontal cortex structure and function. *Nature Reviews. Neuroscience*, *10*(6), 410–422. doi:10.1038/nrn2648.

Arnsten, A. F., and Li, B. M. (2005). Neurobiology of executive functions: Catecholamine influences on prefrontal cortical functions. *Biological Psychiatry*, *57*(11), 1377–1384. doi:10.1016/j.biopsych.2004.08.019.

Aron, A. R., Robbins, T. W., and Poldrack, R. A. (2004). Inhibition and the right inferior frontal cortex. *Trends in Cognitive Sciences*, *8*(4), 170–177.

Aslan, B., and Zech, G. (2005). New test for the multivariate two-sample problem based on the concept of minimum energy. *Journal of Statistical Computation and Simulation*, *75*(2), 109–119.

Astle, D. E., Nobre, A. C., and Scerif, G. (2010). Subliminally presented and stored objects capture spatial attention. *Journal of Neuroscience*, *30*(10), 3567–3571.

Aston-Jones, G., and Cohen, J. D. (2005). An integrative theory of locus coeruleus-norepinephrine function: Adaptive gain and optimal performance. *Annual Review of Neuroscience*, *28*, 403–450.

Augustine, J. R. (1996). Circuitry and functional aspects of the insular lobe in primates including humans. *Brain Research. Brain Research Reviews*, *22*(3), 229–244.

Averbeck, B. B., and Seo, M. (2008). The statistical neuroanatomy of frontal networks in the macaque. *PLoS Computational Biology*, *4*(4), e1000050.

Awh, E., Belopolsky, A. V., and Theeuwes, J. (2012). Top-down versus bottom-up attentional control: A failed theoretical dichotomy. *Trends in Cognitive Sciences*, *16*(8), 437–443. doi:10.1016/j.tics.2012.06.010.

Bach, D. R., Hulme, O., Penny, W. D., and Dolan, R. J. (2011). The known unknowns: Neural representation of second-order uncertainty, and ambiguity. *Journal of Neuroscience*, *31*(13), 4811–4820. doi:10.1523/jneurosci.1452-10.2011.

Bach, D. R., Seymour, B., and Dolan, R. J. (2009). Neural activity associated with the passive prediction of ambiguity and risk for aversive events. *Journal of Neuroscience*, *29*(6), 1648–1656.

Bach, D. R., Talmi, D., Hurlemann, R., Patin, A., and Dolan, R. J. (2011). Automatic relevance detection in the absence of a functional amygdala. *Neuropsychologia*, *49*(5), 1302–1305.

Baddeley, A. (2003). Working memory: Looking back and looking forward. *Nature Reviews. Neuroscience*, *4*(10), 829–839.

Badre, D., and D'Esposito, M. (2007). Functional magnetic resonance imaging evidence for a hierarchical organization of the prefrontal cortex. *Journal of Cognitive Neuroscience*, *19*(12), 2082–2099. doi:10.1162/jocn.2007.19.12.2082.

Baldassi, S., and Simoncini, C. (2011). Reward sharpens orientation coding independently of attention. *Frontiers in Neuroscience*, *5*, 13. doi:10.3389/fnins.2011.00013.

Balderston, N. L., Schultz, D. H., and Helmstetter, F. J. (2011). The human amygdala plays a stimulus specific role in the detection of novelty. *NeuroImage*, *55*(4), 1889–1898. doi:10.1016/j.neuroimage.2011.01.034.

Baldo, B. A., and Kelley, A. E. (2007). Discrete neurochemical coding of distinguishable motivational processes: Insights from nucleus accumbens control of feeding. *Psychopharmacology*, *191*(3), 439–459. doi:10.1007/s00213-007-0741-z.

Baluch, F., and Itti, L. (2011). Mechanisms of top-down attention. *Trends in Neurosciences*, *34*(4), 210–224. doi:10.1016/j.tins.2011.02.003.

Bandler, R., and Shipley, M. T. (1994). Columnar organization in the midbrain periaqueductal gray: Modules for emotional expression? *Trends in Neurosciences*, *17*(9), 379–389.

Banks, S. J., Eddy, K. T., Angstadt, M., Nathan, P. J., and Phan, K. L. (2007). Amygdala-frontal connectivity during emotion regulation. *Social Cognitive and Affective Neuroscience*, *2*(4), 303–312. doi:10.1093/scan/nsm029.

Bar, M. (2003). A cortical mechanism for triggering top-down facilitation in visual object recognition. *Journal of Cognitive Neuroscience*, *15*(4), 600–609.

Bar, M., and Neta, M. (2006). Humans prefer curved visual objects. *Psychological Science*, *17*(8), 645–648.

Barbas, H. (1995). Anatomic basis of cognitive-emotional interactions in the primate prefrontal cortex. *Neuroscience and Biobehavioral Reviews*, *19*(3), 449–510.

Barbas, H., Henion, T. H., and Dermon, C. R. (1991). Diverse thalamic projections to the prefrontal cortex in the rhesus monkey. *Journal of Comparative Neurology*, *313*(1), 65–94.

Barbas, H., and Pandya, D. N. (1989). Architecture and intrinsic connections of the prefrontal cortex in the rhesus monkey. *Journal of Comparative Neurology*, *286*(3), 353–375.

Barbas, H., Saha, S., Rempel-Clower, N., and Ghashghaei, T. (2003). Serial pathways from primate prefrontal cortex to autonomic areas may influence emotional expression. *BMC Neuroscience*, *4*, 25.

Bard, P. (1928). A diencephalic mechanism for the expression of rage with special reference to the sympathetic nervous system. *American Journal of Physiology*, *84*(490), 490–515.

Bard, P. (1929). The central representation of the sympathetic system: As indicated by certain physiologic observations. *Archives of Neurology and Psychiatry*, *22*(2), 230–246. doi:10.1001/archneurpsyc.1929.02220020046002.

Bard, P. (1934). On emotional expression after decortication with some remarks on certain theoretical views: Part I. *Psychological Review*, *41*(4), 309–329.

Bargh, J. A., and Morsella, E. (2008). The unconscious mind. *Perspectives on Psychological Science*, *3*(1), 73–79.

Baron, R. M., and Kenny, D. A. (1986). The moderator-mediator variable distinction in social psychological research: Conceptual, strategic, and statistical considerations. *Journal of Personality and Social Psychology*, *51*, 1173–1182.

Barrett, K. C., and Campos, J. J. (1987). Perspectives on emotional development: II. A functionalist approach to emotions. In J. D. Osofsky, ed., *Handbook of Infant Development* (2nd ed.), 558–578. New York: Wiley.

Barrett, L. F., and Bar, M. (2009). See it with feeling: Affective predictions during object perception. *Philosophical Transactions of the Royal Society of London. Series B, Biological Sciences*, *364*(1521), 1325–1334.

Barrett, L. F., and Bliss-Moreau, E. (2009). Affect as a psychological primitive. *Advances in Experimental Social Psychology*, *41*, 167–218.

Basser, P. J., Pajevic, S., Pierpaoli, C., Duda, J., and Aldroubi, A. (2000). In vivo fiber tractography using DT-MRI data. *Magnetic Resonance in Medicine*, *44*(4), 625–632.

Bassett, D. S., Wymbs, N. F., Porter, M. A., Mucha, P. J., Carlson, J. M., and Grafton, S. T. (2011). Dynamic reconfiguration of human brain networks during learning. *Proceedings of the National Academy of Sciences of the United States of America*, *108*(18), 7641–7646. doi:10.1073/pnas.1018985108.

Basten, U., Stelzel, C., and Fiebach, C. J. (2011). Trait anxiety modulates the neural efficiency of inhibitory control. *Journal of Cognitive Neuroscience*, *23*(10), 3132–3145. doi:10.1162/jocn_a_00003.

Bauer, R. M. (1982). Visual hypoemotionality as a symptom of visual-limbic disconnection in man. *Archives of Neurology*, *39*(11), 702–708.

Bauer, R. M. (1984). Autonomic recognition of names and faces in prosopagnosia: A neuropsychological application of the Guilty Knowledge Test. *Neuropsychologia*, *22*(4), 457–469. doi:10.1016/0028-3932(84)90040-X.

Bayes, T. (1763). An essay towards solving a problem in the doctrine of chances. *Philosophical Transactions of the Royal Society of London*, *53*, 370–418.

Bechara, A., Damasio, H., and Damasio, A. R. (2000). Emotion, decision making and the orbitofrontal cortex. *Cerebral Cortex*, *10*(3), 295–307.

Bechara, A., Damasio, H., Damasio, A. R., and Lee, G. P. (1999). Different contributions of the human amygdala and ventromedial prefrontal cortex to decision-making. *Journal of Neuroscience*, *19*(13), 5473–5481.

Bechtel, W., and Richardson, R. C. (1993). *Discovering Complexity: Decomposition and Localization as Strategies in Scientific Research*. Princeton, NJ: Princeton University Press.

Bechtel, W., and Richardson, R. C. (2010). *Discovering Complexity: Decomposition and Localization as Strategies in Scientific Research* (2nd ed.). Cambridge, MA: MIT Press.

Beck, S. M., Locke, H. S., Savine, A. C., Jimura, K., and Braver, T. S. (2010). Primary and secondary rewards differentially modulate neural activity dynamics during working memory. *PLoS ONE*, *5*(2), e9251. doi:10.1371/journal.pone.0009251.

Belova, M. A., Paton, J. J., Morrison, S. E., and Salzman, C. D. (2007). Expectation modulates neural responses to pleasant and aversive stimuli in primate amygdala. *Neuron*, *55*(6), 970–984.

Belova, M. A., Paton, J. J., and Salzman, C. D. (2008). Moment-to-moment tracking of state value in the amygdala. *Journal of Neuroscience*, *28*(40), 10023–10030.

Bender, D. B. (1983). Visual activation of neurons in the primate pulvinar depends on cortex but not colliculus. *Brain Research*, *279*(1–2), 258–261.

Benevento, L. A., and Port, J. D. (1995). Single neurons with both form/color differential responses and saccade-related responses in the nonretinotopic pulvinar of the behaving macaque monkey. *Visual Neuroscience*, *12*(3), 523–544.

Benuzzi, F., Meletti, S., Zamboni, G., Calandra-Buonaura, G., Serafini, M., Lui, F., et al. (2004). Impaired fear processing in right mesial temporal sclerosis: An fMRI study. *Brain Research Bulletin*, *63*(4), 269–281.

Berger, B., and Gaspar, P. (1994). Comparative anatomy of the catecholaminergic innervation of rat and primate cerebral cortex. In W. J. A. J. Smeets and A. Reiner, eds., *Phylogeny and Development of Catecholamine Systems in the CNS of Vertebrates*, 293–325. Cambridge: Cambridge University Press.

Berger, B., Trottier, S., Verney, C., Gaspar, P., and Alvarez, C. (1988). Regional and laminar distribution of the dopamine and serotonin innervation in the macaque cerebral cortex: A radioautographic study. *Journal of Comparative Neurology*, *273*(1), 99–119. doi:10.1002/cne.902730109.

Berkman, E. T., and Lieberman, M. D. (2009). Using neuroscience to broaden emotion regulation: Theoretical and methodological considerations. *Social and Personality Psychology Compass*, *3*, 2–19.

Berman, R. A., and Wurtz, R. H. (2010). Functional identification of a pulvinar path from superior colliculus to cortical area MT. *Journal of Neuroscience*, *30*(18), 6342–6354.

Berns, G. S., McClure, S. M., Pagnoni, G., and Montague, P. R. (2001). Predictability modulates human brain response to reward. *Journal of Neuroscience*, *21*(8), 2793–2798.

Berridge, C. W., and Waterhouse, B. D. (2003). The locus coeruleus-noradrenergic system: Modulation of behavioral state and state-dependent cognitive processes. *Brain Research. Brain Research Reviews*, *42*(1), 33–84.

Berridge, K. C., and Robinson, T. E. (1998). What is the role of dopamine in reward: Hedonic impact, reward learning, or incentive salience? *Brain Research. Brain Research Reviews*, *28*(3), 309–369.

Bilder, R. M., Sabb, F. W., Parker, D. S., Kalar, D., Chu, W. W., Fox, J., et al. (2009). Cognitive ontologies for neuropsychiatric phenomics research. *Cognitive Neuropsychiatry*, *14*(4–5), 419–450. doi:10.1080/13546800902787180.

Bishop, S. (2007). Neurocognitive mechanisms of anxiety: An integrative account. *Trends in Cognitive Sciences*, *11*(7), 307–316.

Bishop, S. (2009). Trait anxiety and impoverished prefrontal control of attention. *Nature Neuroscience*, *12*(1), 92–98.

Bishop, S., Duncan, J., Brett, M., and Lawrence, A. D. (2004). Prefrontal cortical function and anxiety: Controlling attention to threat-related stimuli. *Nature Neuroscience*, *7*(2), 184–188.

Bishop, S., Jenkins, R., and Lawrence, A. D. (2007). Neural processing of fearful faces: Effects of anxiety are gated by perceptual capacity limitations. *Cerebral Cortex*, *17*(7), 1595–1603.

Bisley, J. W., and Goldberg, M. E. (2010). Attention, intention, and priority in the parietal lobe. *Annual Review of Neuroscience*, *33*, 1–21. doi:10.1146/annurev-neuro-060909-152823.

Biswal, B., Yetkin, F. Z., Haughton, V. M., and Hyde, J. S. (1995). Functional connectivity in the motor cortex of resting human brain using echo-planar MRI. *Magnetic Resonance in Medicine*, *34*(4), 537–541.

Blackford, J. U., Buckholtz, J. W., Avery, S. N., and Zald, D. H. (2010). A unique role for the human amygdala in novelty detection. *NeuroImage*, *50*(3), 1188–1193. doi:10.1016/j.neuroimage.2009.12.083.

Blair, K. S., Smith, B. W., Mitchell, D. G., Morton, J., Vythilingam, M., Pessoa, L., et al. (2007). Modulation of emotion by cognition and cognition by emotion. *NeuroImage*, *35*(1), 430–440.

Boehler, C. N., Hopf, J. M., Stoppel, C. M., and Krebs, R. M. (2012). Motivating inhibition–reward prospect speeds up response cancellation. *Cognition*, *125*(3), 498–503.

Boehnke, S. E., and Munoz, D. P. (2008). On the importance of the transient visual response in the superior colliculus. *Current Opinion in Neurobiology*, *18*(6), 544–551.

Bonda, E., Petrides, M., Ostry, D., and Evans, A. (1996). Specific involvement of human parietal systems and the amygdala in the perception of biological motion. *Journal of Neuroscience*, *16*(11), 3737–3744.

Borgatti, S. P. (2005). Centrality and network flow. *Social Networks*, *27*(1), 55–71.

Botvinick, M. M. (2007). Conflict monitoring and decision making: Reconciling two perspectives on anterior cingulate function. *Cognitive, Affective and Behavioral Neuroscience*, *7*(4), 356–366.

Botvinick, M. M., Braver, T. S., Barch, D. M., Carter, C. S., and Cohen, J. D. (2001). Conflict monitoring and cognitive control. *Psychological Review*, *108*(3), 624–652.

Bowers, R. M. (1984). On being unconsciously influenced and informed. In K. S. Bowers and D. Meichelbaum, eds., *The Unconscious Reconsidered,* 227–272. New York: John Wiley and Sons.

Boyer, J. L., Harrison, S., and Ro, T. (2005). Unconscious processing of orientation and color without primary visual cortex. *Proceedings of the National Academy of Sciences of the United States of America*, *102*(46), 16875–16879.

Bradley, M. M., Cuthbert, B. N., and Lang, P. J. (1996). Picture media and emotion: Effects of a sustained affective context. *Psychophysiology*, *33*(6), 662–670.

Brand, M., Grabenhorst, F., Starcke, K., Vandekerckhove, M. M., and Markowitsch, H. J. (2007). Role of the amygdala in decisions under ambiguity and decisions under risk: Evidence from patients with Urbach-Wiethe disease. *Neuropsychologia*, *45*(6), 1305–1317.

Brandão, M. L., Cardoso, S. H., Melo, L. L., Motta, V., and Coimbra, N. C. (1994). Neural substrate of defensive behavior in the midbrain tectum. *Neuroscience and Biobehavioral Reviews*, *18*(3), 339–346.

Brassen, S., Gamer, M., Rose, M., and Buchel, C. (2010). The influence of directed covert attention on emotional face processing. *NeuroImage*, *50*(2), 545–551.

Braver, T. S. (2012). The variable nature of cognitive control: A dual mechanisms framework. *Trends in Cognitive Sciences*, *16*(2), 106–113. doi:10.1016/j.tics.2011.12.010.

Braver, T. S., Gray, J. R., and Burgess, G. C. (2007). Explaining the many varieties of working memory variation: Dual mechanisms of cognitive control. In A. R. A. Conway, C. Jarrold, M. J. Kane, A. Miyake, and J. N. Towse, eds., *Variation in Working Memory*, 76–106. Oxford: Oxford University Press.

Brazdil, M., Roman, R., Urbanek, T., Chladek, J., Spok, D., Marecek, R., et al. (2009). Neural correlates of affective picture processing: A depth ERP study. *NeuroImage*, *47*(1), 376–383. doi:10.1016/j.neuroimage.2009.03.081.

Bressler, S. L., and Menon, V. (2010). Large-scale brain networks in cognition: Emerging methods and principles. *Trends in Cognitive Sciences*, *14*(6), 277–290.

Bridge, H., Thomas, O., Jbabdi, S., and Cowey, A. (2008). Changes in connectivity after visual cortical brain damage underlie altered visual function. *Brain*, *131*(Pt 6), 1433–1444.

Broca, P. (1861). Remarques sur le siège de la faculté du langage articulé, suivies d'une observation d'aphémie (perte de la parole). *Bulletin de la Société Anatomique*, *6*, 330–357.

Brodal, A. (1981). *Neurological Anatomy in Relation to Clinical Medicine*. New York: Oxford University Press.

Bromberg-Martin, E. S., Matsumoto, M., and Hikosaka, O. (2010). Dopamine in motivational control: Rewarding, aversive, and alerting. *Neuron*, *68*(5), 815–834.

Brosch, T., and Wieser, M. J. (2011). The (non)automaticity of amygdala responses to threat: On the issue of fast signals and slow measures. *Journal of Neuroscience*, *31*(41), 14451–14452. doi:10.1523/jneurosci.4089-11.2011.

Brothers, L. (1997). *Friday's Footprint: How Society Shapes the Human Mind*. New York: Oxford University Press.

Buchel, C., Morris, J., Dolan, R. J., and Friston, K. J. (1998). Brain systems mediating aversive conditioning: An event-related fMRI study. *Neuron*, *20*(5), 947–957.

Buckner, R. L., Sepulcre, J., Talukdar, T., Krienen, F. M., Liu, H., Hedden, T., et al. (2009). Cortical hubs revealed by intrinsic functional connectivity: Mapping, assessment of stability, and relation to Alzheimer's disease. *Journal of Neuroscience*, *29*(6), 1860–1873. doi:10.1523/jneurosci.5062-08.2009.

Buffalo, E. A., Fries, P., Landman, R., Liang, H., and Desimone, R. (2010). A backward progression of attentional effects in the ventral stream. *Proceedings of the National Academy of Sciences of the United States of America*, *107*(1), 361–365. doi:10.1073/pnas.0907658106.

Bullier, J. (2001). Integrated model of visual processing. *Brain Research Reviews*, *36*(2–3), 96–107.

Bullier, J. (2003). Cortical connections and functional interactions between visual cortical areas. In M. Fahle and M. Greenlee, eds., *Neuropsychology of Vision*, 23–67. Oxford: Oxford University Press.

Bullier, J., and Kennedy, H. (1983). Projection of the lateral geniculate nucleus onto cortical area V2 in the macaque monkey. *Experimental Brain Research*, *53*(1), 168–172.

Bullmore, E., and Sporns, O. (2009). Complex brain networks: Graph theoretical analysis of structural and functional systems. *Nature Reviews. Neuroscience*, *10*(3), 186–198. doi:10.1038/nrn2575.

Buodo, G., Sarlo, M., and Palomba, D. (2002). Attentional resources measured by reaction times highlight differences within pleasant and unpleasant, high arousing stimuli. *Motivation and Emotion*, *26*, 123–138.

Buschman, T. J., and Miller, E. K. (2007). Top-down versus bottom-up control of attention in the prefrontal and posterior parietal cortices. *Science*, *315*(5820), 1860–1862.

Buschman, T. J., Siegel, M., Roy, J. E., and Miller, E. K. (2011). Neural substrates of cognitive capacity limitations. *Proceedings of the National Academy of Sciences of the United States of America*, *108*(27), 11252–11255. doi:10.1073/pnas.1104666108.

Bush, G., Luu, P., and Posner, M. I. (2000). Cognitive and emotional influences in anterior cingulate cortex. *Trends in Cognitive Sciences*, *4*(6), 215–222.

Bush, G., Whalen, P. J., Rosen, B. R., Jenike, M. A., McInerney, S. C., and Rauch, S. L. (1998). The counting Stroop: An interference task specialized for functional neuroimaging–validation study with functional MRI. *Human Brain Mapping*, *6*(4), 270–282.

Butler, A. B., and Hodos, W. (2005). *Comparative Vertebrate Neuroanatomy: Evolution and Adaptation*. Wiley Online Library.

Buzsáki, G. (2006). *Rhythms of the Brain*. New York: Oxford University Press.

Cacioppo, J. T., and Tassinary, L. G. (1990). Inferring psychological significance from physiological signals. *American Psychologist*, *45*(1), 16–28.

Callaway, E., and Thompson, S. V. (1953). Sympathetic activity and perception; an approach to the relationships between autonomic activity and personality. *Psychosomatic Medicine*, *15*, 443–455.

Campeau, S., and Davis, M. (1995). Involvement of subcortical and cortical afferents to the lateral nucleus of the amygdala in fear conditioning measured with fear-potentiated startle in rats trained concurrently with auditory and visual conditioned stimuli. *Journal of Neuroscience*, *15*(3 Pt 2), 2312–2327.

Cannon, W. B. (1929). *Bodily Changes in Pain, Hunger, Fear and Rage* (2nd ed.). New York: Appleton.

Cannon, W. B., and Britton, S.W. (1925). Studies on the conditions of activity in endocrine glands: XV. Pseudaffective medulliadrenal secretion. *American Journal of Physiology—Legacy Content, 72*(2), 283–294.

Capalbo, M., Postma, E., and Goebel, R. (2008). Combining structural connectivity and response latencies to model the structure of the visual system. *PLoS Computational Biology*, *4*(8), e1000159.

Capgras, J., and Reboul-Lachaux, J. (1923). L'illusion des "sosies" dans un délire systématisé chronique. *Bulletin de la Société Clinique de Médecine Mentale*, *2*, 6–16.

Card, J. P., Swanson, L. W., and Moore, R. Y. (2003). The hypothalamus: An overview of regulatory systems. In L. R. Squire, F. E. Bloom, S. K. McConnell, J. L. Roberts, N .C. Spitzer, and M. J. Zigmond, eds., *Fundamental Neuroscience* (2nd ed.), 897–909. San Diego: Academic Press.

Cardinal, R. N., Parkinson, J. A., Hall, J., and Everitt, B. J. (2002). Emotion and motivation: The role of the amygdala, ventral striatum, and prefrontal cortex. *Neuroscience and Biobehavioral Reviews*, *26*(3), 321–352.

Carmichael, S. T., and Price, J. L. (1994). Architectonic subdivision of the orbital and medial prefrontal cortex in the macaque monkey. *Journal of Comparative Neurology*, *346*(3), 366–402. doi:10.1002/cne.903460305.

Carmichael, S. T., and Price, J. L. (1996). Connectional networks within the orbital and medial prefrontal cortex of macaque monkeys. *Journal of Comparative Neurology*, *371*(2), 179–207.

Carroll, S. B. (1995). Homeotic genes and the evolution of arthropods and chordates. *Nature*, *376*(6540), 479–485. doi:10.1038/376479a0.

Carver, C. S., and White, T. L. (1994). Behavioral inhibition, behavioral activation, and affective responses to impending reward and punishment: The BIS/BAS scales. *Journal of Personality and Social Psychology*, *67*, 319–333.

Catani, M., Jones, D. K., Donato, R., and Ffytche, D. H. (2003). Occipito-temporal connections in the human brain. *Brain*, *126*(Pt 9), 2093–2107. doi:10.1093/brain/awg203.

Cavada, C., Company, T., Tejedor, J., Cruz-Rizzolo, R. J., and Reinoso-Suarez, F. (2000). The anatomical connections of the macaque monkey orbitofrontal cortex: A review. *Cerebral Cortex, 10*(3), 220–242.

Cave, K. R. (1999). The FeatureGate model of visual selection. *Psychological Research, 62*(2–3), 182–194.

Cave, K. R., and Batty, M. J. (2006). From searching for features to searching for threat: Drawing the boundary between preattentive and attentive vision. *Visual Cognition, 14*(4/5/6/7/8), 629–646.

Cechetto, D. F., and Chen, S. J. (1990). Subcortical sites mediating sympathetic responses from insular cortex in rats. *American Journal of Physiology, 258*(1 Pt 2), R245–R255.

Cechetto, D. F., and Saper, C. B. (1990). Role of the central cortex in autonomic function. In A. D. Loewy and K. M. Spyer, eds., *Central Regulation of Autonomic Functions*, 208–223. New York: Oxford University Press.

Chajut, E., and Algom, D. (2003). Selective attention improves under stress: Implications for theories of social cognition. *Journal of Personality and Social Psychology, 85*(2), 231–248.

Chalfin, B. P., Cheung, D. T., Muniz, J. A., de Lima Silveira, L. C., and Finlay, B. L. (2007). Scaling of neuron number and volume of the pulvinar complex in New World primates: Comparisons with humans, other primates, and mammals. *Journal of Comparative Neurology, 504*(3), 265–274. doi:10.1002/cne.21406.

Chao, A., and Shen, T. J. (2003). Nonparametric estimation of Shannon's index of diversity when there are unseen species in sample. *Environmental and Ecological Statistics, 10*(4), 429–443.

Chelazzi, L., Perlato, A., Santandrea, E., and Libera, C. D. (2013). Rewards teach visual selective attention. *Vision Research.*

Chen, C. M., Lakatos, P., Shah, A. S., Mehta, A. D., Givre, S. J., Javitt, D. C., et al. (2007). Functional anatomy and interaction of fast and slow visual pathways in macaque monkeys. *Cerebral Cortex, 17*(7), 1561–1569.

Choi, J. M., Padmala, S., and Pessoa, L. (2012). Impact of state anxiety on the interaction between threat monitoring and cognition. *NeuroImage, 59*(2), 1912–1923. doi:10.1016/j.neuroimage.2011.08.102.

Choi, J. M., Padmala, S., Spechler, P., and Pessoa, L. (2013). Pervasive competition between threat and reward in the brain. *Social Cognitive and Affective Neuroscience.*

Chomsung, R. D., Petry, H. M., and Bickford, M. E. (2008). Ultrastructural examination of diffuse and specific tectopulvinar projections in the tree shrew. *Journal of Comparative Neurology, 510*(1), 24–46. doi:10.1002/cne.21763.

Chun, M. M., Golomb, J. D., and Turk-Browne, N. B. (2011). A taxonomy of external and internal attention. *Annual Review of Psychology, 62*, 73–101. doi:10.1146/annurev.psych.093008.100427.

Churchland, P. S., Ramachandran, V. S., and Sejnowski, T. J. (1994). A critique of pure vision. In C. Koch and J. L. Davis, eds., *Large-Scale Neuronal Theories of the Brain*, 23–60. Oxford: Oxford University Press.

Cisek, P., and Kalaska, J. F. (2010). Neural mechanisms for interacting with a world full of action choices. *Annual Review of Neuroscience, 33*, 269–298. doi:10.1146/annurev.neuro.051508.135409.

Clauset, A., Moore, C., and Newman, M. E. (2008). Hierarchical structure and the prediction of missing links in networks. *Nature, 453*(7191), 98–101.

Cohen, J. D. (2005). The vulcanization of the human brain: A neural perspective on interactions between cognition and emotion. *Journal of Economic Perspectives, 19*(4), 3–24.

Cohen, J. D., and Castro-Alamancos, M. A. (2007). Early sensory pathways for detection of fearful conditioned stimuli: Tectal and thalamic relays. *Journal of Neuroscience, 27*(29), 7762–7776. doi:10.1523/jneurosci.1124-07.2007.

Cohen, J. D., and Castro-Alamancos, M. A. (2010). Neural correlates of active avoidance behavior in superior colliculus. *Journal of Neuroscience*, *30*(25), 8502–8511. doi:10.1523/jneurosci.1497-10.2010.

Cohen, M. A., Alvarez, G. A., and Nakayama, K. (2011). Natural-scene perception requires attention. *Psychological Science*, *22*(9), 1165–1172.

Cohen, M. A., Cavanagh, P., Chun, M. M., and Nakayama, K. (2012). The attentional requirements of consciousness. *Trends in Cognitive Sciences*, *16*(8), 411–417. doi:10.1016/j.tics.2012.06.013.

Cohen, N., and Henik, A. (2012). Do irrelevant emotional stimuli impair or improve executive control? *Frontiers in Integrative Neuroscience*, *6*, 33. doi:10.3389/fnint.2012.00033.

Cohen, N., Henik, A., and Mor, N. (2011). Can emotion modulate attention? Evidence for reciprocal links in the attentional network test. *Experimental Psychology*, *58*(3), 171–179. doi:10.1027/1618-3169/a000083.

Cohen, N., Henik, A., and Moyal, N. (2012). Executive control attenuates emotional effects: For high reappraisers only? *Emotion*, *12*(5), 970–979. doi:10.1037/a0026890.

Colizza, V., Flammini, A., Serrano, M. A., and Vespignani, A. (2006). Detecting rich-club ordering in complex networks. *Nature Physics*, *2*(2), 110–115.

Comoli, E., Coizet, V., Boyes, J., Bolam, J. P., Canteras, N. S., Quirk, R. H., et al. (2003). A direct projection from superior colliculus to substantia nigra for detecting salient visual events. *Nature Neuroscience*, *6*(9), 974–980. doi:10.1038/nn1113.

Comoli, E., Das Neves Favaro, P., Vautrelle, N., Leriche, M., Overton, P. G., and Redgrave, P. (2012). Segregated anatomical input to sub-regions of the rodent superior colliculus associated with approach and defense. *Frontiers in Neuroanatomy*, *6*, 9. doi:10.3389/fnana.2012.00009.

Compton, R. J., Banich, M. T., Mohanty, A., Milham, M. P., Herrington, J., Miller, G. A., et al. (2003). Paying attention to emotion: An fMRI investigation of cognitive and emotional Stroop tasks. *Cognitive, Affective and Behavioral Neuroscience*, *3*(2), 81–96.

Corbetta, M., Miezin, F. M., Dobmeyer, S., Shuiman, G. L., and Peterson, S. E. (1991). Selective and divided attention during visual discrimnination of shape, color, and speed: Functional anatomy by positron emission tomography. *Journal of Neuroscience*, *11*(8), 2383–2402.

Corbetta, M., Patel, G., and Shulman, G. L. (2008). The reorienting system of the human brain: From environment to theory of mind. *Neuron*, *58*(3), 306–324.

Corbetta, M., and Shulman, G. L. (2002). Control of goal-directed and stimulus-driven attention in the brain. *Nature Reviews. Neuroscience*, *3*(3), 201–215.

Coricelli, G., Critchley, H. D., Joffily, M., O'Doherty, J. P., Sirigu, A., and Dolan, R. J. (2005). Regret and its avoidance: A neuroimaging study of choice behavior. *Nature Neuroscience*, *8*(9), 1255–1262.

Cornsweet, D. M. (1969). Use of cues in the visual periphery under conditions of arousal. *Journal of Experimental Psychology*, *80*(1), 14–18.

Cornwell, B. R., Alvarez, R. P., Lissek, S., Kaplan, R., Ernst, M., and Grillon, C. (2011). Anxiety overrides the blocking effects of high perceptual load on amygdala reactivity to threat-related distractors. *Neuropsychologia*, *49*(5), 1363–1368. doi:10.1016/j.neuropsychologia.2011.02.049.

Cowell, R. A., Bussey, T. J., and Saksida, L. M. (2010). Functional dissociations within the ventral object processing pathway: Cognitive modules or a hierarchical continuum? *Journal of Cognitive Neuroscience*, *22*(11), 2460–2479.

Cowey, A. (2004). The 30th Sir Frederick Bartlett lecture. Fact, artefact, and myth about blindsight. *Quarterly Journal of Experimental Psychology. A, Human Experimental Psychology*, *57*(4), 577–609.

Craig, A. D. (2002). How do you feel? Interoception: The sense of the physiological condition of the body. *Nature Reviews. Neuroscience*, *3*(8), 655–666.

Craig, A. D. (2009). How do you feel—now? The anterior insula and human awareness. *Nature Reviews. Neuroscience*, *10*(1), 59–70.

Crick, F. (1984). Function of the thalamic reticular complex: The searchlight hypothesis. *Proceedings of the National Academy of Sciences of the United States of America*, *81*(14), 4586.

Dalton, K. M., Kalin, N. H., Grist, T. M., and Davidson, R. J. (2005). Neural-cardiac coupling in threat-evoked anxiety. *JFournal of Cognitive Neuroscience*, *17*(6), 969–980. doi:10.1162/0898929054021094.

Damaraju, E., Huang, Y. M., Barrett, L. F., and Pessoa, L. (2009). Affective learning enhances activity and functional connectivity in early visual cortex. *Neuropsychologia*, *47*(12), 2480–2487.

Damasio, A. (1989). The brain binds entities and events by multiregional activation from convergence zones. *Neural Computation*, *1*, 123–132.

Damasio, A. R. (1994). *Descartes' Error: Emotion, Reason, and the Human Brain*. New York: Putnam.

Damasio, A. R. (1999). *The Feeling of What Happens: Body and Emotion in the Making of Consciousness*. New York: Harcourt Brace.

Damoiseaux, J. S., and Greicius, M. D. (2009). Greater than the sum of its parts: A review of studies combining structural connectivity and resting-state functional connectivity. *Brain Structure and Function*, *213*(6), 525–533.

Darke, S. (1988). Effects of anxiety on inferential reasoning task performance. *Journal of Personality and Social Psychology*, *55*(3), 499–505.

David, S. V., Fritz, J. B., and Shamma, S. A. (2012). Task reward structure shapes rapid receptive field plasticity in auditory cortex. *Proceedings of the National Academy of Sciences of the United States of America*, *109*(6), 2144–2149. doi:10.1073/pnas.1117717109.

Davidson, R. J., and Irwin, W. (1999). The functional neuroanatomy of emotion and affective style. *Trends in Cognitive Sciences*, *3*(1), 11–21.

Davis, M., Walker, D. L., Miles, L., and Grillon, C. (2010). Phasic vs. sustained fear in rats and humans: Role of the extended amygdala in fear vs. anxiety. *Neuropsychopharmacology*, *35*(1), 105–135. doi:10.1038/npp.2009.109.

Davis, M., and Whalen, P. J. (2001). The amygdala: Vigilance and emotion. *Molecular Psychiatry*, *6*(1), 13–34.

Day-Brown, J. D., Wei, H., Chomsung, R. D., Petry, H. M., and Bickford, M. E. (2010). Pulvinar projections to the striatum and amygdala in the tree shrew. *Frontiers in Neuroanatomy*, *4*, 143. doi:10.3389/fnana.2010.00143.

Dean, P., Redgrave, P., and Westby, G. W. (1989). Event or emergency? Two response systems in the mammalian superior colliculus. *Trends in Neurosciences*, *12*(4), 137–147.

De Cesarei, A., Codispoti, M., and Schupp, H. T. (2009). Peripheral vision and preferential emotion processing. *Neuroreport*, *20*(16), 1439–1443.

Deen, B., Pitskel, N. B., and Pelphrey, K. A. (2011). Three systems of insular functional connectivity identified with cluster analysis. *Cerebral Cortex*, *21*(7), 1498–1506. doi:10.1093/cercor/bhq186.

de Gelder, B., Vroomen, J., Pourtois, G., and Weiskrantz, L. (1999). Non-conscious recognition of affect in the absence of striate cortex. *Neuroreport*, *10*(18), 3759–3763.

Delgado, M. R. (2007). Reward related responses in the human striatum. *Annals of the New York Academy of Sciences*, *1104*(1), 70–88.

Della Libera, C., and Chelazzi, L. (2006). Visual selective attention and the effects of monetary rewards. *Psychological Science*, *17*(3), 222–227.

Della Libera, C., and Chelazzi, L. (2009). Learning to attend and to ignore is a matter of gains and losses. *Psychological Science*, *20*(6), 778–784. doi:10.1111/j.1467-9280.2009.02360.x.

De Martino, B., Kumaran, D., Seymour, B., and Dolan, R. J. (2006). Frames, biases, and rational decision-making in the human brain. *Science*, *313*(5787), 684–687.

Derakshan, N., and Eysenck, M. W. (1998). Working memory capacity in high trait-anxious and repressor groups. *Cognition and Emotion*, *12*(5), 697–713.

DesJardin, J. T., Holmes, A. L., Forcelli, P. A., Cole, C. E., Gale, J. T., Wellman, L. L., et andal. (2013). Defense-like behaviors evoked by pharmacological disinhibition of the superior colliculus in the primate. *Journal of Neuroscience*, *33*(1), 150–155.

Desimone, R. (1996). Neural mechanisms for visual memory and their role in attention. *Proceedings of the National Academy of Sciences of the United States of America*, *93*(24), 13494–13499.

Desimone, R., and Duncan, J. (1995). Neural mechanisms of selective attention. *Annual Review of Neuroscience*, *18*, 193–222.

Desimone, R., Wessinger, M., Thomas, L., and Schneider, W. (1990). Attentional control of visual perception: Cortical and subcortical mechanisms. *Cold Spring Harbor Symposia on Quantitative Biology*, *55*, 963–971.

Deutch, A. Y., and Roth, R. H. (1990). The determinants of stress-induced activation of the prefrontal cortical dopamine system. *Progress in Brain Research, 85*, 367–402; discussion 402–363.

Devinsky, O., Morrell, M. J., and Vogt, B. A. (1995). Contributions of anterior cingulate cortex to behaviour. *Brain*, *118* (Pt 1), 279–306.

Diamond, I. T., and Hall, W. C. (1969). Evolution of neocortex. *Science*, *164*(3877), 251–262.

Di Lollo, V., Kawahara, J., Zuvic, S. M., and Visser, T. A. (2001). The preattentive emperor has no clothes: A dynamic redressing. *Journal of Experimental Psychology. General*, *130*(3), 479–492.

Dima, D., Stephan, K. E., Roiser, J. P., Friston, K. J., and Frangou, S. (2011). Effective connectivity during processing of facial affect: Evidence for multiple parallel pathways. *Journal of Neuroscience*, *31*(40), 14378–14385. doi:10.1523/jneurosci.2400-11.2011.

Doallo, S., Patai, Z., and Nobre, A. C. (2013). Reward associations magnify memory-based biases on perception. *Journal of Cognitive Neuroscience*, *25*(2), 145–257.

Dolan, R. J., and Vuilleumier, P. (2003). Amygdala automaticity in emotional processing. *Annals of the New York Academy of Sciences*, *985*, 348–355.

Dolcos, F., Iordan, A. D., and Dolcos, S. (2011). Neural correlates of emotion-cognition interactions: A review of evidence from brain imaging investigations. *Journal of Cognitive Psychology*, *23*(6), 669–694. doi:10.1080/20445911.2011.594433.

Dolcos, F., and McCarthy, G. (2006). Brain systems mediating cognitive interference by emotional distraction. *Journal of Neuroscience*, *26*(7), 2072–2079.

Dommett, E., Coizet, V., Blaha, C. D., Martindale, J., Lefebvre, V., Walton, N., et al. (2005). How visual stimuli activate dopaminergic neurons at short latency. *Science*, *307*(5714), 1476–1479. doi:10.1126/science.1107026.

Dosenbach, N. U., Fair, D. A., Cohen, A. L., Schlaggar, B. L., and Petersen, S. E. (2008). A dual-networks architecture of top-down control. *Trends in Cognitive Sciences*, *12*(3), 99–105.

Dosenbach, N. U., Fair, D. A., Miezin, F. M., Cohen, A. L., Wenger, K. K., Dosenbach, R. A., et al. (2007). Distinct brain networks for adaptive and stable task control in humans. *Proceedings of the National Academy of Sciences of the United States of America*, *104*(26), 11073–11078. doi:10.1073/pnas.0704320104.

Dosenbach, N. U., Visscher, K. M., Palmer, E. D., Miezin, F. M., Wenger, K. K., Kang, H. C., et al. (2006). A core system for the implementation of task sets. *Neuron*, *50*(5), 799–812.

Douglas, R. J., and Martin, K. A. (2004). Neuronal circuits of the neocortex. *Annual Review of Neuroscience*, *27*, 419–451. doi:10.1146/annurev.neuro.27.070203.144152.

Drevets, W. C., and Raichle, M. E. (1998). Reciprocal suppression of regional cerebral blood flow during emotional versus higher cognitive processes: Implications for interactions between emotion and cognition. *Cognition and Emotion*, *12*(3), 353–385.

Driver, J. (2001). A selective review of selective attention research from the past century. *British Journal of Psychology*, *92*(Part 1), 53–78.

Duffy, E. (1962). *Activation and Behavior*. New York: Wiley.

Duncan, J., Emslie, H., Williams, P., Johnson, R., and Freer, C. (1996). Intelligence and the frontal lobe: The organization of goal-directed behavior. *Cognitive Psychology*, *30*(3), 257–303.

Duncan, J., and Owen, A. M. (2000). Common regions of the human frontal lobe recruited by diverse cognitive demands. *Trends in Neurosciences*, *23*(10), 475–483.

Duncan, S., and Barrett, L. F. (2007). Affect is a form of cognition: A neurobiological analysis. *Cognition and Emotion*, *21*(6), 1184–1211.

Dupont, S., Bouilleret, V., Hasboun, D., Semah, F., and Baulac, M. (2003). Functional anatomy of the insula: New insights from imaging. *Surgical and Radiologic Anatomy*, *25*(2), 113–119. doi:10.1007/s00276-003-0103-4.

Dykes, R. W. (1997). Mechanisms controlling neuronal plasticity in somatosensory cortex. *Canadian Journal of Physiology and Pharmacology*, *75*(5), 535–545.

Edelman, G. M. (1978). Group selection and phasic reentrant signalling: A theory of higher brain function. In G. M. Edelman and V. B. Mountcastle, eds., *The Mindful Brain*, 51–100. Cambridge, MA: MIT Press.

Edelman, G. M., and Gally, J. A. (2001). Degeneracy and complexity in biological systems. *Proceedings of the National Academy of Sciences of the United States of America*, *98*(24), 13763–13768.

Eger, E., Jedynak, A., Iwaki, T., and Skrandies, W. (2003). Rapid extraction of emotional expression: Evidence from evoked potential fields during brief presentation of face stimuli. *Neuropsychologia*, *41*(7), 808–817.

Egner, T., Etkin, A., Gale, S., and Hirsch, J. (2008). Dissociable neural systems resolve conflict from emotional versus nonemotional distracters. *Cerebral Cortex*, *18*(6), 1475–1484.

Egner, T., and Hirsch, J. (2005). Cognitive control mechanisms resolve conflict through cortical amplification of task-relevant information. *Nature Neuroscience*, *8*(12), 1784–1790.

Eimer, M. (1996). The N2pc component as an indicator of attentional selectivity. *Electroencephalography and Clinical Neurophysiology*, *99*(3), 225–234.

Eippert, F., Veit, R., Weiskopf, N., Erb, M., Birbaumer, N., and Anders, S. (2007). Regulation of emotional responses elicited by threat-related stimuli. *Human Brain Mapping*, *28*(5), 409–423. doi:10.1002/hbm.20291.

Ekman, P. (1992). An argument for basic emotions. *Cognition and Emotion*, *6*, 169–200.

Ellis, H. D., and Young, A. W. (1990). Accounting for delusional misidentifications. *British Journal of Psychiatry*, *157*, 239–248. doi:10.1192/bjp.157.2.239.

Engelmann, J. B., Damaraju, E. C., Padmala, S., and Pessoa, L. (2009). Combined effects of attention and motivation on visual task performance: Transient and sustained motivational effects. *Frontiers in Human Neuroscience*, *3*(4). doi:10.3389/neuro.3309.3004.2009.

Engelmann, J. B., and Pessoa, L. (2007a). Behavioral and neural interactions between reward and attention. Paper presented at the Society for Neuroscience 2007, San Diego.

Engelmann, J. B., and Pessoa, L. (2007b). Motivation sharpens exogenous spatial attention. *Emotion*, *7*(3), 668–674.

Enns, J. T., and Rensink, R. A. (1990). Influence of scene-based properties on visual search. *Science*, *247*(4943), 721–723.

Epstein, R., and Kanwisher, N. (1998). A cortical representation of the local visual environment. *Nature*, *392*(6676), 598–601.

Eriksen, B. A., and Eriksen, C. W. (1974). Effects of noise letters upon the identification of a target letter in a nonsearch task. *Attention, Perception and Psychophysics*, *16*(1), 143–149. doi:10.3758/BF03203267.

Eriksen, C. W. (1960). Discrimination and learning without awareness: A methodological survey and evaluation. *Psychological Review*, *67*, 279–300.

Erk, S., Kleczar, A., and Walter, H. (2007). Valence-specific regulation effects in a working memory task with emotional context. *NeuroImage*, *37*(2), 623–632.

Estrada, E., and Hatano, N. (2008). Communicability in complex networks. *Physical Review E: Statistical, Nonlinear, and Soft Matter Physics*, *77*(3), 036111.

Etkin, A., Egner, T., and Kalisch, R. (2011). Emotional processing in anterior cingulate and medial prefrontal cortex. *Trends in Cognitive Sciences*, *15*(2), 85–93. doi:10.1016/j.tics.2010.11.004.

Etkin, A., Egner, T., Peraza, D. M., Kandel, E. R., and Hirsch, J. (2006). Resolving emotional conflict: A role for the rostral anterior cingulate cortex in modulating activity in the amygdala. *Neuron*, *51*(6), 871–882.

Etkin, A., and Wager, T. D. (2007). Functional neuroimaging of anxiety: A meta-analysis of emotional processing in PTSD, social anxiety disorder, and specific phobia. *American Journal of Psychiatry*, *164*(10), 1476–1488. doi:10.1176/appi.ajp.2007.07030504.

Evans, J. S. B. T. (2008). Dual-processing accounts of reasoning, judgment, and social cognition. *Annual Review of Psychology*, *59*(1), 255–278.

Evans, K. K., and Treisman, A. (2005). Perception of objects in natural scenes: Is it really attention free? *Journal of Experimental Psychology. Human Perception and Performance*, *31*(6), 1476–1492. doi:10.1037/0096-1523.31.6.1476.

Everitt, B. J., Cardinal, R. N., Parkinson, J. A., and Robbins, T. W. (2003). Appetitive behavior: Impact of amygdala-dependent mechanisms of emotional learning. *Annals of the New York Academy of Sciences*, *985*, 233–250.

Everitt, B. J., and Robbins, T. W. (1992). Amygdala-ventral striatal interactions and reward-related processes. In J. P. Aggleton, ed., *The Amygdala: Neurological Aspects of Emotion, Memory, and Mental Dysfunction*, 401–429. New York: Wiley.

Everitt, B. J., and Robbins, T. W. (1997). Central cholinergic systems and cognition. *Annual Review of Psychology*, *48*, 649–684.

Eysenck, M. W., and Calvo, M. G. (1992). Anxiety and performance: The processing efficiency theory. *Cognition and Emotion*, *6*(6), 409–434.

Eysenck, M. W., and Derakshan, N. (2011). New perspectives in attentional control theory. *Personality and Individual Differences*, *50*(7), 955–960.

Eysenck, M. W., Derakshan, N., Santos, R., and Calvo, M. G. (2007). Anxiety and cognitive performance: Attentional control theory. *Emotion*, *7*(2), 336–353.

Fabre-Thorpe, M. (2011). The characteristics and limits of rapid visual categorization. *Frontiers in Psychology*, *2*, 243. doi:10.3389/fpsyg.2011.00243.

Fales, C. L., Barch, D. M., Burgess, G. C., Schaefer, A., Mennin, D. S., Gray, J. R., et al. (2008). Anxiety and cognitive efficiency: Differential modulation of transient and sustained neural activity during a working memory task. *Cognitive, Affective and Behavioral Neuroscience*, *8*(3), 239–253.

Fecteau, J. H., and Munoz, D. P. (2006). Salience, relevance, and firing: A priority map for target selection. *Trends in Cognitive Sciences*, *10*(8), 382–390.

Feinstein, J. S., Adolphs, R., Damasio, A., and Tranel, D. (2011). The human amygdala and the induction and experience of fear. *Current Biology*, *21*(1), 34–38. doi:10.1016/j.cub.2010.11.042.

Felleman, D. J., and Van Essen, D. C. (1991). Retinotopic organization in human visual cortex and the spatial precision of functional MRI. *Cerebral Cortex*, *1*, 1–47.

Fenker, D. B., Heipertz, D., Boehler, C. N., Schoenfeld, M. A., Noesselt, T., Heinze, H. J., et al. (2010). Mandatory processing of irrelevant fearful face features in visual search. *Journal of Cognitive Neuroscience*, *22*(12), 2926–2938. doi:10.1162/jocn.2009.21340.

Ferrarini, L., Veer, I. M., Baerends, E., van Tol, M. J., Renken, R. J., van der Wee, N. J., et al. (2009). Hierarchical functional modularity in the resting-state human brain. *Human Brain Mapping*, *30*(7), 2220–2231.

Finger, S. (1994). *Origins of Neuroscience: A History of Explorations into Brain Function*. New York: Oxford University Press.

Floresco, S. B., and Ghods-Sharifi, S. (2007). Amygdala-prefrontal cortical circuitry regulates effort-based decision making. *Cerebral Cortex*, *17*(2), 251–260.

Floresco, S. B., St. Onge, J. R., Ghods-Sharifi, S., and Winstanley, C. A. (2008). Cortico-limbic-striatal circuits subserving different forms of cost-benefit decision making. *Cognitive, Affective and Behavioral Neuroscience*, *8*(4), 375–389.

Folk, C. L., Remington, R. W., and Johnston, J. C. (1992). Involuntary covert orienting is contingent on attentional control settings. *Journal of Experimental Psychology. Human Perception and Performance*, *18*(4), 1030–1044.

Fox, E., Russo, R., Bowles, R., and Dutton, K. (2001). Do threatening stimuli draw or hold visual attention in subclinical anxiety? *Journal of Experimental Psychology. General*, *130*(4), 681–700.

Fox, E., Russo, R., and Dutton, K. (2002). Attentional bias for threat: Evidence for delayed disengagement from emotional faces. *Cognition and Emotion*, *16*(3), 355–379.

Fox, E., Russo, R., and Georgiou, G. A. (2005). Anxiety modulates the degree of attentive resources required to process emotional faces. *Cognitive, Affective and Behavioral Neuroscience*, *5*(4), 396–404.

Fox, E., Yates, A., and Ashwin, C. (2012). Trait anxiety and perceptual load as determinants of emotion processing in a fear conditioning paradigm. *Emotion*, *12*(2), 236–249. doi:10.1037/a0025321.

Fox, M. D., and Raichle, M. E. (2007). Spontaneous fluctuations in brain activity observed with functional magnetic resonance imaging. *Nature Reviews. Neuroscience*, *8*(9), 700–711.

Fox, M. D., Snyder, A. Z., Vincent, J. L., Corbetta, M., Van Essen, D. C., and Raichle, M. E. (2005). The human brain is intrinsically organized into dynamic, anticorrelated functional networks. *Proceedings of the National Academy of Sciences of the United States of America*, *102*(27), 9673–9678. doi:10.1073/pnas.0504136102.

Freese, J. L., and Amaral, D. G. (2005). The organization of projections from the amygdala to visual cortical areas TE and V1 in the macaque monkey. *Journal of Comparative Neurology*, *486*(4), 295–317.

Friston, K. J., Buechel, C., Fink, G. R., Morris, J., Rolls, E., and Dolan, R. J. (1997). Psychophysiological and modulatory interactions in neuroimaging. *NeuroImage*, *6*(3), 218–229.

Friston, K. J., Frith, C. D., Liddle, P. F., and Frackowiak, R. S. (1993). Functional connectivity: The principal-component analysis of large (PET) data sets. *Journal of Cerebral Blood Flow and Metabolism*, *13*(1), 5–14.

Friston, K. J., Harrison, L., and Penny, W. (2003). Dynamic causal modelling. *NeuroImage*, *19*(4), 1273–1302.

Froemke, R. C., Carcea, I., Barker, A. J., Yuan, K., Seybold, B. A., Martins, A. R. O., et al. (2012). Long-term modification of cortical synapses improves sensory perception. *Nature Neuroscience*, *16*(1), 79–88.

Funahashi, S., Bruce, S. J., and Goldman-Rakic, P. S. (1989). Mnemonic coding of visual space in the monkey's dorsolateral prefrontal cortex. *Journal of Neurophysiology*, *61*, 1–19.

Furey, M. L., Pietrini, P., and Haxby, J. V. (2000). Cholinergic enhancement and increased selectivity of perceptual processing during working memory. *Science*, *290*(5500), 2315–2319. doi:10.1126/science.290.5500.2315.

Furey, M. L., Pietrini, P., Haxby, J. V., and Drevets, W. C. (2008). Selective effects of cholinergic modulation on task performance during selective attention. *Neuropsychopharmacology*, *33*(4), 913–923. doi:10.1038/sj.npp.1301461.

Fuster, J. M. (1973). Unit activity in prefrontal cortex during delayed-response performance: Neuronal correlates of transient memory. *Journal of Neurophysiology*, *36*(1), 61–78.

Fuster, J. M. (1997). *The Prefrontal Cortex: Anatomy, Physiology, and Neuropsychology of the Frontal Lobes*. New York: Raven.

Fuster, J. M. (2008). *The Prefrontal Cortex* (4th ed.). San Diego, CA: Academic Press.

Gaither, N. S., and Stein, B. E. (1979). Reptiles and mammals use similar sensory organizations in the midbrain. *Science*, *205*(4406), 595–597.

Gamer, M., and Buchel, C. (2009). Amygdala activation predicts gaze toward fearful eyes. *Journal of Neuroscience*, *29*(28), 9123–9126. doi:10.1523/jneurosci.1883-09.2009.

Gamer, M., Zurowski, B., and Buchel, C. (2010). Different amygdala subregions mediate valence-related and attentional effects of oxytocin in humans. *Proceedings of the National Academy of Sciences of the United States of America*, *107*(20), 9400–9405.

Germana, J. (1969). Central efferent processes and autonomic-behavioral integration. *Psychophysiology*, *6*(1), 78–90.

Gerstein, G. L., and Perkel, D. H. (1969). Simultaneously recorded trains of action potentials: Analysis and functional interpretation. *Science*, *164*(3881), 828–830.

Ghashghaei, H. T., Hilgetag, C. C., and Barbas, H. (2007). Sequence of information processing for emotions based on the anatomic dialogue between prefrontal cortex and amygdala. *NeuroImage*, *34*(3), 905–923.

Ghods-Sharifi, S., St. Onge, J. R., and Floresco, S. B. (2009). Fundamental contribution by the basolateral amygdala to different forms of decision making. *Journal of Neuroscience*, *29*(16), 5251–5259.

Gibson, B. S., and Kelsey, E. M. (1998). Stimulus-driven attentional capture is contingent on attentional set for displaywide visual features. *Journal of Experimental Psychology. Human Perception and Performance*, *24*(3), 699–706.

Gilaie-Dotan, S., Perry, A., Bonneh, Y., Malach, R., and Bentin, S. (2009). Seeing with profoundly deactivated mid-level visual areas: Non-hierarchical functioning in the human visual cortex. *Cerebral Cortex*, *19*(7), 1687–1703.

Gilbert, A. M., and Fiez, J. A. (2004). Integrating rewards and cognition in the frontal cortex. *Cognitive, Affective and Behavioral Neuroscience*, *4*(4), 540–552.

Girard, P., and Bullier, J. (1989). Visual activity in area V2 during reversible inactivation of area 17 in the macaque monkey. *Journal of Neurophysiology*, *62*(6), 1287–1302.

Girvan, M., and Newman, M. E. (2002). Community structure in social and biological networks. *Proceedings of the National Academy of Sciences of the United States of America*, *99*(12), 7821–7826.

Glasser, M. F., and Van Essen, D. C. (2011). Mapping human cortical areas in vivo based on myelin content as revealed by T1- and T2-weighted MRI. *Journal of Neuroscience*, *31*(32), 11597–11616. doi:10.1523/jneurosci.2180-11.2011.

Goard, M., and Dan, Y. (2009). Basal forebrain activation enhances cortical coding of natural scenes. *Nature Neuroscience*, *12*(11), 1444–1449.

Goel, V., and Dolan, R. J. (2003). Reciprocal neural response within lateral and ventral medial prefrontal cortex during hot and cold reasoning. *NeuroImage*, *20*(4), 2314–2321.

Goldberg, M. E., Bisley, J., Powell, K. D., Gottlieb, J., and Kusunoki, M. (2002). The role of the lateral intraparietal area of the monkey in the generation of saccades and visuospatial attention. *Annals of the New York Academy of Sciences, 956*, 205–215. doi:10.1111/j.1749-6632.2002.tb02820.x.

Goldman-Rakic, P. S. (1988). Topography of cognition: Parallel distributed networks in primate association cortex. *Annual Review of Neuroscience, 11*, 137–156.

Goldman-Rakic, P. S., Leranth, C., Williams, S. M., Mons, N., and Geffard, M. (1989). Dopamine synaptic complex with pyramidal neurons in primate cerebral cortex. *Proceedings of the National Academy of Sciences of the United States of America, 86*(22), 9015–9019.

Goldstein, M., Brendel, G., Tuescher, O., Pan, H., Epstein, J., Beutel, M., et al. (2007). Neural substrates of the interaction of emotional stimulus processing and motor inhibitory control: An emotional linguistic go/no-go fMRI study. *NeuroImage, 36*(3), 1026–1040.

Good, B. H., de Montjoye, Y. A., and Clauset, A. (2010). Performance of modularity maximization in practical contexts. *Physical Review E: Statistical, Nonlinear, and Soft Matter Physics, 81*(4 Pt 2), 046106. doi:10.1103/PhysRevE.81.046106.

Gothard, K. M., Battaglia, F. P., Erickson, C. A., Spitler, K. M., and Amaral, D. G. (2007). Neural responses to facial expression and face identity in the monkey amygdala. *Journal of Neurophysiology, 97*(2), 1671–1683.

Gottfried, J. A., O'Doherty, J., and Dolan, R. J. (2003). Encoding predictive reward value in human amygdala and orbitofrontal cortex. *Science, 301*(5636), 1104–1107.

Grabenhorst, F., Hernádi, I., and Schultz, W. (2012). Prediction of economic choice by primate amygdala neurons. *Proceedings of the National Academy of Sciences of the United States of America, 109*(46), 18950–18955.

Gray, J. A. (1990). Brain systems that mediate both emotion and cognition. *Cognition and Emotion, 4*(3), 269–288.

Gray, J. R., Braver, T. S., and Raichle, M. E. (2002). Integration of emotion and cognition in the lateral prefrontal cortex. *Proceedings of the National Academy of Sciences of the United States of America, 99*(6), 4115–4120.

Gray, J. R., Burgess, G. C., Schaefer, A., Yarkoni, T., Larsen, R. J., and Braver, T. S. (2005). Affective personality differences in neural processing efficiency confirmed using fMRI. *Cognitive, Affective and Behavioral Neuroscience, 5*(2), 182–190.

Gray, K. L., Adams, W. J., Hedger, N., Newton, K. E., and Garner, M. (2013). Faces and awareness: Low-level, not emotional factors determine perceptual dominance. *Emotion*. Epub ahead of print.

Green, D. M., and Swets, J. A. (1966). *Signal Detection Theory and Psychophysics*. New York: Wiley.

Greene, M. R., and Oliva, A. (2009). The briefest of glances: The time course of natural scene understanding. *Psychological Science, 20*(4), 464–472.

Greicius, M. D., Krasnow, B., Reiss, A. L., and Menon, V. (2003). Functional connectivity in the resting brain: A network analysis of the default mode hypothesis. *Proceedings of the National Academy of Sciences of the United States of America, 100*(1), 253–258.

Greicius, M. D., Supekar, K., Menon, V., and Dougherty, R. F. (2009). Resting-state functional connectivity reflects structural connectivity in the default mode network. *Cerebral Cortex, 19*(1), 72–78.

Grieve, K. L., Acuña, C., and Cudeiro, J. (2000). The primate pulvinar nuclei: Vision and action. *Trends in Neurosciences, 23*(1), 35–39.

Gross, J. J., ed. (2007). *Handbook of Emotion Regulation*. New York: Guilford Press.

Grossberg, S. (1964). *The Theory of Embedding Fields with Applications to Psychology and Neurophysiology*. New York: Rockefeller Institute for Medical Research.

Grossberg, S. (1971). Embedding fields: Underlying philosophy, mathematics, and applications to psychology, physiology, and anatomy. *Journal of Cybernetics*, *1*(1), 28–50.

Grossberg, S. (1980). How does a brain build a cognitive code? *Psychological Review*, *87*(1), 1–51.

Grossberg, S., and Levine, D. S. (1987). Neural dynamics of attentionally modulated Pavlovian conditioning: Blocking, interstimulus interval, and secondary reinforcement. *Applied Optics*, *26*(23), 5015–5030.

Grossberg, S., Mingolla, E., and Ross, W. D. (1994). A neural theory of attentive visual search: Interactions of boundary, surface, spatial, and object representations. *Psychological Review*, *101*(3), 470–489.

Gruber, A. J., Dayan, P., Gutkin, B. S., and Solla, S. A. (2006). Dopamine modulation in the basal ganglia locks the gate to working memory. *Journal of Computational Neuroscience*, *20*(2), 153–166. doi:10.1007/s10827-005-5705-x.

Gu, X., Liu, X., Van Dam, N. T., Hof, P. R., and Fan, J. (2012). Cognition-emotion integration in the anterior insular cortex. *Cerebral Cortex*. doi:10.1093/cercor/bhr367.

Guillery, R. W. (1995). Anatomical evidence concerning the role of the thalamus in corticocortical communication: A brief review. *Journal of Anatomy*, *187*(Pt 3), 583–592.

Guimerà, R., and Nunes Amaral, L. A. (2005). Functional cartography of complex metabolic networks. *Nature*, *433*(7028), 895–900.

Guimerà, R., Sales-Pardo, M., and Amaral, L. A. N. (2007). Classes of complex networks defined by role-to-role connectivity profiles. *Nature Physics*, *3*, 63–69.

Haber, S. N., and Knutson, B. (2010). The reward circuit: Linking primate anatomy and human imaging. *Neuropsychopharmacology*, *35*(1), 4–26. doi:10.1038/npp.2009.129.

Habib, M. (1986). Visual hypoemotionality and prosopagnosia associated with right temporal lobe isolation. *Neuropsychologia*, *24*(4), 577–582.

Haidt, J. (2003). The moral emotions. In R. J. Davidson, K. R. Scherer, and H. H. Goldsmith, eds., *Handbook of Affective Sciences*, 852–870. Oxford: Oxford University Press.

Halgren, E., Raij, T., Marinkovic, K., Jousmaki, V., and Hari, R. (2000). Cognitive response profile of the human fusiform face area as determined by MEG. *Cerebral Cortex*, *10*(1), 69–81.

Hamm, A. O., Weike, A. I., Schupp, H. T., Treig, T., Dressel, A., and Kessler, C. (2003). Affective blindsight: Intact fear conditioning to a visual cue in a cortically blind patient. *Brain*, *126*(Pt 2), 267–275.

Hampton, A. N., Adolphs, R., Tyszka, M. J., and O'Doherty, J. P. (2007). Contributions of the amygdala to reward expectancy and choice signals in human prefrontal cortex. *Neuron*, *55*(4), 545–555.

Harsay, H. A., Cohen, M. X., Oosterhof, N. N., Forstmann, B. U., Mars, R. B., and Ridderinkhof, K. R. (2011). Functional connectivity of the striatum links motivation to action control in humans. *Journal of Neuroscience*, *31*(29), 10701–10711. doi:10.1523/jneurosci.5415-10.2011.

Hart, S. J., Green, S. R., Casp, M., and Belger, A. (2010). Emotional priming effects during Stroop task performance. *NeuroImage*, *49*(3), 2662–2670. doi:10.1016/j.neuroimage.2009.10.076.

Hatfield, T., Han, J. S., Conley, M., Gallagher, M., and Holland, P. (1996). Neurotoxic lesions of basolateral, but not central, amygdala interfere with Pavlovian second-order conditioning and reinforcer devaluation effects. *Journal of Neuroscience*, *16*(16), 5256–5265.

Hawellek, D. J., Hipp, J. F., Lewis, C. M., Corbetta, M., and Engel, A. K. (2011). Increased functional connectivity indicates the severity of cognitive impairment in multiple sclerosis. *Proceedings of the National Academy of Sciences of the United States of America*, *108*(47), 19066–19071. doi:10.1073/pnas.1110024108.

Hawkey, D. J., Amitay, S., and Moore, D. R. (2004). Early and rapid perceptual learning. *Nature Neuroscience*, *7*(10), 1055–1056. doi:10.1038/nn1315.

Haynes, J. D., and Rees, G. (2006). Decoding mental states from brain activity in humans. *Nature Reviews. Neuroscience*, *7*(7), 523–534.

He, B. J., Snyder, A. Z., Vincent, J. L., Epstein, A., Shulman, G. L., and Corbetta, M. (2007). Breakdown of functional connectivity in frontoparietal networks underlies behavioral deficits in spatial neglect. *Neuron*, *53*(6), 905–918.

Heimer, L., Van Hoesen, G. W., Trimble, M., and Zahm, D. S. (2008). *Anatomy of Neuropsychiatry: The New Anatomy of the Basal Forebrain and its Implications for Neuropsychiatric Illness*. Burlington, MA: Academic Press.

Hermans, E. J., van Marle, H. J., Ossewaarde, L., Henckens, M. J., Qin, S., van Kesteren, M. T., et al. (2011). Stress-related noradrenergic activity prompts large-scale neural network reconfiguration. *Science*, *334*(6059), 1151–1153. doi:10.1126/science.1209603.

Herrero, J. L., Roberts, M. J., Delicato, L. S., Gieselmann, M. A., Dayan, P., and Thiele, A. (2008). Acetylcholine contributes through muscarinic receptors to attentional modulation in V1. *Nature*, *454*(7208), 1110–1114. doi:10.1038/nature07141.

Herry, C., Bach, D. R., Esposito, F., Di Salle, F., Perrig, W. J., Scheffler, K., et al. (2007). Processing of temporal unpredictability in human and animal amygdala. *Journal of Neuroscience*, *27*(22), 5958–5966.

Hickey, C., Chelazzi, L., and Theeuwes, J. (2010). Reward changes salience in human vision via the anterior cingulate. *Journal of Neuroscience*, *30*(33), 11096–11103. doi:10.1523/jneurosci.1026-10.2010.

Hilgetag, C. C., O'Neill, M. A., and Young, M. P. (1996). Indeterminate organization of the visual system. *Science*, *271*(5250), 776–777.

Hillyard, S. A., Vogel, E. K., and Luck, S. J. (1998). Sensory gain control (amplification) as a mechanism of selective attention: Electrophysiological and neuroimaging evidence. *Philosophical Transactions of the Royal Society of London. Series B, Biological Sciences*, *353*(1373), 1257–1270. doi:10.1098/rstb.1998.0281.

Hindi Attar, C., Muller, M. M., Andersen, S. K., Buchel, C., and Rose, M. (2010). Emotional processing in a salient motion context: Integration of motion and emotion in both V5/hMT+ and the amygdala. *Journal of Neuroscience*, *30*(15), 5204–5210. doi:10.1523/jneurosci.5029-09.2010.

Hirano, Y., Stefanovic, B., and Silva, A. C. (2011). Spatiotemporal evolution of the functional magnetic resonance imaging response to ultrashort stimuli. *Journal of Neuroscience*, *31*(4), 1440–1447. doi:10.1523/JNEUROSCI.3986-10.2011.

Hirstein, W., and Ramachandran, V. S. (1997). Capgras syndrome: A novel probe for understanding the neural representation of the identity and familiarity of persons. *Proceedings of the Royal Sciety Society B*, *264*(1380), 437–444. doi:10.1098/rspb.1997.0062.

Holland, P. C. (2007). Disconnection of the amygdala central nucleus and the substantia innominata/nucleus basalis magnocellularis disrupts performance in a sustained attention task. *Behavioral Neuroscience*, *121*(1), 80–89. doi:10.1037/0735-7044.121.1.80.

Holland, P. C., and Gallagher, M. (1993). Effects of amygdala central nucleus lesions on blocking and unblocking. *Behavioral Neuroscience*, *107*(2), 235–245.

Holland, P. C., and Gallagher, M. (1999). Amygdala circuitry in attentional and representational processes. *Trends in Cognitive Sciences*, *3*(2), 65–73.

Holland, P. C., and Gallagher, M. (2006). Different roles for amygdala central nucleus and substantia innominata in the surprise-induced enhancement of learning. *Journal of Neuroscience*, *26*(14), 3791–3797.

Holland, P. C., Han, J. S., and Gallagher, M. (2000). Lesions of the amygdala central nucleus alter performance on a selective attention task. *Journal of Neuroscience*, *20*(17), 6701–6706.

Holmes, A., Nielsen, M. K., and Green, S. (2008). Effects of anxiety on the processing of fearful and happy faces: An event-related potential study. *Biological Psychology*, *77*(2), 159–173. doi:10.1016/j.biopsycho.2007.10.003.

Holroyd, C. B., and Coles, M. G. (2002). The neural basis of human error processing: Reinforcement learning, dopamine, and the error-related negativity. *Psychological Review*, *109*(4), 679–709.

Holstege, G. (1991). Descending pathways from the periaqueductal gray and adjacent areas. In A. Depaulis and R. Bandler, eds., *The Midbrain Periaqueductal Gray Matter. Functional, Anatomical and Neurochemical Organization*, 239–265. New York: Plenum Press.

Honey, C. J., Kotter, R., Breakspear, M., and Sporns, O. (2007). Network structure of cerebral cortex shapes functional connectivity on multiple time scales. *Proceedings of the National Academy of Sciences of the United States of America*, *104*(24), 10240–10245.

Horvitz, J. C. (2000). Mesolimbocortical and nigrostriatal dopamine responses to salient non-reward events. *Neuroscience*, *96*(4), 651–656.

Hsu, M., Bhatt, M., Adolphs, R., Tranel, D., and Camerer, C. F. (2005). Neural systems responding to degrees of uncertainty in human decision-making. *Science*, *310*(5754), 1680–1683.

Hsu, S. M., Hetrick, W. P., and Pessoa, L. (2008). Depth of facial expression processing depends on stimulus visibility: Behavioral and electrophysiological evidence of priming effects. *Cognitive, Affective and Behavioral Neuroscience*, *8*(3), 282–292.

Hsu, S. M., and Pessoa, L. (2007). Dissociable effects of bottom-up and top-down factors on the processing of unattended fearful faces. *Neuropsychologia*, *45*(13), 3075–3086.

Hu, K., Bauer, A., Padmala, S., and Pessoa, L. (2011). Threat of bodily harm has opposing effects on cognition. *Emotion*, *12*(1), 28–32. doi:10.1037/a0024345.

Hübner, R., and Schlösser, J. (2010). Monetary reward increases attentional effort in the flanker task. *Psychonomic Bulletin and Review*, *17*(6), 821–826. doi:10.3758/PBR.17.6.821.

Huerta, M. F., and Harting, J. K. (1984). The mammalian superior colliculus: Studies of its morphology and connections. In H. Vanegas, ed., *Comparative Neurology of the Optic Tectum*, 687–773. New York: Plenum Press.

Hull, C. L. (1943). *Principles of Behavior: An Introduction to Behavior Theory*. New York: Appleton-Century-Crofts.

Hung, C. P., Kreiman, G., Poggio, T., and DiCarlo, J. J. (2005). Fast readout of object identity from macaque inferior temporal cortex. *Science*, *310*(5749), 863–866.

Hupe, J. M., James, A. C., Girard, P., and Bullier, J. (2001). Response modulations by static texture surround in area V1 of the macaque monkey do not depend on feedback connections from V2. *Journal of Neurophysiology*, *85*(1), 146–163.

Ikemoto, S., and Panksepp, J. (1999). The role of nucleus accumbens dopamine in motivated behavior: A unifying interpretation with special reference to reward-seeking. *Brain Research. Brain Research Reviews*, *31*(1), 6–41.

Imura, K., and Rockland, K. S. (2006). Long-range interneurons within the medial pulvinar nucleus of macaque monkeys. *Journal of Comparative Neurology*, *498*(5), 649–666.

Inagaki, M., and Fujita, I. (2011). Reference frames for spatial frequency in face representation differ in the temporal visual cortex and amygdala. *Journal of Neuroscience*, *31*(28), 10371–10379. doi:10.1523/jneurosci.1114-11.2011.

Ioannides, A. A., Liu, M. J., Liu, L. C., Bamidis, P. D., Hellstrand, E., and Stephan, K. M. (1995). Magnetic field tomography of cortical and deep processes: Examples of "real-time mapping" of averaged and single trial MEG signals. *International Journal of Psychophysiology*, *20*(3), 161–175.

Iordan, A., Molitor, R., Llewellyn, N., Novak, J., Eettickal, S., Sandescu, R., et al. (2011). Behavioral and brain imaging investigation of the impact of positive and negative distraction on cognitive performance. Paper presented at the Society for Neuroscience 2011, Washington, D.C.

Ishai, A., Pessoa, L., Bikle, P. C., and Ungerleider, L. G. (2004). Repetition suppression of faces is modulated by emotion. *Proceedings of the National Academy of Sciences of the United States of America*, *101*, 9827–9832.

Izhikevich, E. M. (2007). Solving the distal reward problem through linkage of STDP and dopamine signaling. *Cerebral Cortex*, *17*(10), 2443. doi:10.1093/cercor/bhl152.

Jagadeesh, B., Chelazzi, L., Mishkin, M., and Desimone, R. (2001). Learning increases stimulus salience in anterior inferior temporal cortex of the macaque. *Journal of Neurophysiology*, *86*(1), 290–303.

James, W. (1890). *Principles of Psychology* (Vol. 1). New York: Holt.

Japee, S., Crocker, L., Carver, F., Pessoa, L., and Ungerleider, L. G. (2009). Individual differences in valence modulation of face-selective M170 response. *Emotion*, *9*(1), 59–69.

Jenison, R. L., Rangel, A., Oya, H., Kawasaki, H., and Howard, M. A. (2011). Value encoding in single neurons in the human amygdala during decision making. *Journal of Neuroscience*, *31*(1), 331–338. doi:10.1523/jneurosci.4461-10.2011.

Jimura, K., Locke, H. S., and Braver, T. S. (2010). Prefrontal cortex mediation of cognitive enhancement in rewarding motivational contexts. *Proceedings of the National Academy of Sciences of the United States of America*, *107*(19), 8871–8876. doi:10.1073/pnas.1002007107.

Joels, M., Pu, Z., Wiegert, O., Oitzl, M. S., and Krugers, H. J. (2006). Learning under stress: How does it work? *Trends in Cognitive Sciences*, *10*(4), 152–158. doi:10.1016/j.tics.2006.02.002.

Johnson, D. N., McGrath, A., and McNeil, C. (2002). Cuing interacts with perceptual load in visual search. *Psychological Science*, *13*(3), 284–287.

Jones, E. G., and Burton, H. (1976). A projection from the medial pulvinar to the amygdala in primates. *Brain Research*, *104*(1), 142–147.

Jonides, J., and Yantis, S. (1988). Uniqueness of abrupt visual onset in capturing attention. *Perception and Psychophysics*, *43*(4), 346–354.

Joseph, J. S., Chun, M. M., and Nakayama, K. (1997). Attentional requirements in a preattentive feature search task. *Nature*, *387*(6635), 805–807.

Kaada, B. R. (1951). Somato-motor, autonomic and electrocorticographic responses to electrical stimulation of rhinencephalic and other structures in primates, cat, and dog:A study of responses from the limbic, subcallosal, orbito-insular, piriform and temporal cortex, hippocampus-fornix and amygdala. *Acta Physiologica Scandinavica. Supplementum*, *24*(83), 1–262.

Kaada, B. R. (1960). Cingulate, posterior orbital, anterior insular and temporal pole cortext. In J. Field, H. W. Magoun, and V. E., eds., *Handbook of Physiology*. Section 1. *Neurophysiology* 2:1345–1371. Washington, DC: American Physiological Society.

Kaada, B. R. (1972). Stimulation and regional ablation of the amygdaloid complex with reference to functional representations. In B. E. Eleftheriou, ed., *The Neurobiology of the Amygdala*, 205–281). New York: Plenum.

Kaas, J. H. (2004). Evolution of somatosensory and motor cortex in primates. *Anatomical Record. Part A, Discoveries in Molecular, Cellular, and Evolutionary Biology*, *281*(1), 1148–1156. doi:10.1002/ar.a.20120.

Kahneman, D. (1973). *Attention and Effort*. Englewood Cliffs, NJ: Prentice-Hall.

Kahneman, D., and Tversky, A. (1979). Prospect theory: An analysis of decision under risk. *Econometrica*, *47*, 263–291.

Kaiser, M., and Hilgetag, C. C. (2006). Nonoptimal component placement, but short processing paths, due to long-distance projections in neural systems. *PLoS Computational Biology*, *2*(7), e95.

Kang, M. S., Blake, R., and Woodman, G. F. (2011). Semantic analysis does not occur in the absence of awareness induced by interocular suppression. *Journal of Neuroscience*, *31*(38), 13535–13545. doi:10.1523/jneurosci.1691-11.2011.

Kanwisher, N. (2010). Functional specificity in the human brain: A window into the functional architecture of the mind. *Proceedings of the National Academy of Sciences of the United States of America*, *107*(25), 11163–11170.

Kanwisher, N., McDermott, J., and Chun, M. M. (1997). The fusiform face area: A module in human extrastriate cortex specialized for face perception. *Journal of Neuroscience*, *17*(11), 4302–4311.

Kapp, B. S., Supple, W. F., Jr., and Whalen, P. J. (1994). Effects of electrical stimulation of the amygdaloid central nucleus on neocortical arousal in the rabbit. *Behavioral Neuroscience*, *108*(1), 81–93.

Karnath, H. O., Himmelbach, M., and Rorden, C. (2002). The subcortical anatomy of human spatial neglect: Putamen, caudate nucleus and pulvinar. *Brain*, *125*(Pt. 2), 350–360.

Kastner, S., O'Connor, D. H., Fukui, M. M., Fehd, H. M., Herwig, U., and Pinsk, M. A. (2004). Functional imaging of the human lateral geniculate nucleus and pulvinar. *Journal of Neurophysiology*, *91*(1), 438–448.

Kastner, S., and Pinsk, M. A. (2004). Visual attention as a multilevel selection process. *Cognitive, Affective and Behavioral Neuroscience*, *4*(4), 483–500.

Kastner, S., and Ungerleider, L. G. (2000). Mechanisms of visual attention in the human cortex. *Annual Review of Neuroscience*, *23*, 315–341.

Katsuki, F., and Constantinidis, C. (2012). Early involvement of prefrontal cortex in visual bottom-up attention. *Nature Neuroscience*, *15*(8), 1160–1166. doi:10.1038/nn.3164.

Kelso, J. A., and Engstrøm, D. A. (2006). *The Complementary Nature*. Cambridge, MA: MIT Press.

Kennedy, D. P., and Adolphs, R. (2010). Impaired fixation to eyes following amygdala damage arises from abnormal bottom-up attention. *Neuropsychologia*, *48*(12), 3392–3398. doi:10.1016/j.neuropsychologia.2010.06.025.

Kennerley, S. W., Walton, M. E., Behrens, T. E., Buckley, M. J., and Rushworth, M. F. (2006). Optimal decision making and the anterior cingulate cortex. *Nature Neuroscience*, *9*(7), 940–947.

Keren, G., and Schul, Y. (2009). Two is not always better than one A critical evaluation of two-system theories. *Perspectives on Psychological Science*, *4*(6), 533–550.

Kilgard, M. P., and Merzenich, M. M. (1998). Cortical map reorganization enabled by nucleus basalis activity. *Science*, *279*(5357), 1714–1718.

Kim, S. H., and Hamann, S. (2007). Neural correlates of positive and negative emotion regulation. *Journal of Cognitive Neuroscience*, *19*(5), 776–798. doi:10.1162/jocn.2007.19.5.776.

King, A. J., and Nelken, I. (2009). Unraveling the principles of auditory cortical processing: Can we learn from the visual system? *Nature Neuroscience*, *12*(6), 698–701.

Kinnison, J., Padmala, S., Choi, J. M., and Pessoa, L. (2012). Network analysis reveals increased integration during emotional and motivational processing. *Journal of Neuroscience*, *32*(24), 8361–8372. doi:10.1523/jneurosci.0821-12.2012.

Kiss, M., Driver, J., and Eimer, M. (2009). Reward priority of visual target singletons modulates event-related potential signatures of attentional selection. *Psychological Science*, *20*(2), 245–251. doi:10.1111/j.1467-9280.2009.02281.x.

Kitsak, M., Gallos, L. K., Havlin, S., Liljeros, F., Muchnik, L., Stanley, H. E., et al. (2010). Identification of influential spreaders in complex networks. *Nature Physics*, *6*(11), 888–893.

Kitzbichler, M. G., Henson, R. N., Smith, M. L., Nathan, P. J., and Bullmore, E. T. (2011). Cognitive effort drives workspace configuration of human brain functional networks. *Journal of Neuroscience*, *31*(22), 8259–8270. doi:10.1523/jneurosci.0440-11.2011.

Kleffner, D. A., and Ramachandran, V. S. (1992). On the perception of shape from shading. *Perception and Psychophysics*, *52*(1), 18–36.

Klüver, H., and Bucy, P. C. (1939). Preliminary analysis of functions of the temporal lobes in monkeys. *Archives of Neurology and Psychiatry*, *42*(6), 979–1000. doi:10.1001/archneurpsyc.1939.02270240017001.

Knutson, B., and Greer, S. M. (2008). Anticipatory affect: Neural correlates and consequences for choice. *Philosophical Transactions of the Royal Society of London. Series B, Biological Sciences*, *363*(1511), 3771–3786. doi:10.1098/rstb.2008.0155.

Kobayashi, S., Kawagoe, R., Takikawa, Y., Koizumi, M., Sakagami, M., and Hikosaka, O. (2007). Functional differences between macaque prefrontal cortex and caudate nucleus during eye movements with and without reward. *Experimental Brain Research*, *176*, 341–355.

Kobayashi, S., Lauwereyns, J., Koizumi, M., Sakagami, M., and Hikosaka, O. (2002). Influence of reward expectation on visuospatial processing in macaque lateral prefrontal cortex. *Journal of Neurophysiology*, *87*(3), 1488–1498.

Kober, H., Barrett, L. F., Joseph, J., Bliss-Moreau, E., Lindquist, K., and Wager, T. D. (2008). Functional grouping and cortical-subcortical interactions in emotion: A meta-analysis of neuroimaging studies. *NeuroImage*, *42*(2), 998–1031.

Koch, C., and Tsuchiya, N. (2007). Attention and consciousness: Two distinct brain processes. *Trends in Cognitive Sciences*, *11*(1), 16–22.

Koch, M. A., Norris, D. G., and Hund-Georgiadis, M. (2002). An investigation of functional and anatomical connectivity using magnetic resonance imaging. *NeuroImage*, *16*(1), 241–250.

Konorski, J. (1967). *Intergrative Activity of the Brain: An Interdisciplinary Approach.* Chicago: University of Chicago Press.

Kouider, S., and Dehaene, S. (2007). Levels of processing during non-conscious perception: A critical review of visual masking. *Philosophical Transactions of the Royal Society of London. Series B, Biological Sciences*, *362*(1481), 857–875. doi:10.1098/rstb.2007.2093.

Kouneiher, F., Charron, S., and Koechlin, E. (2009). Motivation and cognitive control in the human prefrontal cortex. *Nature Neuroscience*, *12*(7), 939–945. doi:10.1038/nn.2321.

Kravitz, D. J., Saleem, K. S., Baker, C. I., Ungerleider, L. G., and Mishkin, M. (2013). The ventral visual pathway: an expanded neural framework for the processing of object quality. *Trends in Cognitive Sciences*, *17*(1), 26–49.

Krebs, R. M., Boehler, C. N., Egner, T., and Woldorff, M. G. (2011). The neural underpinnings of how reward associations can both guide and misguide attention. *Journal of Neuroscience*, *31*(26), 9752–9759. doi:10.1523/jneuroscience.0732-11.2011.

Krebs, R. M., Boehler, C. N., and Woldorff, M. G. (2010). The influence of reward associations on conflict processing in the Stroop task. *Cognition*, *117*(3), 341–347. doi:10.1016/j.cognition.2010.08.018.

Kreiman, G., Koch, C., and Fried, I. (2000a). Category-specific visual responses of single neurons in the human medial temporal lobe. *Nature Neuroscience*, *3*(9), 946–953.

Kreiman, G., Koch, C., and Fried, I. (2000b). Imagery neurons in the human brain. *Nature*, *408*(6810), 357–361.

Kristjansson, A., Sigurjonsdottir, O., and Driver, J. (2010). Fortune and reversals of fortune in visual search: Reward contingencies for pop-out targets affect search efficiency and target repetition effects. *Attention, Perception and Psychophysics*, *72*(5), 1229–1236. doi:10.3758/APP.72.5.1229.

Krolak-Salmon, P., Henaff, M. A., Vighetto, A., Bertrand, O., and Mauguière, F. (2004). Early amygdala reaction to fear spreading in occipital, temporal, and frontal cortex: A depth electrode ERP study in human. *Neuron*, *42*(4), 665–676.

Krubitzer, L. (2009). In search of a unifying theory of complex brain evolution. *Annals of the New York Academy of Sciences*, *1156*, 44–67.

Kruglanski, A. W., Belanger, J. J., Chen, X., Kopetz, C., Pierro, A., and Mannetti, L. (2012). The energetics of motivated cognition: A force-field analysis. *Psychological Review*, *119*(1), 1–20. doi:10.1037/a0025488.

Kruglanski, A. W., Erbs, H. P., Pierro, A., Mannetti, L., and Chun, W. Y. (2006). On parametric continuities in the world of binary either ors. *Psychological Inquiry*, *17*, 153–165.

Kruschke, J. K. (2003). Attention in learning. *Current Directions in Psychological Science*, *12*, 171–175.

Kuhnen, C. M., and Knutson, B. (2005). The neural basis of financial risk taking. *Neuron*, *47*(5), 763–770. doi:10.1016/j.neuron.2005.08.008.

Kunimoto, C., Miller, J., and Pashler, H. (2001). Confidence and accuracy of near-threshold discrimination responses. *Consciousness and Cognition*, *10*(3), 294–340.

Kuraoka, K., and Nakamura, K. (2007). Responses of single neurons in monkey amygdala to facial and vocal emotions. *Journal of Neurophysiology*, *97*(2), 1379–1387.

Kveraga, K., Boshyan, J., and Bar, M. (2007). Magnocellular projections as the trigger of top-down facilitation in recognition. *Journal of Neuroscience*, *27*(48), 13232–13240.

LaBar, K. S., Gatenby, J. C., Gore, J. C., LeDoux, J. E., and Phelps, E. A. (1998). Human amygdala activation during conditioned fear acquisition and extinction: A mixed-trial fMRI study. *Neuron*, *20*(5), 937–945.

LaBerge, D., and Buchsbaum, M. S. (1990). Positron emission tomographic measurements of pulvinar activity during an attention task. *Journal of Neuroscience*, *10*(2), 613–619.

Laird, A. R., Lancaster, J. L., and Fox, P. T. (2005). BrainMap: The social evolution of a human brain mapping database. *Neuroinformatics*, *3*(1), 65–78.

Lamme, V. A. (2003). Why visual attention and awareness are different. *Trends in Cognitive Sciences*, *7*(1), 12–18.

Lamme, V. A., and Roelfsema, P. R. (2000). The distinct modes of vision offered by feedforward and recurrent processing. *Trends in Neurosciences*, *23*(11), 571–579.

Lang, P. J., and Davis, M. (2006). Emotion, motivation, and the brain: Reflex foundations in animal and human research. *Progress in Brain Research*, *156*, 3–29.

Lang, P. J., Davis, M., and Öhman, A. (2000). Fear and anxiety: Animal models and human cognitive psychophysiology. *Journal of Affective Disorders*, *61*(3), 137–159.

Lashley, K. S. (1930). Basic neural mechanisms in behavior. *Psychological Review*, *37*, 1–24.

Lavie, N. (1995). Perceptual load as a necessary condition for selective attention. *Journal of Experimental Psychology. Human Perception and Performance*, *21*(3), 451–468.

Lavie, N., and de Fockert, J. W. (2003). Contrasting effects of sensory limits and capacity limits in visual selective attention. *Perception and Psychophysics*, *65*(2), 202–212.

Lavie, N., Hirst, A., de Fockert, J. W., and Viding, E. (2004). Load theory of selective attention and cognitive control. *Journal of Experimental Psychology. General*, *133*(3), 339–354.

Lavric, A., Rippon, G., and Gray, J. R. (2003). Threat-evoked anxiety disrupts spatial working memory performance: An attentional account. *Cognitive Therapy and Research*, *27*(5), 489–504.

LeDoux, J. E. (1996). *The Emotional Brain*. New York: Simon & Schuster.

LeDoux, J. E. (2000). Emotion circuits in the brain. *Annual Review of Neuroscience*, *23*(1), 155–184.

Lee, L., Harrison, L. M., and Mechelli, A. (2003). A report of the functional connectivity workshop, Dusseldorf 2002. *NeuroImage*, *19*(2 Pt 1), 457–465.

Leh, S. E., Chakravarty, M. M., and Ptito, A. (2008). The connectivity of the human pulvinar: A diffusion tensor imaging tractography study. *International Journal of Biomedical Imaging*, *2008*, 789539.

Leon, M. I., and Shadlen, M. N. (1999). Effect of expected reward magnitude on the response of neurons in the dorsolateral prefrontal cortex of the macaque. *Neuron*, *24*(2), 415–425.

Leonard, C. M., Rolls, E. T., Wilson, F. A., and Baylis, G. C. (1985). Neurons in the amygdala of the monkey with responses selective for faces. *Behavioural Brain Research*, *15*(2), 159–176.

Lewis, M. D. (2005). Bridging emotion theory and neurobiology through dynamic systems modeling. *Behavioral and Brain Sciences*, *28*(2), 169–194, discussion 194–245.

Lewis, M. D., and Todd, R. M. (2007). The self-regulating brain: Cortical-subcortical feedback and the development of intelligent action. *Cognitive Development*, *22*, 406–430.

Li, F. F., VanRullen, R., Koch, C., and Perona, P. (2002). Rapid natural scene categorization in the near absence of attention. *Proceedings of the National Academy of Sciences of the United States of America*, *99*(14), 9596–9601.

Lim, S. L., and Pessoa, L. (2008). Affective learning increases sensitivity to graded emotional faces. *Emotion*, *8*(1), 96–103.

Lim, S. L., Padmala, S., and Pessoa, L. (2008). Affective learning modulates spatial competition during low-load attentional conditions. *Neuropsychologia*, *46*(5), 1267–1278.

Lim, S. L., Padmala, S., and Pessoa, L. (2009). Segregating the significant from the mundane on a moment-to-moment basis via direct and indirect amygdala contributions. *Proceedings of the National Academy of Sciences of the United States of America*, *106*(39), 16841–16846.

Lin, S. C., and Nicolelis, M. A. (2008). Neuronal ensemble bursting in the basal forebrain encodes salience irrespective of valence. *Neuron*, *59*(1), 138–149.

Lindstrom, B. R., and Bohlin, G. (2012). Threat-relevance impairs executive functions: Negative impact on working memory and response inhibition. *Emotion*, *12*(2), 384–393. doi:10.1037/a0027305.

Linke, R., De Lima, A. D., Schwegler, H., and Pape, H. C. (1999). Direct synaptic connections of axons from superior colliculus with identified thalamo-amygdaloid projection neurons in the rat: Possible substrates of a subcortical visual pathway to the amygdala. *Journal of Comparative Neurology*, *403*(2), 158–170.

Liu, H., Agam, Y., Madsen, J. R., and Kreiman, G. (2009). Timing, timing, timing: Fast decoding of object information from intracranial field potentials in human visual cortex. *Neuron*, *62*(2), 281–290.

Liu, X., Hairston, J., Schrier, M., and Fan, J. (2011). Common and distinct networks underlying reward valence and processing stages: A meta-analysis of functional neuroimaging studies. *Neuroscience and Biobehavioral Reviews*, *35*(5), 1219–1236. doi:10.1016/j.neubiorev.2010.12.012.

Liu, Y. Y., Slotine, J. J., and Barabasi, A. L. (2011). Controllability of complex networks. *Nature*, *473*(7346), 167–173. doi:10.1038/nature10011.

Locke, H. S., and Braver, T. S. (2008). Motivational influences on cognitive control: Behavior, brain activation, and individual differences. *Cognitive, Affective and Behavioral Neuroscience*, *8*(1), 99–112.

Loftus, E. F., and Klinger, M. R. (1992). Is the unconscious smart or dumb? *American Psychologist*, *47*(6), 761–765.

Logan, G. D. (1988). Automaticity, resources, and memory: Theoretical controversies and practical implications. *Human Factors*, *30*(5), 583–598.

Longe, O., Senior, C., and Rippon, G. (2009). The lateral and ventromedial prefrontal cortex work as a dynamic integrated system: Evidence from fMRI connectivity analysis. *Journal of Cognitive Neuroscience*, *21*(1), 141–154. doi:10.1162/jocn.2009.21012.

Lowe, M. J., Mock, B. J., and Sorenson, J. A. (1998). Functional connectivity in single and multislice echoplanar imaging using resting-state fluctuations. *NeuroImage*, *7*(2), 119–132.

Ludmer, R., Dudai, Y., and Rubin, N. (2011). Uncovering camouflage: Amygdala activation predicts long-term memory of induced perceptual insight. *Neuron*, *69*(5), 1002–1014. doi:10.1016/j.neuron.2011.02.013.

Luksch, H., Khanbabaie, R., and Wessel, R. (2004). Synaptic dynamics mediate sensitivity to motion independent of stimulus details. *Nature Neuroscience*, *7*(4), 380–388. doi:10.1038/nn1204.

Luo, A. H., Tahsili-Fahadan, P., Wise, R. A., Lupica, C. R., and Aston-Jones, G. (2011). Linking context with reward: A functional circuit from hippocampal CA3 to ventral tegmental area. *Science*, *333*(6040), 353–357. doi:10.1126/science.1204622.

Luo, Q., Holroyd, T., Majestic, C., Cheng, X., Schechter, J., and Blair, R. J. (2010). Emotional automaticity is a matter of timing. *Journal of Neuroscience*, *30*(17), 5825–5829.

Luria, A. R. (1966). *Higher Cortical Functions in Man*. Oxford: Basic Books.

Lyon, D. C., Nassi, J. J., and Callaway, E. M. (2010). A disynaptic relay from superior colliculus to dorsal stream visual cortex in macaque monkey. *Neuron*, *65*(2), 270–279.

Ma, T. P., Lynch, J. C., Donahoe, D. K., Attallah, H., and Rafols, J. A. (1998). Organization of the medial pulvinar nucleus in the macaque. *Anatomical Record*, *250*(2), 220–237.

Mack, A., and Clarke, J. (2012). Gist perception requires attention. *Visual Cognition*, *20*(3), 300–327.

MacLean, P. D. (1949). Psychosomatic disease and the "visceral brain": Recent developments bearing on the Papez theory of emotion. *Psychosomatic Medicine*, *11*, 338–353.

MacLean, P. D. (1952). Some psychiatric implications of physiological studies on frontotemporal portion of limbic system (visceral brain). *Electroencephalography and Clinical Neurophysiology*, *4*(4), 407–418.

MacLean, P. D. (1970). The triune brain, emotion, and scientific bias. In F. O. Schmitt, ed., *The Neurosciences Second Study Program*, 336–349). New York: Rockefeller University Press.

MacLean, P. D. (1990). *The Triune Brain in Evolution: Role in Paleocerebral Functions*. New York: Plenum.

MacLeod, C., Mathews, A., and Tata, P. (1986). Attentional bias in emotional disorders. *Journal of Abnormal Psychology*, *95*(1), 15–20.

MacLeod, C., Rutherford, E., Campbell, L., Ebsworthy, G., and Holker, L. (2002). Selective attention and emotional vulnerability: Assessing the causal basis of their association through the experimental manipulation of attentional bias. *Journal of Abnormal Psychology*, *111*(1), 107–123.

Maddux, J. M., Kerfoot, E. C., Chatterjee, S., and Holland, P. C. (2007). Dissociation of attention in learning and action: Effects of lesions of the amygdala central nucleus, medial prefrontal cortex, and posterior parietal cortex. *Behavioral Neuroscience*, *121*(1), 63–79. doi:10.1037/0735-7044.121.1.63.

Magurran, A. E. (2004). *Measuring Biological Diversity*. Oxford, United Kingdom: Blackwell.

Malkova, L., Gaffan, D., and Murray, E. A. (1997). Excitotoxic lesions of the amygdala fail to produce impairment in visual learning for auditory secondary reinforcement but interfere with reinforcer devaluation effects in rhesus monkeys. *Journal of Neuroscience*, *17*(15), 6011–6020.

Mantini, D., Gerits, A., Nelissen, K., Durand, J. B., Joly, O., Simone, L., et al. (2011). Default mode of brain function in monkeys. *Journal of Neuroscience*, *31*(36), 12954–12962. doi:10.1523/jneurosci.2318-11.2011.

Marcel, A. J. (1983). Conscious and unconscious perception: Experiments on visual masking and word recognition. *Cognitive Psychology*, *15*(2), 197–237.

Markov, N. T., and Kennedy, H. (2013). The importance of being hierarchical. *Current Opinion in Neurobiology*. Epub ahead of print.

Markov, N. T., Misery, P., Falchier, A., Lamy, C., Vezoli, J., Quilodran, R., et al. (2011). Weight consistency specifies regularities of macaque cortical networks. *Cerebral Cortex*, *21*(6), 1254–1272. doi:10.1093/cercor/bhq201.

Marois, R., Chun, M. M., and Gore, J. C. (2000). Neural correlates of the attentional blink. *Neuron*, *28*(1), 299–308.

Marois, R., and Ivanoff, J. (2005). Capacity limits of information processing in the brain. *Trends in Cognitive Sciences*, *9*(6), 296–305.

Martínez-García, F., Novejarque, A., and Lanuza, E. (2007). Evolution of the amygdala in vertebrates. In J. Kaas andT. H. Bullock, eds., *Evolution of Nervous Systems: A Comprehensive Reference. Vol. 2: Non-mammalian Vertebrates*, 255–334. Amsterdam: Academic Press-Elsevier.

Masland, R. H., and Martin, P. R. (2007). The unsolved mystery of vision. *Current Biology*, *17*(15), R577–R582. doi:10.1016/j.cub.2007.05.040.

Mather, M., and Sutherland, M. R. (2011). Arousal-biased competition in perception and memory. *Perspectives on Psychological Science*, *6*(2), 114–133. doi:10.1177/1745691611400234.

Mather, M., Mitchell, K. J., Raye, C. L., Novak, D. L., Greene, E. J., and Johnson, M. K. (2006). Emotional arousal can impair feature binding in working memory. *Journal of Cognitive Neuroscience*, *18*(4), 614–625. doi:10.1162/jocn.2006.18.4.614.

Mathews, A., and Mackintosh, B. (1998). A cognitive model of selective processing in anxiety. *Cognitive Therapy and Research*, *22*(6), 539–560.

Mathews, A., and MacLeod, C. (1985). Selective processing of threat cues in anxiety states. *Behaviour Research and Therapy*, *23*(5), 563–569.

Maunsell, J. H. (2004). Neuronal representations of cognitive state: Reward or attention? *Trends in Cognitive Sciences*, *8*(6), 261–265.

May, P. J. (2006). The mammalian superior colliculus: Laminar structure and connections. *Progress in Brain Research*, *151*, 321–378. doi:10.1016/s0079-6123(05)51011-2.

Mayberg, H. S., Brannan, S. K., Mahurin, R. K., Jerabek, P. A., Brickman, J. S., Tekell, J. L., et al. (1997). Cingulate function in depression: A potential predictor of treatment response. *Neuroreport*, *8*(4), 1057–1061.

Mayberg, H. S., Liotti, M., Brannan, S. K., McGinnis, S., Mahurin, R. K., Jerabek, P. A., et al. (1999). Reciprocal limbic-cortical function and negative mood: Converging PET findings in depression and normal sadness. *American Journal of Psychiatry*, *156*(5), 675–682.

McAlonan, K., Cavanaugh, J., and Wurtz, R. H. (2008). Guarding the gateway to cortex with attention in visual thalamus. *Nature*, *456*(7220), 391–394. doi:10.1038/nature07382.

McGinty, V. B., Hayden, B. Y., Heilbronner, S. R., Dumont, E. C., Graves, S. M., Mirrione, M. M., et al. (2011). Emerging, reemerging, and forgotten brain areas of the reward circuit: Notes from the 2010 Motivational and Neural Networks conference. *Behavioural Brain Research*, *225*, 348–357.

McHaffie, J. G., Stanford, T. R., Stein, B. E., Coizet, V., and Redgrave, P. (2005). Subcortical loops through the basal ganglia. *Trends in Neurosciences*, *28*(8), 401–407. doi:10.1016/j.tins.2005.06.006.

McIntosh, A. R. (2000). Towards a network theory of cognition. *Neural Networks*, *13*(8–9), 861–870.

McKenna, F. P. (1986). Effects of unattended emotional stimuli on color-naming performance. *Current Psychological Research and Reviews*, *5*, 3–9.

McNally, G. P., Johansen, J. P., and Blair, H. T. (2011). Placing prediction into the fear circuit. *Trends in Neurosciences*, *34*(6), 283–292. doi:10.1016/j.tins.2011.03.005.

McTighe, S. M., Cowell, R. A., Winters, B. D., Bussey, T. J., and Saksida, L. M. (2010). Paradoxical false memory for objects after brain damage. *Science*, *330*(6009), 1408–1410. doi:10.1126/science.1194780.

Mechias, M. L., Etkin, A., and Kalisch, R. (2010). A meta-analysis of instructed fear studies: Implications for conscious appraisal of threat. *NeuroImage*, *49*(2), 1760–1768. doi:10.1016/j.neuroimage.2009.09.040.

Mehta, A. D., Ulbert, I., and Schroeder, C. E. (2000). Intermodal selective attention in monkeys: I. Distribution and timing of effects across visual areas. *Cerebral Cortex*, *10*(4), 343–358.

Merikle, P. M., Smilek, D., Eastwood, J. D. (2001). Perception without awareness: Perspectives from cognitive psychology. *Cognition*, *79*(1-2), 115–134.

Messinger, A., Winkle, C. C., Tootell, R. B., and Ungerleider, L. G. (2011). Using electrical stimulation and fMRI to map connections of the amygdala in the monkey. Paper presented at the Society for Neuroscience 2011, Washington, D.C.

Mesulam, M.-M. (1981). A cortical network for directed attention and unilateral neglect. *Annals of Neurology*, *10*, 309–325.

Mesulam, M.-M. (1990). Large-scale neurocognitive networks and distributed processing for attention, language, and memory. *Annals of Neurology*, *28*, 597–613.

Mesulam, M.-M. (1998). From sensation to cognition. *Brain*, *121*(6), 1013–1052.

Mesulam, M.-M. (2000). Behavioral neuroanatomy: Large-scale networks, association cortex, frontal syndromes, the limbic system, and hemispheric specializations. In M.-M. Mesulam, ed., *Principles of Behavioral and Cognitive Neurology* (2nd ed.), 1–120. New York: Oxford University Press.

Mesulam, M-M., Hersh, L. B., Mash, D. C., and Geula, C. (1992). Differential cholinergic innervation within functional subdivisions of the human cerebral cortex: A choline acetyltransferase study. *Journal of Comparative Neurology*, *318*(3), 316–328.

Mesulam, M.-M., and Mufson, E. J. (1982a). Insula of the old world monkey: I. Architectonics in the insulo-orbito-temporal component of the paralimbic brain. *Journal of Comparative Neurology*, *212*(1), 1–22. doi:10.1002/cne.902120102.

Mesulam, M.-M., and Mufson, E. J. (1982b). Insula of the old world monkey: III. Efferent cortical output and comments on function. *Journal of Comparative Neurology*, *212*(1), 38–52. doi:10.1002/cne.902120104.

Milders, M., Sahraie, A., Logan, S., and Donnellon, N. (2006). Awareness of faces is modulated by their emotional meaning. *Emotion*, *6*(1), 10–17.

Milgram, S. (1967). The small world problem. *Psychology Today*, *2*(1), 60–67.

Miller, E. K., and Cohen, J. D. (2001). An integrative theory of prefrontal cortex function. *Annual Review of Neuroscience*, *24*, 167–202.

Miller, E. K., Gochin, P. M., and Gross, C. G. (1991). Habituation-like decrease in the responses of neurons in inferior temporal cortex of the macaque. *Visual Neuroscience*, *7*(4), 357–362.

Mitchell, D. G., Nakic, M., Fridberg, D., Kamel, N., Pine, D. S., and Blair, R. J. (2007). The impact of processing load on emotion. *NeuroImage*, *34*(3), 1299–1309.

Miyake, A., Friedman, N. P., Emerson, M. J., Witzki, A. H., Howerter, A., and Wager, T. D. (2000). The unity and diversity of executive functions and their contributions to complex "Frontal Lobe" tasks: A latent variable analysis. *Cognitive Psychology*, *41*(1), 49–100.

Mizuhiki, T., Richmond, B. J., and Shidara, M. (2012). Encoding of reward expectation by monkey anterior insular neurons. *Journal of Neurophysiology*, *107*(11), 2996–3007. doi:10.1152/jn.00282.2011.

Mobbs, D., Marchant, J. L., Hassabis, D., Seymour, B., Tan, G., Gray, M., et al. (2009). From threat to fear: The neural organization of defensive fear systems in humans. *Journal of Neuroscience*, *29*(39), 12236–12243. doi:10.1523/jneurosci.2378-09.2009.

Mobbs, D., Petrovic, P., Marchant, J. L., Hassabis, D., Weiskopf, N., Seymour, B., et al. (2007). When fear is near: Threat imminence elicits prefrontal-periaqueductal gray shifts in humans. *Science*, *317*(5841), 1079–1083. doi:10.1126/science.1144298.

Mobbs, D., Yu, R., Rowe, J. B., and Eich, H., FeldmanHall, O., and Dalgleish, T. (2010). Neural activity associated with monitoring the oscillating threat value of a tarantula. *Proceedings of the National Academy of Sciences of the United States of America*, *107*(47), 20582–20586. doi: 10.1073/pnas.1009076107

Modha, D. S., and Singh, R. (2010). Network architecture of the long-distance pathways in the macaque brain. *Proceedings of the National Academy of Sciences of the United States of America*, *107*(30), 13485–13490.

Moeller, S., Freiwald, W. A., and Tsao, D. Y. (2008). Patches with links: A unified system for processing faces in the macaque temporal lobe. *Science*, *320*(5881), 1355–1359. doi:10.1126/science.1157436.

Mogg, K., Garner, M., and Bradley, B. P. (2007). Anxiety and orienting of gaze to angry and fearful faces. *Biological Psychology*, *76*(3), 163–169. doi:10.1016/j.biopsycho.2007.07.005.

Mohanty, A., Egner, T., Monti, J. M., and Mesulam, M.-M. (2009). Search for a threatening target triggers limbic guidance of spatial attention. *Journal of Neuroscience*, *29*(34), 10563–10572. doi:10.1523/jneurosci.1170-09.2009.

Mohanty, A., Gitelman, D. R., Small, D. M., and Mesulam, M.-M. (2008). The spatial attention network interacts with limbic and monoaminergic systems to modulate motivation-induced attention shifts. *Cerebral Cortex*, *18*(11), 2604–2613.

Mohr, P. N., Biele, G., and Heekeren, H. R. (2010). Neural processing of risk. *Journal of Neuroscience*, *30*(19), 6613–6619. doi:10.1523/jneurosci.0003-10.2010.

Moll, J., Zahn, R., de Oliveira-Souza, R., Krueger, F., and Grafman, J. (2005). The neural basis of human moral cognition. *Nature Reviews. Neuroscience*, *6*(10), 799–809.

Montague, P. R., Hyman, S. E., and Cohen, J. D. (2004). Computational roles for dopamine in behavioural control. *Nature*, *431*(7010), 760–767. doi:10.1038/nature03015.

Moors, A., and De Houwer, J. (2006). Automaticity: A theoretical and conceptual analysis. *Psychological Bulletin*, *132*(2), 297–326.

Morecraft, R. J., and Tanji, J. (2009). Cingulofrontal interactions and the cingulate motor areas. In B. A. Vogt, ed., *Cingulate Neurobiology and Disease*, 113–144. Oxford: Oxford University Press.

Mormann, F., Dubois, J., Kornblith, S., Milosavljevic, M., Cerf, M., Ison, M., et al. (2011). A category-specific response to animals in the right human amygdala. *Nature Neuroscience*, *14*(10), 1247–1249. doi:10.1038/nn.2899.

Mormann, F., Kornblith, S., Quiroga, R. Q., Kraskov, A., Cerf, M., Fried, I., et al. (2008). Latency and selectivity of single neurons indicate hierarchical processing in the human medial temporal lobe. *Journal of Neuroscience*, *28*(36), 8865–8872.

Morris, J. S., de Gelder, B., Weiskrantz, L., and Dolan, R. J. (2001). Differential extrageniculostriate and amygdala responses to presentation of emotional faces in a cortically blind field. *Brain*, *124*(6), 1241–1252.

Morris, J. S., Friston, K. J., Buchel, C., Frith, C. D., Young, A. W., Calder, A. J., et al. (1998). A neuromodulatory role for the human amygdala in processing emotional facial expressions. *Brain*, *121*(Pt 1), 47–57.

Morris, J. S., Öhman, A., and Dolan, R. J. (1998). Conscious and unconscious emotional learning in the human amygdala. *Nature*, *393*(6684), 467–470.

Morris, J. S., Öhman, A., and Dolan, R. J. (1999). A subcortical pathway to the right amygdala mediating "unseen" fear. *Proceedings of the National Academy of Sciences of the United States of America*, *96*(4), 1680–1685.

Morrison, I., Bjornsdotter, M., and Olausson, H. (2011). Vicarious responses to social touch in posterior insular cortex are tuned to pleasant caressing speeds. *Journal of Neuroscience*, *31*(26), 9554–9562. doi:10.1523/jneurosci.0397-11.2011.

Morrison, S. E., and Salzman, C. D. (2010). Re-valuing the amygdala. *Current Opinion in Neurobiology*, *20*(2), 221–230.

Morrison, S. E., and Salzman, C. D. (2011). Representations of appetitive and aversive information in the primate orbitofrontal cortex. *Annals of the New York Academy of Sciences*, *1239*(1), 59–70.

Moruzzi, G., and Magoun, H. W. (1949). Brain stem reticular formation and activation of the EEG. *Electroencephalography and Clinical Neurophysiology*, *1*(4), 455–473.

Moskowitz, G. B. (2002). Preconscious effects of temporary goals on attention. *Journal of Experimental Social Psychology*, *38*(4), 397–404.

Most, S. B., Chun, M. M., Widders, D. M., and Zald, D. H. (2005). Attentional rubbernecking: Cognitive control and personality in emotion-induced blindness. *Psychonomic Bulletin and Review*, *12*(4), 654–661.

Most, S. B., Scholl, B. J., Clifford, E. R., and Simons, D. J. (2005). What you see is what you set: Sustained inattentional blindness and the capture of awareness. *Psychological Review*, *112*(1), 217–242.

Moussa, M. N., Vechlekar, C. D., Burdette, J. H., Steen, M. R., Hugenschmidt, C. E., and Laurienti, P. J. (2011). Changes in cognitive state alter human functional brain networks. *Frontiers in Human Neuroscience*, *5*, 83. doi:10.3389/fnhum.2011.00083.

Mucha, P. J., Richardson, T., Macon, K., Porter, M. A., and Onnela, J. P. (2010). Community structure in time-dependent, multiscale, and multiplex networks. *Science*, *328*(5980), 876–878.

Muller, M. M., Andersen, S. K., and Keil, A. (2008). Time course of competition for visual processing resources between emotional pictures and foreground task. *Cerebral Cortex*, *18*(8), 1892–1899.

Murphy, K. R., and Myors, B. (2004). *Statistical Power Analysis: A Simple and General Model for Traditional and Modern Hypothesis Tests* (2nd ed.). Mahwaw, NJ: Erlbaum.

Murphy, P. C., and Sillito, A. M. (1991). Cholinergic enhancement of direction selectivity in the visual cortex of the cat. *Neuroscience*, *40*(1), 13–20.

Murray, E. A. (2007). The amygdala, reward and emotion. *Trends in Cognitive Sciences*, *11*(11), 489–497.

Nakamura, H., Gattass, R., Desimone, R., and Ungerleider, L. G. (1993). The modular organization of projections from areas V1 and V2 to areas V4 and TEO in macaques. *Journal of Neuroscience*, *13*(9), 3681–3691.

Nakamura, K., Mikami, A., and Kubota, K. (1992). Activity of single neurons in the monkey amygdala during performance of a visual discrimination task. *Journal of Neurophysiology*, *67*(6), 1447–1463.

Nakayama, K., and Joseph, J. S. (1998). Attention, pattern recognition, and pop-out in visual search. In R. Parasuraman, ed., *The Attentive Brain*, 279–298. Cambridge, MA: MIT Press.

Naqvi, N. H., and Bechara, A. (2009). The hidden island of addiction: The insula. *Trends in Neurosciences*, *32*(1), 56–67. doi:10.1016/j.tins.2008.09.009.

Nassi, J. J., and Callaway, E. M. (2009). Parallel processing strategies of the primate visual system. *Nature Reviews. Neuroscience*, *10*(5), 360–372. doi:10.1038/nrn2619.

Nauta, W. J. H. (1971). The problem of the frontal lobe: A reinterpretation. *Journal of Psychiatric Research*, *8*, 167–187.

Nauta, W. J. [H.] (1972). Neural associations of the frontal cortex. *Acta Neurobiologiae Experimentalis*, *32*(2), 125–140.

Navalpakkam, V., Koch, C., and Perona, P. (2009). Homo economicus in visual search. *Journal of Vision, 9*(1), 31: 1–16. doi: 10.1167/9.1.31

Navon, D. (1984). Resources: A theoretical soup stone? *Psychological Review*, *91*(2), 216–234.

Neafsey, E. J. (1990). Prefrontal cortical control of the autonomic nervous system: Anatomical and physiological observations. *Progress in Brain Research, 85*, 147–165; discussion 165–146.

Neisser, U. (1976). *Cognition and Reality*. San Francisco: Freeman.

Newell, A. (1973). *You Can't Play 20 Questions with Nature and Win: Projective Comments on the Papers of this Symposium*. Pittsburgh: Carnegie Mellon University, Department of Computer Science.

Newman, M. E. J. (2003). Mixing patterns in networks. *Physical Review E: Statistical, Nonlinear, and Soft Matter Physics*, *67*(2 Pt 2), 026126.

Newman, M. E. J. (2005). A measure of betweenness centrality based on random walks. *Social Networks*, *27*(1), 39–54.

Newman, M. E. J. (2010). *Networks: An Introduction*. New York: Oxford University Press.

Newton, J. R., Ellsworth, C., Miyakawa, T., Tonegawa, S., and Sur, M. (2004). Acceleration of visually cued conditioned fear through the auditory pathway. *Nature Neuroscience*, *7*(9), 968–973. doi:10.1038/nn1306.

Nguyen, M. N., Hori, E., Matsumoto, J., Tran, A. H., Ono, T., and Nishijo, H. (2013). Neuronal responses to face-like stimuli in the monkey pulvinar. *European Journal of Neuroscience*, *37*, 35–51.

Nobata, T., Hakoda, Y., and Ninose, Y. (2010). The functional field of view becomes narrower while viewing negative emotional stimuli. *Cognition and Emotion*, *24*(5), 886–891.

Norberg, J., Peira, N., and Wiens, S. (2010). Never mind the spider: Late positive potentials to phobic threat at fixation are unaffected by perceptual load. *Psychophysiology*, *47*(6), 1151–1158. doi:10.1111/j.1469-8986.2010.01019.x.

Norman, D. A., and Bobrow, D. G. (1975). On data-limited and resource-limited processes. *Cognitive Psychology*, *7*, 44–64.

Norman, D. A., and Shallice, T. (1986). Attention to action: Willed and automatic control of behavior. In R. J. Davidson, G. E. Schwartz, and D. Shapiro, eds., *Consciousness and Self-Regulation*, 1–18. New York: Plenum.

Norman, K. A., Polyn, S. M., Detre, G. J., and Haxby, J. V. (2006). Beyond mind-reading: Multi-voxel pattern analysis of fMRI data. *Trends in Cognitive Sciences*, *10*(9), 424–430.

Noudoost, B., and Moore, T. (2011a). Control of visual cortical signals by prefrontal dopamine. *Nature*, *474*(7351), 372–375. doi:10.1038/nature09995.

Noudoost, B., and Moore, T. (2011b). The role of neuromodulators in selective attention. *Trends in Cognitive Sciences*, *15*(12), 585–591. doi:10.1016/j.tics.2011.10.006.

Nowak, L. G., and Bullier, J. (1997). The timing of information transfer in the visual system. In K. Rockland, J. Kass, and A. Peters, eds., *Cerebral Cortex: Extrastriate Cortex in Primate* (Vol. 12), 205–241. New York: Plenum.

Oades, R. D., and Halliday, G. M. (1987). Ventral tegmental (A10) system: Neurobiology: 1. Anatomy and connectivity. *Brain Research. Brain Research Reviews*, *12*(2), 117–165. doi:10.1016/0165-0173(87)90011-7.

Oatley, K., Keltner, D., and Jenkins, J. M. (2006). *Understanding Emotions*. Malden, MA: Blackwell.

Ochsner, K. N., and Gross, J. J. (2005). The cognitive control of emotion. *Trends in Cognitive Sciences*, *9*(5), 242–249.

Ochsner, K. N., and Gross, J. J. (2008). Cognitive emotion regulation: Insights from social cognitive and affective neuroscience. *Current Directions in Psychological Science*, *17*(2), 153.

Ochsner, K. N., Ray, R. D., Cooper, J. C., Robertson, E. R., Chopra, S., Gabrieli, J. D., et al. (2004). For better or for worse: Neural systems supporting the cognitive down- and up-regulation of negative emotion. *NeuroImage*, *23*(2), 483–499. doi:10.1016/j.neuroimage.2004.06.030.

O'Connor, D. H., Fukui, M. M., Pinsk, M. A., and Kastner, S. (2002). Attention modulates responses in the human lateral geniculate nucleus. *Nature Neuroscience*, *5*(11), 1203–1209.

O'Doherty, J. P. (2004). Reward representations and reward-related learning in the human brain: Insights from neuroimaging. *Current Opinion in Neurobiology*, *14*(6), 769–776.

Ogawa, S., Lee, T. M., Stepnoski, R., Chen, W., Zhu, X. H., and Ugurbil, K. (2000). An approach to probe some neural systems interaction by functional MRI at neural time scale down to milliseconds. *Proceedings of the National Academy of Sciences of the United States of America*, *97*(20), 11026–11031. doi:10.1073/pnas.97.20.11026.

Öhman, A. (2005). The role of the amygdala in human fear: Automatic detection of threat. *Psychoneuroendocrinology*, *30*(10), 953–958. doi:10.1016/j.psyneuen.2005.03.019.

Öhman, A., Lundqvist, D., and Esteves, F. (2001). The face in the crowd revisited: A threat advantage with schematic stimuli. *Journal of Personality and Social Psychology*, *80*(3), 381–396.

Öhman, A., and Mineka, S. (2001). Fears, phobias, and preparedness: Toward an evolved module of fear and fear learning. *Psychological Review*, *108*(3), 483–522.

Okon-Singer, H., Tzelgov, J., and Henik, A. (2007). Distinguishing between automaticity and attention in the processing of emotionally significant stimuli. *Emotion*, *7*(1), 147–157. doi:10.1037/1528-3542.7.1.147.

Olivelle, P., trans. (1996). *Upanisads*. Oxford: Oxford University Press.

O'Neill, J., Sobel, T. L., and Vogt, B. A. (2009). Localizing cingulate subregions-of-interest in magnetic resonance images guided by cytological parcellations. In B. A. Vogt, ed., *Cingulate Neurobiology and Disease. Affective Neuroscience: The Foundations of Human and Animal Emotions*, 803–817. New York: Oxford University Press.

Öngür, D., and Price, J. L. (2000). The organization of networks within the orbital and medial prefrontal cortex of rats, monkeys and humans. *Cerebral Cortex*, *10*(3), 206–219.

Ossandon, T., Vidal, J. R., Ciumas, C., Jerbi, K., Hamame, C. M., Dalal, S. S., et al. (2012). Efficient "pop-out" visual search elicits sustained broadband gamma activity in the dorsal attention network. *Journal of Neuroscience*, *32*(10), 3414–3421. doi:10.1523/jneurosci.6048-11.2012.

Otmakhova, N. A., and Lisman, J. E. (1996). D1/D5 dopamine receptor activation increases the magnitude of early long-term potentiation at CA1 hippocampal synapses. *Journal of Neuroscience*, *16*(23), 7478–7486.

Ouellette, B. G., and Casanova, C. (2006). Overlapping visual response latency distributions in visual cortices and LP-pulvinar complex of the cat. *Experimental Brain Research*, *175*(2), 332–341.

Ousdal, O. T., Jensen, J., Server, A., Hariri, A. R., Nakstad, P. H., and Andreassen, O. A. (2008). The human amygdala is involved in general behavioral relevance detection: Evidence from an event-related functional magnetic resonance imaging Go-NoGo task. *Neuroscience*, *156*(3), 450–455.

Ousdal, O. T., Reckless, G. E., Server, A., Andreassen, O. A., and Jensen, J. (2012). Effect of relevance on amygdala activation and association with the ventral striatum. *NeuroImage*, *62*(1), 95–101.

Oya, H., Kawasaki, H., Howard, M. A., III, and Adolphs, R. (2002). Electrophysiological responses in the human amygdala discriminate emotion categories of complex visual stimuli. *Journal of Neuroscience*, *22*(21), 9502–9512.

Padmala, S., Lim, S.-L., and Pessoa, L. (2010). Pulvinar and affective significance: Responses track moment-to-moment visibility. *Frontiers in Human Neuroscience*, *4*, 1–9.

Padmala, S., and Pessoa, L. (2008). Affective learning enhances visual detection and responses in primary visual cortex. *Journal of Neuroscience*, *28*(24), 6202–6210.

Padmala, S., and Pessoa, L. (2010). Interactions between cognition and motivation during response inhibition. *Neuropsychologia*, *48*(2), 558–565.

Padmala, S., and Pessoa, L. (2011). Reward reduces conflict by enhancing attentional control and biasing visual cortical processing. *Journal of Cognitive Neuroscience*, *23*(11), 3419–3432. doi:10.1162/jocn_a_00011.

Palla, G., Derenyi, I., Farkas, I., and Vicsek, T. (2005). Uncovering the overlapping community structure of complex networks in nature and society. *Nature*, *435*(7043), 814–818.

Panksepp, J. (1998). *Affective Neuroscience: The Foundations of Human and Animal Emotions*. New York: Oxford University Press.

Pantoja, J., Ribeiro, S., Wiest, M., Soares, E., Gervasoni, D., Lemos, N. A., et al. (2007). Neuronal activity in the primary somatosensory thalamocortical loop is modulated by reward contingency during tactile discrimination. *Journal of Neuroscience*, *27*(39), 10608–10620.

Papez, J. W. (1937). A proposed mechanism of emotion. *Archives of Neurology and Psychiatry*, *38*, 725–743.

Parikh, V., Kozak, R., Martinez, V., and Sarter, M. (2007). Prefrontal acetylcholine release controls cue detection on multiple timescales. *Neuron*, *56*(1), 141–154.

Parikh, V., and Sarter, M. (2008). Cholinergic mediation of attention: Contributions of phasic and tonic increases in prefrontal cholinergic activity. *Annals of the New York Academy of Sciences*, *1129*, 225–235.

Park, S., Kim, M. S., and Chun, M. M. (2007). Concurrent working memory load can facilitate selective attention: Evidence for specialized load. *Journal of Experimental Psychology. Human Perception and Performance*, *33*(5), 1062–1075.

Park, S. Q., Kahnt, T., Rieskamp, J., and Heekeren, H. R. (2011). Neurobiology of value integration: When value impacts valuation. *Journal of Neuroscience*, *31*(25), 9307–9314. doi:10.1523/jneurosci.4973-10.2011.

Parvizi, J. (2009). Corticocentric myopia: Old bias in new cognitive sciences. *Trends in Cognitive Sciences*, *13*(8), 354–359. doi:10.1016/j.tics.2009.04.008.

Pashler, H. (1994). Dual-task interference in simple tasks: Data and theory. *Psychological Bulletin*, *116*(2), 220–244.

Pashler, H. (1998). *The Psychology of Attention*. Cambridge, MA: MIT Press.

Passarelli, L., Rosa, M. G., Gamberini, M., Bakola, S., Burman, K. J., Fattori, P., et al. (2011). Cortical connections of area V6Av in the macaque: A visual-input node to the eye/hand coordination system. *Journal of Neuroscience*, *31*(5), 1790–1801. doi:10.1523/jneurosci.4784-10.2011.

Passingham, R. E., Stephan, K. E., and Kotter, R. (2002). The anatomical basis of functional localization in the cortex. *Nature Reviews. Neuroscience*, *3*(8), 606–616.

Passingham, R. E., and Wise, S. P. (2012). *The Neurobiology of the Prefrontal Cortex: Anatomy, Evolution, and the Origin of Insight*. Oxford: Oxford University Press.

Paulus, M. P., and Stein, M. B. (2006). An insular view of anxiety. *Biological Psychiatry*, *60*(4), 383–387.

Pavlov, I. P. (1927). *Conditioned Reflexes: An Investigation of the Physiological Activity of the Cerebral Cortex*. G. V. Anrep, trans. Oxford: Oxford University Press.

Pearce, J. M., and Hall, G. (1980). A model for Pavlovian learning: Variations in the effectiveness of conditioned but not of unconditioned stimuli. *Psychological Review*, *87*(6), 532–552.

Peck, C. J., Jangraw, D. C., Suzuki, M., Efem, R., and Gottlieb, J. (2009). Reward modulates attention independently of action value in posterior parietal cortex. *Journal of Neuroscience*, *29*(36), 11182–11191. doi:10.1523/jneurosci.1929-09.2009.

Peck, C. J., Lau, B., and Salzman, C. D. (2011). The role of the primate amygdala in visual spatial attention. Paper presented at the Society for Neuroscience 2011 Washington, D.C.

Peck, C. J., Lau, B., and Salzman, C. D. (2013). The primate amygdala combines information about space and value. *Nature Neuroscience*, *13*(3), 340–348.

Peelen, M. V., Fei-Fei, L., and Kastner, S. (2009). Neural mechanisms of rapid natural scene categorization in human visual cortex. *Nature*, *460*(7251), 94–97.

Pegna, A. J., Khateb, A., Lazeyras, F., and Seghier, M. L. (2005). Discriminating emotional faces without primary visual cortices involves the right amygdala. *Nature Neuroscience*, *8*(1), 24–25.

Pereira, M. G., de Oliveira, L., Erthal, F. S., Joffily, M., Mocaiber, I. F., Volchan, E., et al. (2010). Emotion affects action: Midcingulate cortex as a pivotal node of interaction between negative emotion and motor signals. *Cognitive, Affective and Behavioral Neuroscience*, *10*(1), 94–106. doi:10.3758/CABN.10.1.94.

Pereira, M. G., Volchan, E., de Souza, G. G., Oliveira, L., Campagnoli, R. R., Pinheiro, W. M., et al. (2006). Sustained and transient modulation of performance induced by emotional picture viewing. *Emotion*, *6*(4), 622–634.

Perlstein, W. M., Elbert, T., and Stenger, V. A. (2002). Dissociation in human prefrontal cortex of affective influences on working memory-related activity. *Proceedings of the National Academy of Sciences of the United States of America*, *99*(3), 1736–1741.

Pessoa, L. (2005). To what extent are emotional visual stimuli processed without attention and awareness? *Current Opinion in Neurobiology*, *15*(2), 188–196.

Pessoa, L. (2008). On the relationship between emotion and cognition. *Nature Reviews Neuroscience*, *9*(2), 148–158.

Pessoa, L. (2009). How do emotion and motivation direct executive function? *Trends in Cognitive Sciences*, *13*(4), 160–166.

Pessoa, L. (2010a). Emergent processes in cognitive-emotional interactions. *Dialogues in Clinical Neuroscience*, *12*(4), 433–448.

Pessoa, L. (2010b). Emotion and cognition and the amygdala: From "what is it?" to "what's to be done?" *Neuropsychologia*, *48*(12), 3416–3429. doi:10.1016/j.neuropsychologia.2010.06.038.

Pessoa, L. (2010c). Emotion and attention effects: Is it all a matter of timing? Not yet. *Frontiers in Human Neuroscience, 4*.

Pessoa, L., and Adolphs, R. (2010). Emotion processing and the amygdala: From a "low road" to "many roads" of evaluating biological significance. *Nature Reviews. Neuroscience*, *11*(11), 773–783. doi:10.1038/nrn2920.

Pessoa, L., and Adolphs, R. (in preparation). Subcortical affective processing: A critical analysis and re-evaluation.

Pessoa, L., and Engelmann, J. B. (2010a). Embedding reward signals into perception and cognition. *Frontiers in Neuroscience, 4*: 17. doi:10.3389/fnins.2010.00017.

Pessoa, L., Gutierrez, E., Bandettini, P. B., and Ungerleider, L. G. (2002). Neural correlates of visual working memory: fMRI amplitude predicts task performance. *Neuron*, *35*, 975–987.

Pessoa, L., Japee, S., Sturman, D., and Ungerleider, L. G. (2006). Target visibility and visual awareness modulate amygdala responses to fearful faces. *Cerebral Cortex*, *16*(3), 366–375.

Pessoa, L., Japee, S., and Ungerleider, L. G. (2005). Visual awareness and the detection of fearful faces. *Emotion*, *5*(2), 243–247.

Pessoa, L., Kastner, S., and Ungerleider, L. G. (2002). Attentional control of the processing of neutral and emotional stimuli. *Brain Research. Cognitive Brain Research*, *15*(1), 31–45.

Pessoa, L., McKenna, M., Gutierrez, E., and Ungerleider, L. G. (2002). Neural processing of emotional faces requires attention. *Proceedings of the National Academy of Sciences of the United States of America*, *99*(17), 11458–11463.

Pessoa, L., and Padmala, S. (2007). Decoding near-threshold perception of fear from distributed single-trial brain activation. *Cerebral Cortex*, *17*(3), 691–701.

Pessoa, L., Padmala, S., Kenzer, A., and Bauer, A. (2012). Interactions between cognition and emotion during response inhibition. *Emotion*, *12*(1), 192–197. doi:10.1037/a0024109.

Pessoa, L., Padmala, S., and Morland, T. (2005). Fate of unattended fearful faces in the amygdala is determined by both attentional resources and cognitive modulation. *NeuroImage*, *28*(1), 249–255.

Pessoa, L., and Ungerleider, L. G. (2004a). Neural correlates of change detection and change blindness in a working memory task. *Cerebral Cortex*, *14*(4), 511–520.

Pessoa, L., and Ungerleider, L. G. (2004b). Neuroimaging studies of attention and the processing of emotion-laden stimuli. *Progress in Brain Research*, *144*, 171–182.

Pessoa, L., and Ungerleider, L. G. (2004c). Top-down mechanisms for working memory and attentional processes. In M. S. Gazzaniga, ed., *The New Cognitive Neurosciences* (3rd ed.), 919–930). Cambridge, MA: MIT Press.

Petersen, S. E., Robinson, D. L., and Keys, W. (1985). Pulvinar nuclei of the behaving rhesus monkey: Visual responses and their modulation. *Journal of Neurophysiology*, *54*(4), 867–886.

Petrovich, G. D., Canteras, N. S., and Swanson, L. W. (2001). Combinatorial amygdalar inputs to hippocampal domains and hypothalamic behavior systems. *Brain Research. Brain Research Reviews*, *38*(1–2), 247–289.

Phan, K. L., Fitzgerald, D. A., Nathan, P. J., Moore, G. J., Uhde, T. W., and Tancer, M. E. (2005). Neural substrates for voluntary suppression of negative affect: A functional magnetic resonance imaging study. *Biological Psychiatry*, *57*(3), 210–219. doi:10.1016/j.biopsych.2004.10.030.

Phelps, E. A. (2006). Emotion and cognition: Insights from studies of the human amygdala. *Annual Review of Psychology*, *57*, 27–53.

Pichon, S., de Gelder, B., and Grezes, J. (2012). Threat prompts defensive brain responses independently of attentional control. *Cerebral Cortex*, *22*(2), 274–285. doi:10.1093/cercor/bhr060.

Pickens, C. L., Saddoris, M. P., Setlow, B., Gallagher, M., Holland, P. C., and Schoenbaum, G. (2003). Different roles for orbitofrontal cortex and basolateral amygdala in a reinforcer devaluation task. *Journal of Neuroscience*, *23*(35), 11078–11084.

Piech, R. M., McHugo, M., Smith, S. D., Dukic, M. S., Van Der Meer, J., Abou-Khali, B., et al. (2010). Fear-enhanced visual search persists after amygdala lesions. *Neuropsychologia*, *48*(12), 3430–3435 doi:10.1016/j.neuropsychologia.2010.07.09.

Piech, R. M., McHugo, M., Smith, S. D., Dukic, M. S., Van Der Meer, J., Abou-Khalil, B., et al. (2011). Attentional capture by emotional stimuli is preserved in patients with amygdala lesions. *Neuropsychologia*, *49*(12), 3314–3319. doi:10.1016/j.neuropsychologia.2011.08.004.

Pittel, B., Spencer, J., and Wormald, N. (1996). Sudden emergence of a giant k-core in a random graph. *Journal of Combinatorial Theory Series B*, *67*(1), 111–151.

Pizzagalli, D., Regard, M., and Lehmann, D. (1999). Rapid emotional face processing in the human right and left brain hemispheres: An ERP study. *Neuroreport*, *10*(13), 2691–2698.

Platt, M. L., and Glimcher, P. W. (1999). Neural correlates of decision variables in parietal cortex. *Nature*, *400*(6741), 233–238.

Platt, M. L., and Huettel, S. A. (2008). Risky business: The neuroeconomics of decision making under uncertainty. *Nature Neuroscience*, *11*(4), 398–403. doi:10.1038/nn2062.

Pleger, B., Blankenburg, F., Ruff, C. C., Driver, J., and Dolan, R. J. (2008). Reward facilitates tactile judgments and modulates hemodynamic responses in human primary somatosensory cortex. *Journal of Neuroscience*, *28*(33), 8161–8168. doi:10.1523/jneurosci.1093-08.2008.

Pleger, B., Ruff, C. C., Blankenburg, F., Kloppel, S., Driver, J., and Dolan, R. J. (2009). Influence of dopaminergically mediated reward on somatosensory decision-making. *PLoS Biology*, *7*(7), e1000164. doi:10.1371/journal.pbio.1000164.

Pochon, J.-B., Levy, R., Fossati, P., Lehericy, S., Poline, J. B., Pillon, B., et al. (2002). The neural system that bridges reward and cognition in humans: An fMRI study. *Proceedings of the National Academy of Sciences of the United States of America*, *99*(8), 5669–5674.

Poggio, T., Fahle, M., and Edelman, S. (1992). Fast perceptual learning in visual hyperacuity. *Science*, *256*(5059), 1018–1021. doi:10.1126/science.1589770.

Poldrack, R. A. (2006). Can cognitive processes be inferred from neuroimaging data? *Trends in Cognitive Sciences*, *10*(2), 59–63. doi:10.1016/j.tics.2005.12.004.

Poldrack, R. A. (2011). Inferring mental states from neuroimaging data: From reverse inference to large-scale decoding. *Neuron*, *72*(5), 692–697. doi:10.1016/j.neuron.2011.11.001.

Polk, T. A., Drake, R. M., Jonides, J. J., Smith, M. R., and Smith, E. E. (2008). Attention enhances the neural processing of relevant features and suppresses the processing of irrelevant features in humans: A functional magnetic resonance imaging study of the Stroop task. *Journal of Neuroscience*, *28*(51), 13786–13792.

Ponce, C. R., Hunter, J. N., Pack, C. C., Lomber, S. G., and Born, R. T. (2011). Contributions of indirect pathways to visual response properties in macaque middle temporal area MT. *Journal of Neuroscience*, *31*(10), 3894–3903. doi:10.1523/jneurosci.5362-10.2011.

Ponce, C. R., Lomber, S. G., and Born, R. T. (2008). Integrating motion and depth via parallel pathways. *Nature Neuroscience*, *11*(2), 216–223. doi:10.1038/nn2039.

Poppel, E., Held, R., and Frost, D. (1973). Letter: Residual visual function after brain wounds involving the central visual pathways in man. *Nature*, *243*(5405), 295–296.

Postuma, R. B., and Dagher, A. (2006). Basal ganglia functional connectivity based on a meta-analysis of 126 positron emission tomography and functional magnetic resonance imaging publications. *Cerebral Cortex*, *16*(10), 1508–1521.

Potter, M. C., and Levy, E. I. (1969). Recognition memory for a rapid sequence of pictures. *Journal of Experimental Psychology*, *81*(1), 10–15.

Pourtois, G., Schettino, A., and Vuilleumier, P. (2012). Brain mechanisms for emotional influences on perception and attention: What is magic and what is not. *Biological Psychology*. doi:10.1016/j.biopsycho.2012.02.007.

Pourtois, G., Spinelli, L., Seeck, M., and Vuilleumier, P. (2010). Temporal precedence of emotion over attention modulations in the lateral amygdala: Intracranial ERP evidence from a patient with temporal lobe epilepsy. *Cognitive, Affective and Behavioral Neuroscience*, *10*(1), 83–93.

Preuss, T. M. (1995). Do rats have prefrontal cortex? The Rose-Woolsey-Akert program reconsidered. *Journal of Cognitive Neuroscience*, *7*(1), 1–24.

Preuss, T. M. (2007). Evolutionary specializations of primate brain systems. In M. J. Ravosa and M. Dagosta, eds., *Primate Origins: Evolution and Adaptations*, 625–675. New York: Springer.

Pribram, K. H. (1967a). The limbic systems, efferent control of neural inhibition and behavior. *Progress in Brain Research*, *27*, 318–336.

Pribram, K. H. (1967b). The new neurology and the biology of emotion: A structural approach. *American Psychologist*, *22*(10), 830–838.

Pribram, K. H., and McGuinness, D. (1975). Arousal, activation, and effort in the control of attention. *Psychological Review*, *82*(2), 116–149.

Price, C. J., and Friston, K. J. (2002). Degeneracy and cognitive anatomy. *Trends in Cognitive Sciences*, *6*(10), 416–421.

Price, C. J., and Friston, K. J. (2005). Functional ontologies for cognition: The systematic definition of structure and function. *Cognitive Neuropsychology*, *22*(3/4), 262–275.

Price, J. L. (2006). Architectonic structure of the orbital and medial prefrontal cortex. In D. H. Zald and S. L. Rauch, eds., *The Orbitofrontal Cortex*, 3–17. Oxford: Oxford University Press.

Qin, S., Hermans, E. J., van Marle, H. J., Luo, J., and Fernandez, G. (2009). Acute psychological stress reduces working memory-related activity in the dorsolateral prefrontal cortex. *Biological Psychiatry*, *66*(1), 25–32. doi:10.1016/j.biopsych.2009.03.006.

Raghanti, M. A., Stimpson, C. D., Marcinkiewicz, J. L., Erwin, J. M., Hof, P. R., and Sherwood, C. C. (2008). Cholinergic innervation of the frontal cortex: Differences among humans, chimpanzees, and macaque monkeys. *Journal of Comparative Neurology*, *506*(3), 409–424. doi:10.1002/cne.21546.

Ramachandran, V. S. (2012). *Edge, 2012: What Is Your Favorite Deep, Elegant, or Beautiful Explanation?* Genes, claustrum, and consciousness. Retrieved from http://edge.org/response-detail/2945/what-is-your-favorite-deep-elegant-or-beautiful-explanation

Ramsey, J. D., Hanson, S. J., Hanson, C., Halchenko, Y. O., Poldrack, R. A., and Glymour, C. (2010). Six problems for causal inference from fMRI. *NeuroImage*, *49*(2), 1545–1558. doi:10.1016/j.neuroimage.2009.08.065.

Raymond, J. E., and O'Brien, J. L. (2009). Selective visual attention and motivation: The consequences of value learning in an attentional blink task. *Psychological Science*, *20*(8), 981–988. doi:10.1111/j.1467-9280.2009.02391.x.

Raymond, J. E., Shapiro, K. L., and Arnell, K. M. (1992). Temporary suppression of visual processing in an RSVP task: An attentional blink? *Journal of Experimental Psychology. Human Perception and Performance*, *18*(3), 849–860.

Redgrave, P., and Gurney, K. (2006). The short-latency dopamine signal: A role in discovering novel actions? *Nature Reviews. Neuroscience*, *7*(12), 967–975.

Redgrave, P., Prescott, T. J., and Gurney, K. (1999). Is the short-latency dopamine response too short to signal reward error? *Trends in Neurosciences*, *22*(4), 146–151.

Rempel-Clower, N. L., and Barbas, H. (1998). Topographic organization of connections between the hypothalamus and prefrontal cortex in the rhesus monkey. *Journal of Comparative Neurology*, *398*(3), 393–419.

Rempel-Clower, N. L., and Barbas, H. (2000). The laminar pattern of connections between prefrontal and anterior temporal cortices in the rhesus monkey is related to cortical structure and function. *Cerebral Cortex*, *10*(9), 851–865.

Rescorla, R. A., and Wagner, A. R. (1972). A theory of Pavlovian conditioning: Variations in the effectiveness of reinforcement and non-reinforcement. In A. H. Black and W. F. Prokasy, eds., *Classical Conditioning: II. Current Research and Theory*, 64–99. New York: Appleton Century Crofts.

Reynolds, J. H., Chelazzi, L., and Desimone, R. (1999). Competitive mechanisms subserve attention in macaque areas V2 and V4. *Journal of Neuroscience*, *19*, 1736–1753.

Reynolds, J. H., and Heeger, D. J. (2009). The normalization model of attention. *Neuron*, *61*(2), 168–185. doi:10.1016/j.neuron.2009.01.002.

Richards, A., French, C. C., Keogh, E., and Carter, C. (2000). Test-anxiety, inferential reasoning and working memory load. *Anxiety, Stress, and Coping*, *13*(1), 87–109.

Richardson, F. M., Seghier, M. L., Leff, A. P., Thomas, M. S., and Price, C. J. (2011). Multiple routes from occipital to temporal cortices during reading. *Journal of Neuroscience*, *31*(22), 8239–8247. doi:10.1523/jneurosci.6519-10.2011.

Ridderinkhof, K. R., Ullsperger, M., Crone, E. A., and Nieuwenhuis, S. (2004). The role of the medial frontal cortex in cognitive control. *Science*, *306*(5695), 443–447. doi:10.1126/science.1100301.

Risold, P. Y., Thompson, R. H., and Swanson, L. W. (1997). The structural organization of connections between hypothalamus and cerebral cortex. *Brain Research. Brain Research Reviews*, *24*(2–3), 197–254.

Rissman, J., Gazzaley, A., and D'Esposito, M. (2004). Measuring functional connectivity during distinct stages of a cognitive task. *NeuroImage*, *23*(2), 752–763.

Robbins, T. W., and Everitt, B. J. (1992). Functions of dopamine in the dorsal and ventral striatum. *Seminars in Neuroscience*, *4*(2), 119–127. doi:10.101/1044-5765(92)90010-Y.

Robbins, T. W., and Everitt, B. J. (1996). Neurobehavioural mechanisms of reward and motivation. *Current Opinion in Neurobiology*, *6*(2), 228–236.

Robbins, T. W., and Everitt, B. J. (2007). A role for mesencephalic dopamine in activation: Commentary on Berridge (2006). *Psychopharmacology*, *191*(3), 433–437. doi:10.1007/s00213-006-0528-7.

Robinson, D. L., and Cowie, R. J. (1997). The primate pulvinar: Structural, functional, and behavioral components of visual salience. In D. McCormick, ed., *The Thalamus*, 53–92. New York: Elsevier.

Robinson, O. J., Letkiewicz, A. M., Overstreet, C., Ernst, M., and Grillon, C. (2011). The effect of induced anxiety on cognition: Threat of shock enhances aversive processing in healthy individuals. *Cognitive, Affective and Behavioral Neuroscience*, *11*(2), 217–227. doi:10.3758/s13415-011-0030-5.

Roesch, M. R., Calu, D. J., Esber, G. R., and Schoenbaum, G. (2010). Neural correlates of variations in event processing during learning in basolateral amygdala. *Journal of Neuroscience*, *30*(7), 2464–2471. doi:10.1523/jneurosci.5781-09.2010.

Roland, P. E. (2010). Six principles of visual cortical dynamics. *Frontiers in Systems Neuroscience*, *4*, 28. doi:10.3389/fnsys.2010.00028.

Rolls, E. T. (2005). *Emotion Explained*. Oxford: Oxford University Press.

Romanski, L. M., Giguere, M., Bates, J. F., and Goldman-Rakic, P. S. (1997). Topographic organization of medial pulvinar connections with the prefrontal cortex in the rhesus monkey. *Journal of Comparative Neurology*, *379*(3), 313–332.

Romanski, L. M., and LeDoux, J. E. (1992). Equipotentiality of thalamo-amygdala and thalamo-cortico-amygdala circuits in auditory fear conditioning. *Journal of Neuroscience*, *12*(11), 4501–4509.

Rosen, B. R., Buckner, R. L., and Dale, A. M. (1998). Event-related functional MRI: Past, present, and future. *Proceedings of the National Academy of Sciences of the United States of America*, *95*(3), 773–780.

Rosenberg, D. S., Mauguière, F., Catenoix, H., Faillenot, I., and Magnin, M. (2009). Reciprocal thalamocortical connectivity of the medial pulvinar: A depth stimulation and evoked potential study in human brain. *Cerebral Cortex*, *19*(6), 1462–1473.

Rotshtein, P., Richardson, M. P., Winston, J. S., Kiebel, S. J., Vuilleumier, P., Eimer, M., et al. (2010). Amygdala damage affects event-related potentials for fearful faces at specific time windows. *Human Brain Mapping*, *31*(7), 1089–1105.

Roy, A. K., Shehzad, Z., Margulies, D. S., Kelly, A. M., Uddin, L. Q., Gotimer, K., et al. (2009). Functional connectivity of the human amygdala using resting state fMRI. *NeuroImage*, *45*(2), 614–626. doi:10.1016/j.neuroimage.2008.11.030.

Rubia, K., Smith, A. B., Brammer, M. J., and Taylor, E. (2003). Right inferior prefrontal cortex mediates response inhibition while mesial prefrontal cortex is responsible for error detection. *NeuroImage*, *20*(1), 351–358.

Rubinov, M., and Sporns, O. (2010). Complex network measures of brain connectivity: Uses and interpretations. *NeuroImage*, *52*(3), 1059–1069. doi:10.1016/j.neuroimage.2009.10.003.

Rudrauf, D., David, O., Lachaux, J. P., Kovach, C. K., Martinerie, J., Renault, B., et al. (2008). Rapid interactions between the ventral visual stream and emotion-related structures rely on a two-pathway architecture. *Journal of Neuroscience*, *28*(11), 2793–2803.

Rumelhart, D. E., and McClelland, J. L. (1986). *Parallel Distributed Processing: Explorations in the Microstructure of Cognition* (Vol. 1). Cambridge, MA: MIT Press.

Rushworth, M. F., and Behrens, T. E. (2008). Choice, uncertainty and value in prefrontal and cingulate cortex. *Nature Neuroscience*, *11*(4), 389–397. doi:10.1038/nn2066.

Rutishauser, U., Tudusciuc, O., Neumann, D., Mamelak, A. N., Heller, A. C., Ross, I. B., et al. (2011). Single-unit responses selective for whole faces in the human amygdala. *Current Biology*, *21*(19), 1654–1660. doi:10.1016/j.cub.2011.08.035.

Saalmann, Y. B., Pinsk, M. A., Wang, L., Li, X., and Kastner, S. (2012). The pulvinar regulates information transmission between cortical areas based on attention demands. *Science*, *337*(6095), 753–756.

Sadaghiani, S., Hesselmann, G., Friston, K. J., and Kleinschmidt, A. (2010). The relation of ongoing brain activity, evoked neural responses, and cognition. *Frontiers in System Neuroscience*, *4*, 20.

Sagan, C. (1977). *The Dragons of Eden: Speculations on the Evolution of Human Intelligence*. New York: Ballantine.

Sah, P., Faber, E. S. L., De Armentia, M. L., and Power, J. (2003). The amygdaloid complex: Anatomy and physiology. *Physiological Reviews*, *83*(3), 803–834.

Salamone, J. D. (1994). The involvement of nucleus accumbens dopamine in appetitive and aversive motivation. *Behavioural Brain Research*, *61*(2), 117–133.

Salamone, J. D., Correa, M., Farrar, A. M., Nunes, E. J., and Pardo, M. (2009). Dopamine, behavioral economics, and effort. *Frontiers in Behavioral Neuroscience*, *3*, 13. doi:10.3389/neuro.08.013.2009.

Saleem, K. S., Kondo, H., and Price, J. L. (2008). Complementary circuits connecting the orbital and medial prefrontal networks with the temporal, insular, and opercular cortex in the macaque monkey. *Journal of Comparative Neurology*, *506*(4), 659–693. doi:10.1002/cne.21577.

Sallet, J., Quilodran, R., Rothe, M., Vezoli, J., Joseph, J. P., and Procyk, E. (2007). Expectations, gains, and losses in the anterior cingulate cortex. *Cognitive, Affective and Behavioral Neuroscience*, *7*(4), 327–336.

Salvador, R., Suckling, J., Coleman, M. R., Pickard, J. D., Menon, D., and Bullmore, E. (2005). Neurophysiological architecture of functional magnetic resonance images of human brain. *Cerebral Cortex*, *15*(9), 1332–1342.

Salzman, C. D., and Fusi, S. (2010). Emotion, cognition, and mental state representation in amygdala and prefrontal cortex. *Annual Review of Neuroscience*, *33*, 173–202.

Salzman, C. D., Paton, J. J., Belova, M. A., and Morrison, S. E. (2007). Flexible neural representations of value in the primate brain. *Annals of the New York Academy of Sciences*, *1121*, 336–354.

Samanez-Larkin, G. R., Gibbs, S. E., Khanna, K., Nielsen, L., Carstensen, L. L., and Knutson, B. (2007). Anticipation of monetary gain but not loss in healthy older adults. *Nature Neuroscience*, *10*(6), 787–791. doi:10.1038/nn1894.

Sander, D., Grafman, J., and Zalla, T. (2003). The human amygdala: An evolved system for relevance detection. *Reviews in the Neurosciences*, *14*(4), 303–316.

Sanger, J., and Wascher, E. (2011). The influence of extrinsic motivation on competition-based selection. *Behavioural Brain Research*, *224*(1), 58–64. doi:10.1016/j.bbr.2011.05.015.

Santos, A., Mier, D., Kirsch, P., and Meyer-Lindenberg, A. (2011). Evidence for a general face salience signal in human amygdala. *NeuroImage*, *54*(4), 3111–3116. doi:10.1016/j.neuroimage.2010.11.024.

Saper, C. B., and Loewy, A. D. (1980). Efferent connections of the parabrachial nucleus in the rat. *Brain Research*, *197*(2), 291–317.

Sarter, M., Berntson, G. G., and Cacioppo, J. T. (1996). Brain imaging and cognitive neuroscience. Toward strong inference in attributing function to structure. *American Psychologist*, *51*(1), 13–21.

Sarter, M., and Bruno, J. P. (1999). Abnormal regulation of corticopetal cholinergic neurons and impaired information processing in neuropsychiatric disorders. *Trends in Neurosciences*, *22*(2), 67–74.

Sarter, M., and Bruno, J. P. (2000). Cortical cholinergic inputs mediating arousal, attentional processing and dreaming: Differential afferent regulation of the basal forebrain by telencephalic and brainstem afferents. *Neuroscience*, *95*(4), 933–952.

Sarter, M., Bruno, J. P., and Turchi, J. (1999). Basal forebrain afferent projections modulating cortical acetylcholine, attention, and implications for neuropsychiatric disorders. *Annals of the New York Academy of Sciences*, *877*, 368–382.

Sarter, M., Gehring, W. J., and Kozak, R. (2006). More attention must be paid: The neurobiology of attentional effort. *Brain Research. Brain Research Reviews*, *51*(2), 145–160. doi:10.1016/j.brainresrev.2005.11.002.

Sarter, M., Hasselmo, M. E., Bruno, J. P., and Givens, B. (2005). Unraveling the attentional functions of cortical cholinergic inputs: Interactions between signal-driven and cognitive modulation of signal detection. *Brain Research. Brain Research Reviews*, *48*(1), 98–111. doi:10.1016/j.brainresrev.2004.08.006.

Sasaki, Y., Náñez, J. E., and Watanabe, T. (2010). Advances in visual perceptual learning and plasticity. *Nature Reviews. Neuroscience*, *11*(1), 53–60. doi:10.1038/nrn2737.

Sato, H., Hata, Y., Masui, H., and Tsumoto, T. (1987). A functional role of cholinergic innervation to neurons in the cat visual cortex. *Journal of Neurophysiology*, *58*(4), 765–780.

Savoy, R. L., Bandettini, P. A., O'Craven, K. M., Kwong, K. K., Davis, T. L., Baker, J. R., Weisskopf, R. M., and Rosen, B. R. (1995). Pushing the temporal resolution of fMRI: Studies of very brief visual stimuli, onset variability and asynchrony, and stimulus-correlated changes in noise. Paper presented at the 15th Annual Meeting of the Society of Magnetic Resonance in Medicine.

Sawaguchi, T., and Matsumura, M. (1985). Laminar distributions of neurons sensitive to acetylcholine, noradrenaline and dopamine in the dorsolateral prefrontal cortex of the monkey. *Neuroscience Research*, *2*(4), 255–273.

Saygin, Z. M., Osher, D. E., Augustinack, J., Fischl, B., and Gabrieli, J. D. (2011). Connectivity-based segmentation of human amygdala nuclei using probabilistic tractography. *NeuroImage*, *56*(3), 1353–1361. doi:10.1016/j.neuroimage.2011.03.006.

Schaefer, A., Braver, T. S., Reynolds, J. R., Burgess, G. C., Yarkoni, T., and Gray, J. R. (2006). Individual differences in amygdala activity predict response speed during working memory. *Journal of Neuroscience*, *26*(40), 10120–10128.

Schenberg, L. C., Póvoa, R. M. F., Costa, A. L. P., Caldellas, A. V., Tufik, S., and Bittencourt, A. S. (2005). Functional specializations within the tectum defense systems of the rat. *Neuroscience and Biobehavioral Reviews*, *29*(8), 1279–1298.

Schiller, P. H., and Koerner, F. (1971). Discharge characteristics of single units in superior colliculus of the alert rhesus monkey. *Journal of Neurophysiology*, *34*(5), 920–936.

Schmid, M. C., Mrowka, S. W., Turchi, J., Saunders, R. C., Wilke, M., Peters, A. J., et al. (2010). Blindsight depends on the lateral geniculate nucleus. *Nature*, *466*(7304), 373–377.

Schmid, M. C., Panagiotaropoulos, T., Augath, M. A., Logothetis, N. K., and Smirnakis, S. M. (2009). Visually driven activation in macaque areas V2 and V3 without input from the primary visual cortex. *PLoS ONE*, *4*(5), e5527.

Schmidt, L., Lebreton, M., Clery-Melin, M. L., Daunizeau, J., and Pessiglione, M. (2012). Neural mechanisms underlying motivation of mental versus physical effort. *PLoS Biology*, *10*(2), e1001266. doi:10.1371/journal.pbio.1001266.

Schmolesky, M. T., Wang, Y., Hanes, D. P., Thompson, K. G., Leutgeb, S., Schall, J. D., et al. (1998). Signal timing across the macaque visual sysschneider tem. *Journal of Neurophysiology*, *79*(6), 3272–3278.

Schneider, W., Dumais, S. T., and Shiffrin, R. (1984). Autonomic and control processing and attention. In R. S. Parasuraman and D. R. Davies, eds., *Varieties of Attention*, 1–27. New York: Academic Press.

Schultz, W. (2000). Multiple reward signals in the brain. *Nature Reviews. Neuroscience*, *1*(3), 199–207.

Schultz, W. (2002). Getting formal with dopamine and reward. *Neuron*, *36*(2), 241–263.

Schultz, W. (2006). Behavioral theories and the neurophysiology of reward. *Annual Review of Psychology*, *57*, 87–115. doi:10.1146/annurev.psych.56.091103.070229.

Schultz, W., Apicella, P., Scarnati, E., and Ljungberg, T. (1992). Neuronal activity in monkey ventral striatum related to the expectation of reward. *Journal of Neuroscience*, *12*(12), 4595–4610.

Schumann, C. M., Bauman, M. D., and Amaral, D. G. (2011). Abnormal structure or function of the amygdala is a common component of neurodevelopmental disorders. *Neuropsychologia*, *49*(4), 745–759. doi:10.1016/j.neuropsychologia.2010.09.028.

Schupp, H. T., Junghofer, M., Weike, A. I., and Hamm, A. O. (2003a). Attention and emotion: An ERP analysis of facilitated emotional stimulus processing. *Neuroreport*, *14*(8), 1107–1110.

Schupp, H. T., Junghofer, M., Weike, A. I., and Hamm, A. O. (2003b). Emotional facilitation of sensory processing in the visual cortex. *Psychological Science*, *14*(1), 7–13.

Schupp, H. T., Stockburger, J., Bublatzky, F., Junghofer, M., Weike, A. I., and Hamm, A. O. (2007). Explicit attention interferes with selective emotion processing in human extrastriate cortex. *BMC Neuroscience*, *8*, 16.

Schwabe, L., Merz, C. J., Walter, B., Vaitl, D., Wolf, O. T., and Stark, R. (2011). Emotional modulation of the attentional blink: The neural structures involved in capturing and holding attention. *Neuropsychologia*, *49*(3), 416–425. doi:10.1016/j.neuropsychologia.2010.12.037.

Semba, K. (2000). Multiple output pathways of the basal forebrain: Organization, chemical heterogeneity, and roles in vigilance. *Behavioural Brain Research*, *115*(2), 117–141.

Serences, J. T. (2008). Value-based modulations in human visual cortex. *Neuron*, *60*(6), 1169–1181. doi:10.1016/j.neuron.2008.10.051.

Serences, J. T., and Yantis, S. (2006). Selective visual attention and perceptual coherence. *Trends in Cognitive Sciences*, *10*(1), 38–45.

Serre, T., Kreiman, G., Kouh, M., Cadieu, C., Knoblich, U., and Poggio, T. (2007). A quantitative theory of immediate visual recognition. *Progress in Brain Research*, *165*, 33–56.

Serre, T., Oliva, A., and Poggio, T. (2007). A feedforward architecture accounts for rapid categorization. *Proceedings of the National Academy of Sciences of the United States of America*, *104*(15), 6424–6429.

Seth, A. K., Dienes, Z., Cleeremans, A., Overgaard, M., and Pessoa, L. (2008). Measuring consciousness: Relating behavioural and neurophysiological approaches. *Trends in Cognitive Sciences*, *12*(8), 314–321

Seymour, B., and Dolan, R. (2008). Emotion, decision making, and the amygdala. *Neuron*, *58*(5), 662–671.

Seymour, B., Singer, T., and Dolan, R. (2007). The neurobiology of punishment. *Nature Reviews. Neuroscience*, *8*(4), 300–311.

Shackman, A. J., Salomons, T. V., Slagter, H. A., Fox, A. S., Winter, J. J., and Davidson, R. J. (2011). The integration of negative affect, pain and cognitive control in the cingulate cortex. *Nature Reviews. Neuroscience*, *12*(3), 154–167. doi:10.1038/nrn2994.

Shackman, A. J., Sarinopoulos, I., Maxwell, J. S., Pizzagalli, D. A., Lavric, A., and Davidson, R. J. (2006). Anxiety selectively disrupts visuospatial working memory. *Emotion*, *6*(1), 40–61. doi:10.1037/1528-3542.6.1.40.

Shafer, A. T., Matveychuk, D., Penney, T., O'Hare, A. J., Stokes, J., and Dolcos, F. (2012). Processing of emotional distraction is both automatic and modulated by attention: Evidence from an event-related fMRI investigation. *Journal of Cognitive Neuroscience*, *24*(5), 1233–1252. doi:10.1162/jocn_a_00206.

Shallice, T. (1988). *From Neuropsychology to Mental Structure*. New York: Cambridge University Press.

Shallice, T., and Cooper, R. P. (2011). *The Organisation of Mind*. Oxford: Oxford University Press.

Shannon, C. E. (1948). A mathematical theory of communication. *The Bell System Technical Journal, 27*, 379–423 and 623–656.

Shen, Y. J., and Chun, M. M. (2011). Increases in rewards promote flexible behavior. *Attention, Perception and Psychophysics*, *73*(3), 938–952. doi:10.3758/s13414-010-0065-7.

Sherman, S. M. (2007). The thalamus is more than just a relay. *Current Opinion in Neurobiology*, *17*(4), 417–422. doi:10.1016/j.conb.2007.07.003.

Sherman, S. M., and Guillery, R. W. (1996). Functional organization of thalamocortical relays. *Journal of Neurophysiology*, *76*(3), 1367–1395.

Sherman, S. M., and Guillery, R. W. (1998). On the actions that one nerve cell can have on another: Distinguishing "drivers" from "modulators." *Proceedings of the National Academy of Sciences of the United States of America*, *95*(12), 7121–7126.

Shi, C., and Davis, M. (2001). Visual pathways involved in fear conditioning measured with fear-potentiated startle: Behavioral and anatomic studies. *Journal of Neuroscience*, *21*(24), 9844–9855.

Shidara, M., and Richmond, B. J. (2002). Anterior cingulate: Single neuronal signals related to degree of reward expectancy. *Science*, *296*(5573), 1709–1711. doi:10.1126/science.1069504.

Shiffrin, R. M. (1988). Attention. In R. C. Atkinson, R. J. Hernstein, G. Lindzey, and R. D. Luce, eds., *Stevens' Handbook of Experimental Psychology*, 739–811. New York: Wiley.

Shiffrin, R. M., and Schneider, W. (1977). Controlled and automatic human information processing: II. Perceptual learning, automatic attending and a general theory. *Psychological Review*, *84*(2), 127–190.

Shima, K., and Tanji, J. (1998). Role for cingulate motor area cells in voluntary movement selection based on reward. *Science*, *282*(5392), 1335–1338.

Shipp, S. (2003). The functional logic of cortico-pulvinar connections. *Philosophical Transactions of the Royal Society of London. Series B, Biological Sciences*, *358*(1438), 1605–1624.

Shipp, S. (2004). The brain circuitry of attention. *Trends in Cognitive Sciences*, *8*(5), 223–230.

Shohamy, D., and Adcock, R. A. (2010). Dopamine and adaptive memory. *Trends in Cognitive Sciences*, *14*(10), 464–472. doi:10.1016/j.tics.2010.08.002.

Shuler, M. G., and Bear, M. F. (2006). Reward timing in the primary visual cortex. *Science*, *311*(5767), 1606–1609.

Sierra, M., Lopera, F., Lambert, M. V., Phillips, M. L., and David, A. S. (2002). Separating depersonalisation and derealisation: The relevance of the "lesion method." *Journal of Neurology, Neurosurgery, and Psychiatry*, *72*(4), 530–532.

Sillito, A. M., and Kemp, J. A. (1983). Cholinergic modulation of the functional organization of the cat visual cortex. *Brain Research*, *289*(1–2), 143–155.

Silvert, L., Lepsien, J., Fragopanagos, N., Goolsby, B., Kiss, M., Taylor, J. G., et al. (2007). Influence of attentional demands on the processing of emotional facial expressions in the amygdala. *NeuroImage*, *38*(2), 357–366.

Simmons, A., Matthews, S. C., Stein, M. B., and Paulus, M. P. (2004). Anticipation of emotionally aversive visual stimuli activates right insula. *Neuroreport*, *15*(14), 2261–2265.

Simmons, A., Strigo, I., Matthews, S. C., Paulus, M. P., and Stein, M. B. (2006). Anticipation of aversive visual stimuli is associated with increased insula activation in anxiety-prone subjects. *Biological Psychiatry*, *60*(4), 402–409. doi:10.1016/j.biopsych.2006.04.038.

Simon, H. (1969). *The Sciences of the Artificial*. Cambridge, MA: MIT Press.

Simpson, J. R., Ongur, D., Akbudak, E., Conturo, T. E., Ollinger, J. M., Snyder, A. Z., et al. (2000). The emotional modulation of cognitive processing: An fMRI study. *Journal of Cognitive Neuroscience*, *12*(Suppl 2), 157–170.

Sincich, L. C., Park, K. F., Wohlgemuth, M. J., and Horton, J. C. (2004). Bypassing V1: A direct geniculate input to area MT. *Nature Neuroscience*, *7*(10), 1123–1128.

Singer, T., Critchley, H. D., and Preuschoff, K. (2009). A common role of insula in feelings, empathy and uncertainty. *Trends in Cognitive Sciences*, *13*(8), 334–340.

Small, D. M., Gitelman, D., Simmons, K., Bloise, S. M., Parrish, T., and Mesulam, M.-M. (2005). Monetary incentives enhance processing in brain regions mediating top-down control of attention. *Cerebral Cortex*, *15*(12), 1855–1865.

Smith, F. W., and Schyns, P. G. (2009). Smile through your fear and sadness: Transmitting and identifying facial expression signals over a range of viewing distances. *Psychological Science*, *20*(10), 1202–1208.

Smith, M. L., Cottrell, G. W., Gosselin, F., and Schyns, P. G. (2005). Transmitting and decoding facial expressions. *Psychological Science*, *16*(3), 184–189.

Smith, S. M., Miller, K. L., Moeller, S., Xu, J., Auerbach, E. J., Woolrich, M. W., et al. (2012). Temporally independent functional modes of spontaneous brain activity. *Proceedings of the National Academy of Sciences of the United States of America*, *109*(8), 3131–3136. doi:10.1073/pnas.1121329109.

Smith, Y., Surmeier, D. J., Redgrave, P., and Kimura, M. (2011). Thalamic contributions to Basal ganglia-related behavioral switching and reinforcement. *Journal of Neuroscience*, *31*(45), 16102–16106. doi:10.1523/jneurosci.4634-11.2011.

Snodgrass, M., Bernat, E., and Shevrin, H. (2004). Unconscious perception: A model-based approach to method and evidence. *Perception and Psychophysics*, *66*(5), 846–867.

Snyder, M., Killackey, H., and Diamond, I. T. (1969). Color vision in the tree shrew after removal of posterior neocortex. *Journal of Neurophysiology*, *32*(4), 554–563.

Solano-Castiella, E., Anwander, A., Lohmann, G., Weiss, M., Docherty, C., Geyer, S., et al. (2010). Diffusion tensor imaging segments the human amygdala in vivo. *NeuroImage*, *49*(4), 2958–2965.

Solomon, R. L., and Corbit, J. D. (1974). An opponent-process theory of motivation: I. Temporal dynamics of affect. *Psychological Review*, *81*(2), 119.

Somerville, L. H., and Casey, B. J. (2010). Developmental neurobiology of cognitive control and motivational systems. *Current Opinion in Neurobiology*, *20*(2), 236–241. doi:10.1016/j.conb.2010.01.006.

Somerville, L. H., Whalen, P. J., and Kelley, W. M. (2010). Human bed nucleus of the stria terminalis indexes hypervigilant threat monitoring. *Biological Psychiatry*, *68*(5), 416–424. doi:10.1016/j.biopsych.2010.04.002.

Sperling, G., and Dosher, B. A. (1986). Strategy and optimization in human information processing. In K. Boff, L. Kaufman, and J. Thomas, eds., *Handbook of Perception and Performance*, 1:1–65. New York: Wiley.

Sporns, O., Honey, C. J., and Kotter, R. (2007). Identification and classification of hubs in brain networks. *PLoS ONE, 2*(10), e1049. doi:1010.1371/journal.pone.00001049.

Sporns, O., and Zwi, J. D. (2004). The small world of the cerebral cortex. *Neuroinformatics*, *2*(2), 145–162.

Sprenger, C., Eippert, F., Finsterbusch, J., Bingel, U., Rose, M., and Buchel, C. (2012). Attention modulates spinal cord responses to pain. *Current Biology*, *22*(11), 1019–1022. doi:10.1016/j.cub.2012.04.006.

Sripati, A. P., and Olson, C. R. (2009). Representing the forest before the trees: A global advantage effect in monkey inferotemporal cortex. *Journal of Neuroscience*, *29*(24), 7788–7796.

Stalnaker, T. A., Roesch, M. R., Franz, T. M., Calu, D. J., Singh, T., and Schoenbaum, G. (2007). Cocaine-induced decision-making deficits are mediated by miscoding in basolateral amygdala. *Nature Neuroscience*, *10*(8), 949–951.

Stein, B. E. (1981). Organization of the rodent superior colliculus: Some comparisons with other mammals. *Behavioural Brain Research*, *3*(2), 175–188.

Stein, B. E., and Meredith, M. A. (1993). *The Merging of the Senses*. Cambridge, MA: MIT Press.

Stein, T., Peelen, M. V., Funk, J., and Seidl, K. N. (2010). The fearful-face advantage is modulated by task demands: Evidence from the attentional blink. *Emotion*, *10*(1), 136–140.

Stepniewska, I. (2004). The pulvinar complex. In J. H. Kaas and C. E. Collins, eds., *The Primate Visual System*, 53–80. Boca Raton, FL: CRC Press.

Stevenson, I. H., Rebesco, J. M., Miller, L. E., and Kording, K. P. (2008). Inferring functional connections between neurons. *Current Opinion in Neurobiology*, *18*(6), 582–588.

Stock, G., Rupprecht, U., Stumpf, H., and Schlor, K. H. (1981). Cardiovascular changes during arousal elicited by stimulation of amygdala, hypothalamus and locus coeruleus. *Journal of the Autonomic Nervous System*, *3*(2–4), 503–510.

St. Onge, J. R., Stopper, C. M., Zahm, D. S., and Floresco, S. B. (2012). Separate prefrontal-subcortical circuits mediate different components of risk-based decision making. *Journal of Neuroscience*, *32*(8), 2886–2899. doi:10.1523/jneurosci.5625-11.2012.

St. Peters, M., Demeter, E., Lustig, C., Bruno, J. P., and Sarter, M. (2011). Enhanced control of attention by stimulating mesolimbic-corticopetal cholinergic circuitry. *Journal of Neuroscience*, *31*(26), 9760–9771. doi:10.1523/jneurosci.1902-11.2011.

Straube, T., Dietrich, C., Mothes-Lasch, M., Mentzel, H. J., and Miltner, W. H. (2010). The volatility of the amygdala response to masked fearful eyes. *Human Brain Mapping*, *31*(10), 1601–1608. doi:10.1002/hbm.20960.

Straube, T., Schmidt, S., Weiss, T., Mentzel, H. J., and Miltner, W. H. (2009). Dynamic activation of the anterior cingulate cortex during anticipatory anxiety. *NeuroImage*, *44*(3), 975–981. doi:10.1016/j.neuroimage.2008.10.022.

Straube, T., Weiss, T., Mentzel, H. J., and Miltner, W. H. (2007). Time course of amygdala activation during aversive conditioning depends on attention. *NeuroImage*, *34*(1), 462–469.

Streit, M., Dammers, J., Simsek-Kraues, S., Brinkmeyer, J., Wolwer, W., and Ioannides, A. (2003). Time course of regional brain activations during facial emotion recognition in humans. *Neuroscience Letters*, *342*(1–2), 101–104.

Striedter, G. F. (2005). *Principles of Brain Evolution*. Sunderland, MA: Sinauer Associates.

Stuss, D., and Knight, R. T., eds. (2002). *Principles of Frontal Lobe Function*. Oxford: Oxford University Press.

Sugase, Y., Yamane, S., Ueno, S., and Kawano, K. (1999). Global and fine information coded by single neurons in the temporal visual cortex. *Nature*, *400*(6747), 142–147.

Sugase-Miyamoto, Y., and Richmond, B. J. (2005). Neuronal signals in the monkey basolateral amygdala during reward schedules. *Journal of Neuroscience*, *25*(48), 11071–11083.

Summerfield, C., and Koechlin, E. (2009). Decision making and prefrontal executive function. In M. S. Gazzaniga, ed., *The Cognitive Neurosciences*, 1019–1030. Cambridge, MA: MIT Press.

Sutton, R., and Barto, A. (1988). *Reinforcement Learning*. Cambridge, MA: MIT Press.

Swanson, L. W. (2000). Cerebral hemisphere regulation of motivated behavior. *Brain Research*, *886*(1–2), 113–164.

Swanson, L. W. (2003). The amygdala and its place in the cerebral hemisphere. *Annals of the New York Academy of Sciences*, *985*, 174–184.

Swanson, L. W., and Petrovich, G. D. (1998). What is the amygdala? *Trends in Neurosciences*, *21*(8), 323–331.

Sweeny, T. D., Suzuki, S., Grabowecky, M., and Paller, K. A. (2013). Detecting and categorizing fleeting emotions in faces. *Emotion*, *13*(1), 76–91.

Szczepanowski, R., and Pessoa, L. (2007). Fear perception: Can objective and subjective awareness measures be dissociated? *Journal of Vision (Charlottesville, Va.)*, *7*(4), 10.

Szell, M., Lambiotte, R., and Thurner, S. (2010). Multirelational organization of large-scale social networks in an online world. *Proceedings of the National Academy of Sciences of the United States of America*, *107*(31), 13636–13641.

Takaura, K., Yoshida, M., and Isa, T. (2011). Neural substrate of spatial memory in the superior colliculus after damage to the primary visual cortex. *Journal of Neuroscience*, *31*(11), 4233–4241.

Talmi, D., Hurlemann, R., Patin, A., and Dolan, R. J. (2010). Framing effect following bilateral amygdala lesion. *Neuropsychologia*, *48*(6), 1823–1827. doi:10.1016/j.neuropsychologia.2010.03.005.

Tamietto, M., Cauda, F., Corazzini, L. L., Savazzi, S., Marzi, C. A., Goebel, R., et al. (2010). Collicular vision guides nonconscious behavior. *Journal of Cognitive Neuroscience*, *22*(5), 888–902. doi:10.1162/jocn.2009.21225.

Tamietto, M., and de Gelder, B. (2010). Neural bases of the non-conscious perception of emotional signals. *Nature Reviews. Neuroscience*, *11*(10), 697–709.

Tamietto, M., Pullens, P., de Gelder, B., Weiskrantz, L., and Goebel, R. (2012). Subcortical connections to human amygdala and changes following destruction of the visual cortex. *Current Biology*. doi:10.1016/j.cub.2012.06.006.

Taylor, S. F., Welsh, R. C., Wager, T. D., Phan, K. L., Fitzgerald, K. D., and Gehring, W. J. (2004). A functional neuroimaging study of motivation and executive function. *NeuroImage*, *21*(3), 1045–1054. doi:10.1016/j.neuroimage.2003.10.032.

Theyel, B. B., Llano, D. A., and Sherman, S. M. (2010). The corticothalamocortical circuit drives higher-order cortex in the mouse. *Nature Neuroscience*, *13*(1), 84–88. doi:10.1038/nn.2449.

Thielscher, A., and Pessoa, L. (2007). Neural correlates of perceptual choice and decision making during fear-disgust discrimination. *Journal of Neuroscience*, *27*(11), 2908–2917.

Thompson, E. (2007). *Mind in Life: Biology, Phenomenology, and the Sciences of the Mind*. Cambridge, MA: Harvard University Press.

Thompson, E., and Varela, F. J. (2001). Radical embodiment: Neural dynamics and consciousness. *Trends in Cognitive Sciences*, *5*(10), 418–425.

Thompson, K. G., and Bichot, N. P. (2005). A visual salience map in the primate frontal eye field. *Progress in Brain Research*, *147*, 251–262. doi:10.1016/s0079-6123(04)47019-8.

Tigges, J., and Tigges, M. (1985). Subcortical sources of direct projections to visual cortex. *Cerebral Cortex*, *3*, 351–378.

Tombu, M. N., Asplund, C. L., Dux, P. E., Godwin, D., Martin, J. W., and Marois, R. (2011). A unified attentional bottleneck in the human brain. *Proceedings of the National Academy of Sciences of the United States of America*, *108*(33), 13426–13431. doi:10.1073/pnas.1103583108.

Toro, R., Fox, P. T., and Paus, T. (2008). Functional coactivation map of the human brain. *Cerebral Cortex*, *18*(11), 2553–2559.

Tovee, M. J., and Rolls, E. T. (1995). Information encoding in short firing rate epochs by single neurons in the primate temporal visual cortex. *Visual Cognition*, *2*(1), 35–58.

Tranel, D., and Damasio, A. R. (1985). Knowledge without awareness: An autonomic index of facial recognition by prosopagnosics. *Science*, *228*(4706), 1453–1454.

Tranel, D., and Hyman, B. T. (1990). Neuropsychological correlates of bilateral amygdala damage. *Archives of Neurology*, *47*(3), 349–355.

Treisman, A. M., and Gelade, G. (1980). A feature-integration theory of attention. *Cognitive Psychology*, *12*(1), 97–136.

Tsal, Y., and Benoni, H. (2010). Diluting the burden of load: Perceptual load effects are simply dilution effects. *Journal of Experimental Psychology. Human Perception and Performance*, *36*(6), 1645–1656. doi:10.1037/a0018172.

Tsotsos, J. K. (1990). Analyzing vision at the complexity level. *Behavioral and Brain Sciences*, *13*, 423–469.

Tsotsos, J. K. (1997). Limited capacity of any realizable perceptual system is a sufficient reason for attentive behavior. *Consciousness and Cognition*, *6*(2–3), 429–436. doi:10.1006/ccog.1997.0302.

Tsuchiya, N., Moradi, F., Felsen, C., Yamazaki, M., and Adolphs, R. (2009). Intact rapid detection of fearful faces in the absence of the amygdala. *Nature Neuroscience*, *12*(10), 1224–1225.

Tyszka, J. M., Kennedy, D. P., Adolphs, R., and Paul, L. K. (2011). Intact bilateral resting-state networks in the absence of the corpus callosum. *Journal of Neuroscience*, *31*(42), 15154–15162. doi:10.1523/jneurosci.1453-11.2011.

Tzelgov, J. (1997). Specifying the relations between automaticity and consciousness: A theoretical note. *Consciousness and Cognition*, *6*(2–3), 441–451.

Uddin, L., Kinnison, J., Pessoa, L. M., and Anderson, M. L. (2013). Beyond the tripartite cognition-emotion-interoception model of the human insular cortex. Submitted for publication.

Ungerleider, L. G., and Christensen, C. A. (1979). Pulvinar lesions in monkeys produce abnormal scanning of a complex visual array. *Neuropsychologia*, *17*(5), 493–501.

Ungerleider, L. G., and Mishkin, M. (1982). Two cortical visual systems. In D. J. Ingle, M. A. Goodale, and R. J. W. Mansfield, eds., *Analysis of Visual Behavior*, 549–586. Cambridge, MA: MIT Press.

Urry, H. L., van Reekum, C. M., Johnstone, T., Kalin, N. H., Thurow, M. E., Schaefer, H. S., et al. (2006). Amygdala and ventromedial prefrontal cortex are inversely coupled during regulation of negative affect and predict the diurnal pattern of cortisol secretion among older adults. *Journal of Neuroscience*, *26*(16), 4415–4425. doi:10.1523/JNEUROSCI.3215-05.2006.

Ursin, H., and Kaada, B. R. (1960). Functional localization within the amygdaloid complex in the cat. *Electroencephalography and Clinical Neurophysiology*, *12*, 1–20.

Van Bavel, J. J., Packer, D. J., and Cunningham, W. A. (2008). The neural substrates of in-group bias: A functional magnetic resonance imaging investigation. *Psychological Science*, *19*(11), 1131–1139. doi:10.1111/j.1467-9280.2008.02214.x.

van Boxtel, J. J., Tsuchiya, N., and Koch, C. (2010). Opposing effects of attention and consciousness on afterimages. *Proceedings of the National Academy of Sciences of the United States of America*, *107*(19), 8883–8888.

Van den Bussche, E., Hughes, G., Humbeeck, N. V., and Reynvoet, B. (2010). The relation between consciousness and attention: An empirical study using the priming paradigm. *Consciousness and Cognition*, *19*(1), 86–97.

van den Heuvel, M. P., Mandl, R. C., Kahn, R. S., and Hulshoff Pol, H. E. (2009). Functionally linked resting-state networks reflect the underlying structural connectivity architecture of the human brain. *Human Brain Mapping*, *30*(10), 3127–3141. doi:10.1002/hbm.20737.

van den Heuvel, M. P., and Sporns, O. (2011). Rich-club organization of the human connectome. *Journal of Neuroscience*, *31*(44), 15775–15786. doi:10.1523/jneurosci.3539-11.2011.

Van Dillen, L. F., Heslenfeld, D. J., and Koole, S. L. (2009). Tuning down the emotional brain: An fMRI study of the effects of cognitive load on the processing of affective images. *NeuroImage*, *45*(4), 1212–1219. doi:10.1016/j.neuroimage.2009.01.016.

Van Essen, D. C. (1985). Functional organization of primate visual cortex. In A. Peters and E. G. Jones, eds., *Cerebral Cortex*, 3:259–329. New York: Plenum Press.

van Honk, J., Eisenegger, C., Terburg, D., Stein, D. J., and Morgan, B. (2013). Generous economic investments after basolateral amygdala damage. *Proceedings of the National Academy of Sciences of the United States of America*, *110*(7), 2506–2510.

Van Orden, G. C., Pennington, B. F., and Stone, G. O. (2001). What do double dissociations prove? *Cognitive Science*, *25*(1), 111–172.

Van Snellenberg, J. X., and Wager, T. D. (2010). Cognitive and motivational functions of the human prefrontal cortex. In A.-L. Christensen, E. Goldberg, and D. Bougakov, eds., *Luria's Legacy in the 21st Century*, 30–60. Oxford: Oxford University Press.

Varela, F. J., Lachaux, J. P., Rodriguez, E., and Martinerie, J. (2001). The brainweb: Phase synchronization and large-scale integration. *Nature Reviews. Neuroscience*, *2*(4), 229–239.

Varela, F. J., Thompson, E., and Rosch, E. (1992). *The Embodied Mind: Cognitive Science and Human Experience*. Cambridge, MA: MIT Press.

Verbruggen, F., and De Houwer, J. (2007). Do emotional stimuli interfere with response inhibition? Evidence from the stop signal paradigm. *Cognition and Emotion*, *21*(2), 391–403.

Vickery, T. J., Chun, M. M., and Lee, D. (2011). Ubiquity and specificity of reinforcement signals throughout the human brain. *Neuron*, *72*(1), 166–177. doi:10.1016/j.neuron.2011.08.011.

Vlachos, I., Aertsen, A., and Kumar, A. (2012). Beyond statistical significance: Implications of network structure on neuronal activity. *PLoS Computational Biology*, *8*(1), e1002311. doi:10.1371/journal.pcbi.1002311.

Vogel, E. K., Woodman, G. F., and Luck, S. J. (2005). Pushing around the locus of selection: Evidence for the flexible-selection hypothesis. *Journal of Cognitive Neuroscience*, *17*(12), 1907–1922. doi:10.1162/089892905775008599.

Vogt, B. A., ed. (2008). *Cingulate Neurobiology and Disease*. Oxford: Oxford University Press.

Vogt, B. A., Aston-Jones, G., and Vogt, L. J. (2009). Shared norepinephrinergic and cingulate circuits, nociceptive and allostatic interactions, and models of functional pain and stress disorders. In B. A. Vogt, ed., *Cingulate Neurobiology and Disease*, 467–497. Oxford: Oxford University Press.

Vogt, B. A., and Derbyshire, S. W. G. (2009). Visceral circuits and cingulate-mediated functions. In B. A. Vogt, ed., *Cingulate Neurobiology and Disease*, 219–235. Oxford: Oxford University Press.

Vogt, B. A., and Vogt, L. J. (2009). Mu-opioid receptors, placebo map, descending systems, and cingulate-mediated control of vocalization and pain. In B. A. Vogt, ed., *Cingulate Neurobiology and Disease*, 339–364. Oxford: Oxford University Press.

Vogt, J., De Houwer, J., and Crombez, G. (2011). Multiple goal management starts with attention: Goal prioritizing affects the allocation of spatial attention to goal-relevant events. *Experimental Psychology*, *58*(1), 55–61. doi:10.1027/1618-3169/a000066.

Vogt, J., De Houwer, J., Crombez, G., and Van Damme, S. (2012). Competing for attentional priority: Temporary goals versus threats. *Emotion*. doi:10.1037/a0027204.

Vogt, J., De Houwer, J., Moors, A., Van Damme, S., and Crombez, G. (2010). The automatic orienting of attention to goal-relevant stimuli. *Acta Psychologica*, *134*(1), 61–69. doi:10.1016/j.actpsy.2009.12.006.

Vogt, J., Lozo, L., Koster, E. H., and De Houwer, J. (2011). On the role of goal relevance in emotional attention: Disgust evokes early attention to cleanliness. *Cognition and Emotion*, *25*(3), 466–477. doi:10.1080/02699931.2010.532613.

Vuilleumier, P. (2005). How brains beware: Neural mechanisms of emotional attention. *Trends in Cognitive Sciences*, *9*(12), 585–594.

Vuilleumier, P., Armony, J. L., Driver, J., and Dolan, R. J. (2001). Effects of attention and emotion on face processing in the human brain: An event-related fMRI study. *Neuron*, *30*(3), 829–841.

Vuilleumier, P., Armony, J. L., Driver, J., and Dolan, R. J. (2003). Distinct spatial frequency sensitivities for processing faces and emotional expressions. *Nature Neuroscience*, *6*(6), 624–631.

Vuilleumier, P., Richardson, M. P., Armony, J. L., Driver, J., and Dolan, R. J. (2004). Distant influences of amygdala lesion on visual cortical activation during emotional face processing. *Nature Neuroscience*, *7*(11), 1271–1278.

Wager, T. D., Davidson, M. L., Hughes, B. L., Lindquist, M. A., and Ochsner, K. N. (2008). Prefrontal-subcortical pathways mediating successful emotion regulation. *Neuron*, *59*(6), 1037–1050.

Walker, A. E. (1938). *The Primate Thalamus*. Chicago: University of Chicago Press.

Walton, M. E., Rudebeck, P. H., Bannerman, D. M., and Rushworth, M. F. (2007). Calculating the cost of acting in frontal cortex. *Annals of the New York Academy of Sciences*, *1104*, 340–356.

Wang, J., Zuo, X., and He, Y. (2010). Graph-based network analysis of resting-state functional MRI. *Frontiers in Systems Neuroscience*, *4*, 16. doi:10.3389/fnsys.2010.00016.

Ward, R., Calder, A. J., Parker, M., and Arend, I. (2007). Emotion recognition following human pulvinar damage. *Neuropsychologia*, *45*(8), 1973–1978.

Ward, R., Danziger, S., and Bamford, S. (2005). Response to visual threat following damage to the pulvinar. *Current Biology*, *15*(6), 571–573.

Ward, R., Danziger, S., Owen, V., and Rafal, R. (2002). Deficits in spatial coding and feature binding following damage to spatiotopic maps in the human pulvinar. *Nature Neuroscience*, *5*(2), 99–100.

Wardak, C., Vanduffel, W., and Orban, G. A. (2010). Searching for a salient target involves frontal regions. *Cerebral Cortex*, *20*(10), 2464–2477.

Watanabe, M. (1990). Prefrontal unit activity during associative learning in the monkey. *Experimental Brain Research*, *80*, 296–309.

Watanabe, M. (1996). Reward expectancy in primate prefrontal neurons. *Nature*, *382*, 629–632.

Watts, D. J., and Strogatz, S. H. (1998). Collective dynamics of "small-world" networks. *Nature*, *393*(6684), 440–442.

Webster, M. J., Bachevalier, J., and Ungerleider, L. G. (1993). Subcortical connections of inferior temporal areas TE and TEO in macaques. *Journal of Comparative Neurology*, *335*, 73–91.

Weil, R. S., Furl, N., Ruff, C. C., Symmonds, M., Flandin, G., Dolan, R. J., et al. (2010). Rewarding feedback after correct visual discriminations has both general and specific influences on visual cortex. *Journal of Neurophysiology*, *104*(3), 1746–1757. doi:10.1152/jn.00870.2009.

Weinberger, N. M. (1995). Retuning the brain by fear conditioning. In M. S. Gazzaniga, ed., *The Cognitive Neurosciences*, 1071–1089. Cambridge, MA: MIT Press.

Weinberger, N. M. (2003). The nucleus basalis and memory codes: Auditory cortical plasticity and the induction of specific, associative behavioral memory. *Neurobiology of Learning and Memory*, *80*(3), 268–284.

Weiskrantz, L. (1956). Behavioral changes associated with ablation of the amygdaloid complex in monkeys. *Journal of Comparative and Physiological Psychology*, *49*(4), 381–391.

Weiskrantz, L. (1986). *Blindsight: A Case Study and Implications* (Vol. 12). Oxford: Oxford University Press.

Weller, J. A., Levin, I. P., Shiv, B., and Bechara, A. (2007). Neural correlates of adaptive decision making for risky gains and losses. *Psychological Science*, *18*(11), 958–964.

Whalen, P. J. (1998). Fear, vigilance, and ambiguity: Initial neuroimaging studies of the human amygdala. *Current Directions in Psychological Science*, *7*(6), 177–188.

Whalen, P. J., Bush, G., McNally, R. J., Wilhelm, S., McInerney, S. C., Jenike, M. A., et al. (1998). The emotional counting Stroop paradigm: A functional magnetic resonance imaging probe of the anterior cingulate affective division. *Biological Psychiatry*, *44*(12), 1219–1228.

Whalen, P. J., and Phelps, E. A., eds. (2009). *The Human Amygdala*. New York, NY: Guilford Press.

Whalen, P. J., Rauch, S. L., Etcoff, N. L., McInerney, S. C., Lee, M. B., and Jenike, M. A. (1998). Masked presentations of emotional facial expressions modulate amygdala activity without explicit knowledge. *Journal of Neuroscience*, *18*(1), 411–418.

Wiggs, C. L., and Martin, A. (1998). Properties and mechanisms of perceptual priming. *Current Opinion in Neurobiology*, *8*(2), 227–233.

Wilke, M., Mueller, K. M., and Leopold, D. A. (2009). Neural activity in the visual thalamus reflects perceptual suppression. *Proceedings of the National Academy of Sciences of the United States of America*, *106*(23), 9465–9470.

Wilkinson, D., and Halligan, P. (2004). The relevance of behavioural measures for functional-imaging studies of cognition. *Nature Reviews. Neuroscience*, *5*(1), 67–73. doi:10.1038/nrn1302.

Willis, T. (1664). *The Anatomy of the Brain and Nerves*. Montreal: McGill University Press, 1965.

Wilson, D. E., Muroi, M., and MacLeod, C. M. (2011). Dilution, not load, affects distractor processing. *Journal of Experimental Psychology. Human Perception and Performance*, *37*(2), 319–335. doi:10.1037/a0021433.

Winkielman, P., and Schooler, J. W. (2011). Splitting consciousness: Unconscious, conscious, and metaconscious processes in social cognition. *European Review of Social Psychology*, *22*(1), 1–35.

Winstanley, C. A., Theobald, D. E., Cardinal, R. N., and Robbins, T. W. (2004). Contrasting roles of basolateral amygdala and orbitofrontal cortex in impulsive choice. *Journal of Neuroscience*, *24*(20), 4718–4722.

Wolfe, J. M. (1994). Guided search 2.0: A revised model of visual search. *Psychonomic Bulletin and Review*, *1*(2), 202–238.

Wolfe, J. M., Vo, M. L., Evans, K. K., and Greene, M. R. (2011). Visual search in scenes involves selective and nonselective pathways. *Trends in Cognitive Sciences*, *15*(2), 77–84. doi:10.1016/j.tics.2010.12.001.

Wright, P., and Liu, Y. (2006). Neutral faces activate the amygdala during identity matching. *NeuroImage*, *29*(2), 628–636. doi:10.1016/j.neuroimage.2005.07.047.

Wurtz, R. H., and Albano, J. E. (1980). Visual-motor function of the primate superior colliculus. *Annual Review of Neuroscience*, *3*, 189–226.

Wurtz, R. H., McAlonan, K., Cavanaugh, J., and Berman, R. A. (2011). Thalamic pathways for active vision. *Trends in Cognitive Sciences*, *15*(4), 177–184. doi:10.1016/j.tics.2011.02.004.

Yamasaki, H., LaBar, K. S., and McCarthy, G. (2002). Dissociable prefrontal brain systems for attention and emotion. *Proceedings of the National Academy of Sciences of the United States of America*, *99*(17), 215–222.

Yang, E., Zald, D. H., and Blake, R. (2007). Fearful expressions gain preferential access to awareness during continuous flash suppression. *Emotion*, *7*(4), 882–886.

Yantis, S., and Johnston, D. N. (1990). Mechanisms of attentional priority. *Journal of Experimental Psychology. Human Perception and Performance*, *16*(4), 812–825.

Yantis, S., and Jonides, J. (1984). Abrupt visual onsets and selective attention: Evidence from visual search. *Journal of Experimental Psychology. Human Perception and Performance*, *10*(5), 601–621.

Yantis, S., and Jonides, J. (1990). Abrupt visual onsets and selective attention: Voluntary versus automatic allocation. *Journal of Experimental Psychology. Human Perception and Performance*, *16*(1), 121–134.

Yarkoni, T., Poldrack, R. A., Van Essen, D. C., and Wager, T. D. (2010). Cognitive neuroscience 2.0: Building a cumulative science of human brain function. *Trends in Cognitive Sciences*, *14*(11), 489–496. doi:10.1016/j.tics.2010.08.004.

Yasui, Y., Breder, C. D., Saper, C. B., and Cechetto, D. F. (1991). Autonomic responses and efferent pathways from the insular cortex in the rat. *Journal of Comparative Neurology*, *303*(3), 355–374.

Yates, A., Ashwin, C., and Fox, E. (2010). Does emotion processing require attention? The effects of fear conditioning and perceptual load. *Emotion*, *10*(6), 822–830. doi:10.1037/a0020325.

Yeo, B. T., Krienen, F. M., Sepulcre, J., Sabuncu, M. R., Lashkari, D., Hollinshead, M., et al. (2011). The organization of the human cerebral cortex estimated by intrinsic functional connectivity. *Journal of Neurophysiology*, *106*(3), 1125–1165. doi:10.1152/jn.00338.2011.

Yoshor, D., Bosking, W. H., Ghose, G. M., and Maunsell, J. H. (2007). Receptive fields in human visual cortex mapped with surface electrodes. *Cerebral Cortex*, *17*(10), 2293–2302.

Young, A. W., Reid, I., Wright, S., and Hellawell, D. J. (1993). Face-processing impairments and the Capgras delusion. *British Journal of Psychiatry*, *162*, 695–698.

Young, M. P., Scannell, J. W., Burns, G. A. P. C., and Blakemore, C. (1994). Analysis of connectivity: Neural systems in the cerebral cortex. *Reviews in the Neurosciences*, *5*, 227–249.

Yukie, M., and Iwai, E. (1981). Direct projection from the dorsal lateral geniculate nucleus to the prestriate cortex in macaque monkeys. *Journal of Comparative Neurology*, *201*(1), 81–97.

Zaborszky, L. (2002). The modular organization of brain systems: Basal forebrain: The last frontier. *Progress in Brain Research*, *136*, 359–372.

Zaborszky, L., Hoemke, L., Mohlberg, H., Schleicher, A., Amunts, K., and Zilles, K. (2008). Stereotaxic probabilistic maps of the magnocellular cell groups in human basal forebrain. *NeuroImage*, *42*(3), 1127–1141.

Zaborszky, L., Pang, K., Somogyi, J., Nadasdy, Z., and Kallo, I. (1999). The basal forebrain corticopetal system revisited. *Annals of the New York Academy of Sciences*, *877*, 339–367.

Zald, D. H., and Rauch, S. L. (2007). *The Orbitofrontal Cortex*. Oxford: Oxford University Press.

Zamora-Lopez, G., Zhou, C., and Kurths, J. (2010). Cortical hubs form a module for multisensory integration on top of the hierarchy of cortical networks. *Frontiers in Neuroinformatics*, *4*, 1.

Zelazo, P. D., and Cunningham, W. A. (2007). Executive function: Mechanisms underlying emotion regulation. In J. Gross, ed., *Handbook of Emotion Regulation*, 135–158. New York: Guilford.

Zhang, P., Jamison, K., Engel, S., He, B., and He, S. (2011). Binocular rivalry requires visual attention. *Neuron*, *71*(2), 362–369. doi:10.1016/j.neuron.2011.05.035.

Zihl, J., and von Cramon, D. (1979). The contribution of the "second" visual system to directed visual attention in man. *Brain*, *102*(4), 835–856.

Zikopoulos, B., and Barbas, H. (2012). Pathways for emotions and attention converge on the thalamic reticular nucleus in primates. *Journal of Neuroscience*, *32*(15), 5338–5350. doi:10.1523/jneurosci.4793-11.2012.

Zink, C. F., Pagnoni, G., Martin-Skurski, M. E., Chappelow, J. C., and Berns, G. S. (2004). Human striatal responses to monetary reward depend on saliency. *Neuron*, *42*(3), 509–517.

Zuo, X. N., Ehmke, R., Mennes, M., Imperati, D., Castellanos, F. X., Sporns, O., et al. (2012). Network centrality in the human functional connectome. *Cerebral Cortex*, *22*(8), 1862–1875. doi:10.1093/cercor/bhr269.

Index